INSECT LEARNING

INSECT LEARNING
Ecological and Evolutionary
Perspectives

❖

edited by Daniel R. Papaj
and
Alcinda C. Lewis

Chapman & Hall
New York • London

First published in 1993 by
Chapman & Hall
an imprint of
Routledge, Chapman & Hall, Inc.
29 West 35 Street
New York, NY 10001-2291

Published in Great Britain by
Chapman & Hall
2-6 Boundary Row
London SE1 8HN

Library of Congress Cataloging in Publication Data

Insect learning : ecological and evolutionary perspectives / edited by
 Daniel R. Papaj and Alcinda C. Lewis.
     p.   cm.
   Includes bibliographical references and index.
   ISBN 0-412-02561-2
   1. Insects—Behavior. 2. Insects—Psychology. 3. Avoidance
(Psychology) I. Papaj, Daniel Richard, 1956-   . II. Lewis.
Alcinda C., 1949-
QL496.I387 1992
595.7'051–dc20

                                                              92-30107
                                                                 CIP

British Library Cataloguing in Publication Data also available.

Contributors

Elizabeth A. Bernays
Department of Entomology and
Department of Ecology and
Evolutionary Biology
University of Arizona
Tucson, Arizona

Reuven Dukas
Department of Biology
University of California
Riverside, California

James L. Gould
Department of Ecology and
Evolutionary Biology
Princeton University
Princeton, New Jersey

Uwe Greggers
Fachbereich Biologie
Institut für Neurobiologie
Freie Universität Berlin
Berlin, Germany

Martin Hammer
Fachbereich Biologie
Institut für Neurobiologie
Freie Universität Berlin
Berlin, Germany

Alcinda C. Lewis
Department of Environmental,
Population and Organismic
Biology
University of Colorado
Boulder, Colorado

W. Joseph Lewis
U.S. Department of Agriculture
Agricultural Research Service
Tifton, Georgia

Chao Li
Department of Biological Sciences
Simon Fraser University
Burnaby, British Columbia
Canada

Marc Mangel
Department of Zoology and Center
for Population Biology
University of Caifornia, Davis
Davis, California

Randolf Menzel
Fachbereich Biologie
Institut für Neurobiologie
Freie Universität Berlin
Berlin, Germany

Daniel R. Papaj
Department of Ecology and
Evolutionary Biology and Center
for Insect Science
University of Arizona
Tucson, Arizona

Ronald J. Prokopy
Department of Entomology
University of Massachusetts
Amherst, Massachusetts

Leslie A. Real
Department of Biology
University of North Carolina
Chapel Hill, North Carolina

Mary L. Reid
Department of Biological Sciences
Simon Fraser University
Burnaby, British Columbia
Canada

Bernard D. Roitberg
Department of Biological Sciences
Simon Fraser University
Burnaby, British Columbia
Canada

Jay A. Rosenheim
Department of Entomology
University of California, Davis
Davis, California

Brian H. Smith
Department of Entomology
The Ohio State University
Columbus, Ohio

David W. Stephens
School of Biological Sciences
University of Nebraska, Lincoln
Lincoln, Nebraska

James H. Tumlinson
U.S. Department of Agriculture
Agricultural Research Service
Insect Attractants, Behavior and
  Basic Biology Research
  Laboratory
Gainesville, Florida

Ted C. J. Turlings
U.S. Department of Agriculture
Agricultural Research Service
Insect Attractants, Behavior and
  Basic Biology Research
  Laboratory
Gainesville, Florida

Louise E. M. Vet
Department of Entomology
Agricultural University of
  Wageningen
Wageningen, The Netherlands

Felix L. Wäckers
Department of Entomology
Agricultural University of
  Wageningen
Wageningen, The Netherlands

Contents

Preface

In the past decade, awareness has grown of the importance of learning in the life history of insects in numerous taxa. This awareness has come simultaneously to researchers in areas as diverse as neurobiology, ethology, behavioral ecology, population genetics and pest management. Working in what was once the almost-exclusive domain of psychology, these insect scientists are asking novel questions about learning and using new approaches to answer those questions. Their endeavors are not to be construed as a substitute for psychological inquiry, but rather as a useful complement. This book constitutes a survey of those efforts.

The chapters in this book review a broad range of ecological and evolutionary problems in insect learning, usually in the context of new data and original models. Certain questions of context, mechanism and function arise again and again: What exactly is learned in an ecological setting? How is information represented in the nervous system and do such representations have adaptive significance? Under what conditions is information useful to the insect and are those the conditions in which learning is observed? Are differences in learning abilities among species better explained by differences in ecological niche or in taxonomic status? How does learning on the part of one species influence interactions with other species? Taken together, these questions constitute evidence of an abiding interest in the functional significance of learning in nature. The authors in this volume, whether wearing the hats of physiologist, theoretician or field ecologist, are thus united by an evolutionary perspective on issues in insect learning. At the same time, their approaches and points of view are varied and occasionally even in opposition.

Each of the first two chapters review aspects of the physiological ecology of learning. Bernays summarizes research on aversion learning in insects and raises questions on its origin and function. She proposes that aversion learning is part of a general ability to learn to avoid anything with a negative consequence and therefore may be widespread in insects. Bernays further suggests that in generalist insects such as grasshoppers, aversion learning may be most important in selection of high quality mixed diets rather than in avoidance of toxic plants.

Turlings and coauthors evaluate the flurry of recent work on the role of learning in host selection by parasitoids, concentrating on associative learning of olfactory and visual cues. The authors outline a basic problem with which parasitic insects must deal, namely that host-related stimuli that most reliably predict host presence in a particular microhabitat are (presumably by evolutionary design) usually the most difficult to detect. They illustrate by example how learning permits a female parasitoid to make the best of this detectability-reliability tradeoff. Their thorough review makes it abundantly clear that sophisticated learning abilities in insects are not restricted to social or even semi-social groups.

The next three chapters make ample reference to concepts in neurobiology and psychology. Gould places work on learning in honey bees in an ethological context, relating a variety of learning mechanisms to the nature of particular problems faced by that remarkable insect. In his chapter as in his work in general, Gould surveys learning mechanisms originally defined in psychology (often in vertebrate systems), provides evidence of these mechanisms in honey bees, and shows how they are well designed in the honey bee for the function that they serve.

Menzel and coauthors also infer ecological function from design, specifically linking what is currently known about the temporal dynamics of memory formation in honey bees to the time course or floral encounters (and, hence, flower distributions) in the field. They advocate an approach in which analysis of function follows careful, painstaking elucidation of neurobiological and behavioral mechanism (and not vice-versa). In their view, theoretical constructs such as two-armed bandits are not only painfully far from neurological reality, but are likely to lead to inappropriate conclusions about the value of learning in nature.

Like Gould and Menzel et al., Smith argues that knowledge of mechanism is vital to an understanding of the adaptive nature of learning and vice versa. Using several bee species as examples, Smith suggests that stimulus generalization—a learning phenomenon studied almost exclusively in the laboratory—may reflect not a physiological constraint (as traditionally assumed), but a general adaptation for coping with variability in stimuli emitted by biotic resources such as food and mates.

The next three chapters bring advances in modelling developed in behavioral ecology to bear on questions of the function and adaptive evolution of learning. Using dynamic, state-variable models, Mangel addresses how information of two sorts, information about external resources and information about internal state, ought to influence behavior. His learning model (adapted from a model based on learning in fishermen) illustrates how selection might simultaneously favor learning and forgetting. Mangel concludes by appealing for construction of neural networks that would accomplish the Bayesian updating incorporated in the learning model.

Roitberg and coauthors investigate two factors that potentially influence how much information an animal needs to make decisions: the frequency of a particular decision and its effect on the fitness of the animal. Their model predicts conditions under which learning would be expected to be employed in host and mate choice. The conclusion that information can sometimes have value in selection of appropriate mates calls into question traditional views of courtship and mating in insects as hard-wired behavior.

Stephens presents two models: the first considers what conditions make information valuable to an individual and the second (a population genetic model) distinguishes between the dual effects of environmental predictability and environmental unpredictability on learning. Stephens ends with a discussion of how behavioral ecology figures to contribute to the general process view of learning developed in psychology.

These last three chapters generally assume that natural selection acts to shape behavior. The next three chapters discuss, in the context of learning, the notions of constraints on selection, evolutionary trajectory and phylogenetic history respectively. By way of an hypothesis by Darwin on flower handling by bees, Lewis asks how learning and, in particular, constraints on learning might influence the evolution of an insect's biotic resources. She weighs the possibility that flower constancy in generalist pollinators, a behavior that has puzzled biologists for centuries, is a consequence of constraints on memory. Lewis' chapter forces us to recognize that, while we can easily think of learning as a way in which the environment shapes the insect, the insect can (through learning) do much to shape its biotic environment over evolutionary time.

Papaj addresses a question which occupied both Lamarck and Darwin: are instincts derived from learned behavior or vice versa? He critically evaluates observations of early naturalists on the automatism, or consistency, of individual behavior that caused these naturalists to believe that instincts arose through the acquisition of habit. Using his own work on consistency in movement by parasitic wasps, Papaj presents a model showing how changes in consistency might be of adaptive value to the individual and another (genetic) model illustrating how learning of an adapting form might influence the trajectory along which instincts evolve in predictable environments.

Several contributors make mention of the potential power of the comparative approach for answering questions in the evolution of learning. Rosenheim meets the issue headon. At a loss for hard evidence in the literature on even the most basic questions, Rosenheim instead evaluates recent developments in the analysis of comparative data. He suggests analyses that might be useful for insect learning studies and sets forth several hypotheses on insect learning which might be addressed by these analyses. He urges students of insect learning to consider phylogenetic history as a factor in patterns of learning among species.

The last two chapters illustrate how future research in insect learning stands to make a contribution to applied science, on one hand, and basic science, on the other. Prokopy and Lewis consider how learning by pest and beneficial insects impacts on pest control practices. They provide valuable guidelines for future research while, at the same time, reminding us of the great gulf between what might potentially contribute to pest management and what actually will. At chapter's end, the sobering possibility remains that insect flexibility may do more to impede our efforts to suppress pest populations than to facilitate those efforts.

Finally, Dukas and Real provide a framework for future research on insect cognition. They review aspects of cognition in bees and discuss evolutionary and ecological factors that may influence cognitive architecture. Their stance on issues from cognitive maps to the waggle dance as language will provoke thought and even controversy. Dukas and Real illustrate by example how students of insect behavior stand to benefit from increased attention to principles of cognition and how students of cognitive science stand to benefit from increased attention to insects.

An afterword by Papaj integrates the chapters into a general perspective on how learning adapts the insect to its current environment and how such adapting mechanisms figure in the evolution of functional (even optimal) behavior. He raises the possibility that learning is part of a set of optimizing routines that evolved long ago in ancestral taxa, routines which have been commandeered by succeeding taxa to solve novel ecological problems with relatively little genetic change.

The careful reader will note some inconsistencies among chapters. Authors vary greatly, for example, in their definitions of learning. One author's perspective on memory dynamics may be wholly at odds with the dynamics built into another's model. And so on. In general, we refrained from wielding the kind of editorial hand that would have smoothed out these wrinkles. It is our conviction that such incongruities form the core of academic debate and that it is more in the collective interest for editors to foster debate than to feign consensus.

Truly, consensus on matters related to the behavioral ecology of learning is currently beyond our grasp. Such consensus is likely to remain elusive for quite some time. In 1984, Shettleworth noted that the behavioral ecology of learning was still at a rather primitive level. She speculated that this was due to the difficulty of making field observations of learned behavior and the need for controlled laboratory experiments. Insects are ideal organisms for both laboratory and field studies and, in fact, many laboratory experiments have been performed and some field observations have been made in the intervening years by students of insect learning. Yet this book makes clear how much more needs to be done. It should also make clear to all the desirability for dialogue among evolutionary ecologists, psychologists, geneticists, and neurophysiologists. We trust that this book will aid in expanding that dialogue.

We thank those who reviewed various chapters (Steve Buchmann, Hugh Cresswell, Marcel Dicke, Ann Hedrick, John Jaenike, Bill Mitchell, Tom Valone, Bill Wcislo, Aileen Wardle, Nick Waser, various authors, and a number of anonymous reviewers) as well as the editorial staff at Chapman & Hall, Inc. Special thanks is owed to Greg Payne, science editor at Chapman & Hall, for his patience and advice. A. C. Lewis thanks C. H. Lewis for assistance of all kinds.

D. R. Papaj
A. C. Lewis

Literature Cited

Shettleworth, S. J. 1984. Learning and behavioural ecology. In J. R. Krebs and N. B. Davies (eds.), Behavioural Ecology: An Evolutionary Approach. Blackwell Scientific Publications, Oxford, England, pp. 170–194.

1

Aversion Learning and Feeding
Elizabeth A. Bernays

Introduction

It is now well known that, among vertebrates, learned associations develop between the taste of a food and a subsequent nausea or other negative internal effect, and such a food becomes unacceptable. Characteristic of this type of learning is the relatively long delay between the taste and the visceral effect: often many hours. It is for this reason that food aversion learning has often been considered as a special class of learning (Rozin and Kalat, 1971), although more recently, with more species of animals to compare, there appears to be a continuum from food aversion learning as first described to other types of learned negative associations (MacFarland, 1983).

Learned avoidance of nutritionally deficient food was first described for rats about 50 years ago (Harris et al., 1933), while Garcia et al. (1961) first demonstrated the ability of rats to associate a novel food taste with sickness (caused in various ways) occurring up to several hours later. Learned responses to nutrient deficiency or to specific noxious effects of ingested food are now considered as part of the same spectrum of food aversion learning.

Few studies have dealt with food aversion learning in insects and even fewer adhere to experimental designs that allow an unequivocal interpretation of learning. It is clear, however, that food preferences alter with experience and that, on occasion, acceptability of a food markedly declines with experience of it. Such declines in acceptability of a food probably often reflect aversion learning. This review deals with theoretical aspects of aversion learning, the limited studies carried out so far, the variety of possible functions, and the outstanding questions that need most attention. Reference is made to studies with animals other than insects that have been important in the progress made on the subject.

Where Learned Aversions Are Expected

Because there is often a time delay between sensory patterns associated with food intake and the negative consequences of ingestion, it is to be expected that certain patterns of feeding will enhance the likelihood of aversion learning. For example, discrete meals on single food items will allow associations to be made more readily than grazing on a mixture of foods within a meal (Zahorik and Houpt, 1981). Generalist predators such as mantids and carabids are examples of insects that take discrete meals on particular items with long intermeal gaps, while ground-dwelling scavengers such as crickets are species that probably feed on many miscellaneous items over a short period of time. Extended observations of individual grasshoppers in the field indicate that both types of foraging behavior occur even within species. Thus in *Taeniopoda eques* and *Dactylotum variegatus* most meals are small and on small items while a few large meals occur on specific, apparently preferred plants (Lee, 1991; Raubenheimer and Bernays, 1992).

Another feeding habit that may allow learned aversions to form is that of short-term fidelity to a particular resource such that a learned aversion can develop over a series of meals on a single food type, whereupon rejection and movement away may follow if the food is unsuitable. Species that tend to rest on or near their food, as do most plant-feeding insects, are in this category. It may be that many such insects have extensive abilities to learn associations between food characteristics and unsuitability, but establishing this with certainty requires long-term continuous observations.

Inability to move readily from one food resource to another is a constraint in many insect species, and in others, distances between potential alternative foods prohibit any useful learning occurring. For example, many holometabolous larvae have no alternative but to remain on or in their food source and aversion learning has no relevance for them. This is clearly the case with many fly larvae—for example, leaf miners or carcass dwellers; and for many beetle larvae—for example, wood-boring species. In an extreme case, certain homopterans, such as scale insects, are totally immobile and must feed in the position first selected.

It has been suggested that generalists may be more likely than specialists to learn about food qualities and to develop aversions (McFarland, 1983; Daly et al., 1982). The rat, as an example of an extreme generalist, has been shown to be very adept at learning to avoid toxic or deficient foods, and it follows that if such an animal is exploratory and versatile, and tests many foods, then learning which ones to avoid would be an important part of its food selection process. Gelperin and Forsythe (1975) suggest that similarly, among plant-feeding species, polyphagous species would learn more readily than oligophagous or monophagous ones. In other words, species which are hardwired to accept a narrow range of plants and reject all others are less likely to be able to learn to avoid plants through experience.

The experiments of Dethier (1980) and Dethier and Yost (1979), described below, indicate that this may be true in certain caterpillars, but it could be that generalizations concerning host range are less widely justified among insects. While extreme specialists may never ingest food that is toxic to them, some at least move from low- to high-quality plants. For example, Wang (1990) showed that creosote bush grasshoppers *Ligurotettix coquilletti* moved away from bushes they started to develop on and accumulated on bushes known to be of higher quality for development. Similarly, Parker (1984) found that the grasshopper *Hesperotettix viridis* had a shorter tenure time on damaged host plants than on undamaged ones.

At the other end of the spectrum, individuals of many polyphagous species do not individually mix their diets or select any other food than that upon which they hatched. Thus the situation is not really comparable to the case of vertebrates or molluscs which served as the original models for the idea. Among caterpillar species the most common effect of experience reported in the literature is a preference for the food experienced, even if it is unsuitable or the species is polyphagous (Jermy, 1987).

Larger size and greater mobility create a fine-grained environment in which the encounter rate of different substrates is relatively high. Many large insect herbivores including lepidopterous larvae are rather sedentary, however, and it may be that mobility is the most important factor determining the likelihood of aversion learning among insects in nature. If food aversion learning is considered simply part of a general ability to show learned avoidance responses to anything that has a negative influence, there is a good chance that the potential for such learning is widespread or even universal among insects, but that only some species in certain circumstances normally exhibit it.

Experiments on Aversion/Avoidance Learning

Adverse Effects After Feeding and the Evidence of Aversion Learning

Predators

Berenbaum and Militczky (1984) found that mantids refused milkweed bugs after experiencing them and it is likely that the behavior was a learned aversion to them. Since such predators have discrete meals on a prey item, often with a long gap following, they are excellent candidates for a more thorough study. However, no other published work has been found on predators.

A strong suggestion of learned aversions has been found with the vespid wasp *Mischocyttarus flavitarsus*. Experiments were carried out with wasps that nested in various protected sites in a greenhouse in Berkeley. The tests were designed to examine the relative vulnerability of generalist and specialist herbivores to predation by this wasp (Bernays, 1988), but during the course of the work, several brightly colored caterpillar species were used that the wasps often refused. In two cases detailed records were kept on the frequency of aversive responses over time (Bernays, unpublished). For example, with the grape leaf skeletonizer, *Harrisina brillians*, available in the presence of the salt marsh caterpillar, *Estigmene acrea*, fewer and fewer encounters were made over time, after the initial contacts and bites made when the wasps were naive (Table 1.1). *H. brillians* is highly aposematic and avoided by vertebrate predators and apparently is also unacceptable to wasps. Similar patterns appeared to be occurring with five other species of caterpillars that are normally characterized as aposematic, suggesting that such predators may have added to the selective pressure of vertebrates in the evolution of aposematism.

While aversion learning was not specifically tested in this series of experiments, the data do indicate a change following experience, and the situation deserves a thorough study of the role of learning.

Table 1.1. Changes in encounter rate of predatory wasps with caterpillars of the aposematic *Harrisina brillians* and of the brown *Estigmeme acrea*

Successive 30-minute intervals	Encounters as a % of insects present	
	H. brillians	*E. acrea*
1	21	8
2	25	18
3	17	17
4	4	15
5	0	17

Note: Wasps were naive with respect to both species at the start of the experiment and had equal numbers of similar-sized caterpillars to forage upon.

Herbivores

Dethier (1980) provided the first information on aversion learning in phytophagous insects. Using two polyphagous caterpillars, *Diacrisia virginica* and *Estigmene congrue*, he found that *Petunia hybrida* was the most acceptable in a three-way test with two other plant species. However, if caterpillars had only *Petunia* for 24 hours most individuals became ill: they regurgitated extensively and had convulsive movements. After the illness, the insects were left for a period without food and then given a choice of the same three foods including *Petunia*. In both insect species, relative amounts of *Petunia* eaten were significantly less than in the first test. This demonstrates a preference change that may well result from food aversion learning, although acceptability of alternative food may have increased (as opposed to decreased acceptability of *Petunia*), and there is a possibility of a change with age of the caterpillar. In any case, there was a sharp contrast with the results obtained when the test insect was the oligophagous *Manduca sexta* (Dethier and Yost, 1979). In this species, *Petunia* caused severe symptoms of toxicosis, but insects still preferred the *Petunia* afterward. Dethier comments on the fact that the polyphagous species tend to be relatively active foragers, the two features "theoretically placing a premium on possession of a capacity for food-aversion learning."

In a later study, Dethier (1988) demonstrated that when the polyphagous caterpillar *Diacrisia virginica* was allowed to feed in a field environment that was experimentally altered by adding *Petunia* plants, 24 hours of confinement on *Petunia* had no significant effect on subsequent distribution of insects on *Petunia* plants put into the plots. However, 48 hours of confinement on *Petunia* appeared to greatly reduce subsequent occurrence on this plant in the field. The implication is that there has been food aversion learning, albeit after a most extreme forced confinement. Dethier concludes that such aversions may serve as a backup system for innate sensory systems that normally prevent feeding on toxic plants.

It is still too soon, however, to tell whether or not aversion learning is common in lepidopterous larvae. Raffa (1987) found little evidence for it in the noctuid *Spodoptera frugiperda*. In his extensive trials with 15 toxins, two compounds caused a reduction in feeding but this could have been due to a direct effect on motivational state (i.e., general readiness to feed).

There are occasional reports in the literature suggesting that lepidopterous larvae may change their preferences and there is a possibility that aversion learning is involved. Wasserman (1982), for example, showed that larvae of the gypsy moth, *Lymantria dispar*, ranked *Prunus serotina* consistently low in preference tests after rearing on this plant, while insects reared on other plants ranked it relatively high. It should be emphasized that in examining the effects of feeding experience in caterpillars, induced preferences have been repeatedly demonstrated (Jermy, 1987) while induced aversions have not. One explanation could be that many plants are deterrent but not toxic and the initial unacceptability of such plants is overcome rather readily. Recently, however, it has been demonstrated that in caterpillars of *Heliothis virescens*, the induced preference gives way to a preference for novel foods over time if the plant under test is one that does not support moderately good development (D. Champagne, unpublished). This may suggest an aversion learning.

Recent experiments with the polyphagous grasshopper, *Schistocerca americana*, have clearly demonstrated the ability of these insects to associate toxic effects with the tastes (or perhaps other qualities) of particular foods. In the first study (Bernays and Lee, 1988), individuals were given leaves of spinach to eat and after a meal were injected with nicotine hydrogen tartrate (NHT) at a dose just below the level that elicits twitching symptoms. Individuals rejected spinach or ate little of it, but accepted other novel foods (Table 1.2). Uninjected control insects, or controls injected with water only, showed significantly less tendency to reject spinach. In the second study, Lee and Bernays (1990) demonstrated a similar learned aversion with injections of lithium chloride, coumarin, quinine, and ouabain. The controls here also demonstrated that the reduced consumption was not due to direct nonassociative effects of injected compounds on acceptability of the food. The learned association of the food with the effects of the injection was retained for between 2 and 4 days. Insects learned to associate the novel foods spinach and onion with aversive stimuli, but did not learn to eat less blackberry or broccoli leaves, which are both very palatable foods. As discussed by Bernays and Lee, the differences between foods may be due to differences in relative novelty or relative initial acceptability—both of which are important in aversion learning in

Table 1.2. Successive meal lengths on different foods in a test of aversion learning in *Schistocerca americana*

Experiment	Meal 2 food	Meal length (minutes)	
		Meal 2 broccoli or spinach	Meal 3 spinach[a]
NHT	Spinach	4.0	0.7a
	Broccoli	4.6	2.9b
Water	Spinach	4.2	1.4c
	Broccoli	5.8	3.7b
Untouched controls	Spinach	3.4	1.5c
	Broccoli	5.3	3.4b

Following a meal of wheat, individuals were allowed to take a meal of either broccoli or spinach and then were injected with nicotine hydrogen tartrate or water. The meal following this was spinach and if individuals had learned to associate the taste of spinach with the effect of the injection, it should have been eaten significantly less when they had received the NHT injection following a spinach meal, compared with all other treatments (after Bernays and Lee, 1988).

[a]Numbers followed by different letters indicate meal lengths that are significantly different from one another at $p < 0.01$ (Mann-Whitney U-tests).

vertebrates (Etscorn, 1973; Kalat, 1974; Gustavson, 1977; Zahorik and Houpt, 1981; Logue, 1985).

Clearly, at least one species of grasshopper can be shown to associate sensory qualities of foods (probably tastes or odors) with toxicity in the laboratory. However, there is little evidence to date on the occurrence of the phenomenon in nature. Furthermore, several studies indicate that grasshoppers are orders of magnitude more sensitive to secondary compounds in plants at the level of food acceptance, compared to the post-ingestional physiological level (Cottee et al., 1988; Bernays, 1990; Chapman et al., 1988), suggesting that learned aversions to toxins may be relatively uncommon and unnecessary in practice.

Aversions to nutritionally deficient foods are also learned. Lee and Bernays (1988) demonstrated a learned aversion to spinach, *Spinacea oleracea*, in the polyphagous grasshopper, *Schistocerca americana*. The aversion has been demonstrated to be due to a nutritional deficiency rather than any direct toxic effect (Champagne and Bernays, 1991). The sterol profile in spinach consists entirely of compounds unusable by grasshoppers, and since like all insects they must obtain sterols in the diet, spinach is nutritionally deficient for them. Successive meals on spinach in *S. americana* become progressively smaller until it is rejected after simply biting (Lee and Bernays, 1988). If the normal animal sterol, cholesterol, or the common and usable plant sterol, sitosterol, is added to the leaves of spinach the aversion no longer forms (Fig. 1.1). Champagne and Bernays showed that the sterols are not actually tasted, and the sterol-derived insects showed no ability to distinguish food with or food without cholesterol or sitosterol. On the other hand, by 10 minutes after initiation of feeding on a substrate rich in an appropriate sterol, such individuals maintained a fidelity for that substrate and ate for inordinately long periods on it. It would appear that they learned a positive association as a result of satisfying the deficit.

The elegant experiments of Delaney and Gelperin (1986) with slugs showed unequivocally that an aversion developed to a food with one essential amino acid missing. The behavior was reversed if the amino acid was injected. Frain and Newell (1982) also showed that novel foods that were initially eaten by slugs declined in acceptability over time. These experiments suggest that further investigations with insects might demonstrate the widespread occurrence of aversions to nutritionally deficient foods.

In 1971, Rozin and Kalat reviewed the role of aversion learning in rats. They showed that the "specific hungers" for missing nutrients in the diet could be explained by aversions to the deficient food. That is, learned aversion to a particular food made alternates relatively more palatable. It seems highly likely that the "self-selection" demonstrated by insects, whereby individuals choose a mixture of artificial diets, each deficient in a single

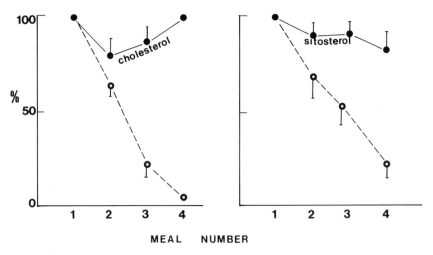

MEAL NUMBER

Figure 1. Aversion learning by nymphs of *Schistocerca americana* on spinach (broken lines—successive meals become progressively shorter), and the lack of such learning when utilizable sterols are added to the spinach (solid lines). Meal lengths are given as percentages of meal 1 in each experiment (after Champagne and Bernays, 1991).

nutrient, is brought about by successive aversion learning to each diet. The ability to "self-select" has been demonstrated in both grasshoppers and caterpillars (Cohen et al., 1987; Schiff et al., 1989; Simpson et al., 1988; Simpson and Simpson, 1990).

In many experiments in which more than one food is made available, such as those on self-selection, or specific hungers mentioned above, it is difficult to distinguish between a decrease in acceptability of an unsuitable food and a short-term increase in acceptability of any novel food. They may be two parts of the same phenomenon as suggested by Bernays and Raubenheimer (1991), or the increase in acceptability of a novel food may simply reflect an increasing readiness to feed ("motivation") due to reduced acceptability of the experienced food resulting in a longer interfeed period. In any case Bernays and Raubenheimer showed with the grasshopper *Schistocerca americana* that following feeding for 4 hours on an inadequate food with a specific flavor, a novel flavor led to rapid acceptance of a nutritionally identical food, and relatively large meals on it. Using two deficient but complementary artificial foods, Bernays and Bright (1991) found that more mixing occurred when unique flavors were added to the two foods. Both of these experiments demonstrate the importance of taste in the aversion learning/neophilia phenomenon.

Learned Avoidance of Food Without Feeding

Blaney and his co-workers (Blaney and Simmonds, 1985, 1987; Blaney et al., 1985) have demonstrated in both grasshoppers and caterpillars that individuals presented with a relatively unfavorable plant will bite and reject the plant but that on subsequent contacts rejection tends to occur after palpation only. On successive contacts a greater proportion of individuals reject at palpation (i.e., before biting) (Fig. 1.2). They presumed that insects learn to associate the superficial taste or smell with the internal constituents of the plant. They demonstrated that the phenomenon was not simply a depression of feeding-related activity, since a second unfavorable plant presented after the initial learning on the first was bitten upon and then in its turn rejected earlier in the behavioral hierarchy. Although technically this process could be sensitization to the particular chemical profile of the plant under test, rather than associative learning, the experience does nonetheless influence the insects in causing avoidance reactions. Recently, Chapman (personal communication) has demonstrated that individual grasshoppers may feed on a plant on first encounter with it, then reject at biting on the next encounter, and later at palpation. There was also an indication that this was followed by total avoidance, with movement away even before an encounter through odor or visual

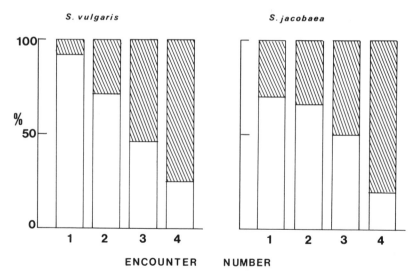

Figure 2. Avoidance responses to foods by *Locusta migratoria* become more extreme as a result of experience. Two species of *Senecio* are unacceptable to all insects, but rejection initially occurs mostly after biting (open bars), while on successive encounters rejection is more likely to occur after palpation only (hatched bars) (after Blaney and Simmonds, 1985).

association. The results of Chapman suggest a continuum between the aversion learning on the deficient spinach demonstrated by Lee and Bernays (1988) and the avoidance resulting from experience without feeding reported by Blaney. There is the added complication, however, that an initial deterrence increases over time for reasons of sensitization—perhaps a separate or even an additional mechanism.

Learned Avoidance in General—the Continuum

It is possible that the variety of learned avoidance reactions demonstrated in insects are of a similar basic nature at the physiological level. Numerous species have been shown to learn to avoid places, positions, visual stimuli, or odors when they are associated with an electric shock (e.g., Alloway, 1972; Tully and Quinn, 1985; Punzo, 1985; Gelperin, 1987; McGuire et al., 1990). There has been considerable work on the ability of an isolated insect leg and associated ganglion to learn to avoid positions associated with an electric shock (reviewed by Eisenstein and Reep, 1985), while in molluscs, a learned suppression of the feeding motor program was demonstrated by an isolated lip-brain preparation (Culligan and Gelperin, 1983). In a mollusc, avoidance of other "dangers" can also be conditioned (Walters et al., 1981). Evidence is accumulating that the variety of different experiments on avoidance/aversion learning may be different more in the nature of the experiment than in the type of learning involved. Similarly, there is evidence that even learning types as apparently distinct as habituation, sensitization, and associative learning have much in common (Duerr and Quinn, 1982). In addition, avoidance learning by fly larvae is controlled by the same genes as avoidance learning in adult flies (Aceves-Pina and Quinn, 1979), perhaps again suggesting some common factor in learning ability.

In experiments designed to demonstrate habituation in insects Jermy et al. (1982) showed that the locust *Schistocerca gregaria* became daily more tolerant of nicotine hydrogen tartrate and had significantly habituated even on the second encounter, when the nicotine was presented on sorghum leaves. On the other hand, when it was presented on sucrose-impregnated filter papers the nicotine became progressively more deterrent. One interpretation is that nicotine became positively associated with a good food and negatively associated with an inadequate food, and that here again, there is a possible gradation from one situation to another. Even in learned preferences by ovipositing females for particular host plants (see Papaj and Prokopy, 1989, for references) there may be a combination of learning positive associations with the preferred plants and negative associations with the nonpreferred ones. There seems to be progressively less reason to consider many of the learning processes, including aversion learning, as uniquely different from one another.

Significance of Aversion Learning

In a general sense, aversion learning can only be seen as a useful ability. If it allows avoidance of toxins that are not normally found to be deterrent, there can be no question about value. There is not yet enough data to determine whether it is a common ability in insects. It is also difficult to determine whether, in species where it can be demonstrated in the laboratory, it is likely to occur naturally. Relevant questions relating to the importance of aversion learning are

1. Receptor sensitivity (as measured by behavioral deterrence) is often orders of magnitude greater than concentrations needed to elicit postingestional toxic symptoms when these occur (Cottee et al., 1988; Bernays, 1990), and it has been suggested that one might expect phytophagous insects to have evolved a capacity to identify and reject toxic foods before ingestion (Papaj and Prokopy, 1989), thus obviating the need for aversion learning.

2. Laboratory findings of aversion learning may be misleading. Confinement in laboratory conditions may result in feeding on diets that are normally rejected and walked away from in the field.

3. If the memory of aversive stimuli is demonstrably short as was shown by Bernays and Lee (1988), the potential functional significance with respect to avoiding toxins in a habitat where the toxic food may be encountered after forgetting is greatly reduced. In this case, the importance of aversion learning may be more in the arena of self-selection of high-quality nutrient mixtures. Indeed, the need to obtain a mixed diet with a balanced nutritional intake may favor the short-term learning so far observed.

4. In a heterogeneous environment or a dynamically changing one, one would expect advantages in short-term learning, since learned details may result in inappropriate decisions in other habitats or at other times. There is, in addition, no information on how much information can be retained at any one time.

Further investigation of generalist predators that take discrete prey items would be worthwhile. Studies of mobile polyphagous herbivores in at least seminatural situations need further study. The challenge is to determine not just a capacity for aversion learning but also its natural occurrence. The pitfalls are many since, in any natural habitat, insects observed are already experienced, and their rejection of food may reflect both innate and learned responses. It may be that, in spite of the difficulty of studying them, newly hatched insects provide the best subjects for study.

The potential value of aversion learning in relation to dietary deficiency has been demonstrated with grasshoppers. It needs examination in a natural setting although it will be difficult to separate from other potential factors causing declining acceptability of foods. Certainly there are changes in chemoreceptor inputs with experience of different foods (Blaney et al., 1986), and some of these relate directly to recent feeding activity (Simpson and Simpson, 1990). Among grasshoppers, a mixed diet is generally better than any single plant for survival and development (Barnes, 1955; Kauffmann, 1965; MacFarlane and Thorsteinson, 1980). Since no food plant is likely to be optimal for all stages of development, it follows that being able to select on the basis of nutritional need would be advantageous. This could be done by direct nutritional feedbacks influencing perception and selection of foods (Bernays and Simpson, 1990) but could also be done by successive aversion learning on a sequence of plants with different nutrient profiles. It will be difficult to distinguish between these possibilities, and indeed, both may occur.

It has been postulated that learning in general in relation to foraging should be of particular value in an unpredictable environment (McFarland, 1983). With respect to mobile generalist insects such as grasshoppers, the habitat would often be described as spatially unpredictable since the scale of the insects and the variety of plants available may mean that the plant species are all encountered infrequently. However, Stephens (1987) has demonstrated that learning can be a disadvantage if unpredictability is too high. Therefore, in complex vegetation a grasshopper may be better off if it tries out the wide variety of available plants, moves after every feed, and relies on whatever direct nutritional feedbacks may be available when it contacts and chooses subsequent plants. This could potentially serve the same purpose as aversion learning in allowing individuals to select an appropriate suite of plants to maximize the nutrient intake and balance. However, the study of insect foraging (other than by bees) has only just begun and one of the greatest needs is to find out just what individuals do in natural habitats as well as in defined arenas with foods of known quality.

Does aversion learning have equal value in habitats of different vegetational complexity? Recent field studies on the rainbow grasshopper *Dactylotum variegatum* (Lee, 1991) indicate that in the complex habitat of this insect, aversion learning will be difficult to detect even if it occurs. This is because individuals continuously move forward, contact numerous different plant species in a short period, and feed for very short times on the majority of them. In a more extensive study involving continuous all-day observations of the polyphagous grasshopper *Taeniopoda eques* (Raubenheimer and Bernays, 1992), it was found that individuals fed on up to 30 different food items in a day and that individual meals usually consisted of bouts of feeding on several different plant species. Under these conditions it is difficult to envisage an important role for learning of any kind, since it would be impossible to associate any taste with feedbacks related to any single food item. Although successive feeding bouts on a single plant species significantly declined, and bouts on newly encountered species were relatively long, these data could be generated by other processes: for example, sensitization to deterrents, or a decision rule to switch food types after a period of feeding. Recent work suggests that in this species the latter processes are most likely (Bernays et al., 1992). However, it will be very important to examine and compare the behavior of other foraging insect species that typically inhabit areas with low plant species diversity or that do not indulge in the excessive locomotor activity found in the only two grasshopper species studied in the field in detail so far.

Since caterpillars generally prefer the plant initially fed upon they may be less rewarding subjects for the study of aversion learning, but cases of diet switching occur in the literature that may be worth further investigation. Are mobile caterpillar species, such as those worked upon by Dethier (1988), the only ones to demonstrate learned aversions to toxins? Is short-term learning in relation to nutrients occurring on a small scale (within plant), or is the self-selection described for caterpillars dependent on other processes (Waldbauer et al., 1984; Simpson et al., 1988)?

Acknowledgments

Part of this work was funded by NSF grant BSR 8705014. Discussions with Reg Chapman resulted in a greatly improved manuscript.

References

Aceves-Pina, E.O., and Quinn, W.G. 1979. Learning in normal and mutant *Drosophila* larvae. Science **206**:93–96.

Alloway, T.M. 1972. Learning and memory in insects. Annu. Rev. Entomol. **17**:43–56.

Barnes, O.L. 1955. Effect of food plants on the lesser migratory grasshopper. J. Econ. Entomol. **48**:119–124.

Berenbaum, M.R., and Militczky, E. 1984. Mantids and milkweed bugs: efficacy of aposematic coloration against invertebrate predators. Am. Midl. Nat. **111**:64–68.

Bernays, E.A. 1988. Host specificity in phytophagous insects: selection pressure from generalist predators. Entomol. Exp. Appl. **49**:131–140.

Bernays, E.A. 1990. Plant secondary compounds deterrent but not toxic to the grass specialist *Locusta migratoria*: implications for the evolution of graminivory. Entomol. Exp. Appl. **54**:53–56.

Bernays, E.A., and Bright, K. 1991. Dietary mixing in grasshoppers: switching induced by nutritional imbalances in foods. Entomol. Exp. Appl. **61**:247–254.

Bernays, E.A., Bright, K.L., Howard, J.J., and Champagne, D. 1992. Variety is the spice of life: the basis of dietary mixing in a polyphagous grasshopper. Anim. Behav., in press.

Bernays, E.A., and Lee, J.C. 1988. Food aversion learning in the polyphagous grasshopper *Schistocerca americana*. Physiol. Entomol. 13:131–137.

Bernays, E.A., and Raubenheimer, D. 1991. Dietary mixing in grasshoppers: changes in acceptability of different plant secondary compounds associated with low levels of dietary protein. J. Insect. Behav. 4:545–556.

Bernays, E.A., and Simpson, S.J. 1990. Nutrition. In R.F. Chapman and A. Joern (eds.), Biology of Grasshoppers. Wiley, New York, pp. 105–128.

Blaney, W.M., Schoonhoven, L.M., and Simmonds, M.S.J. 1986. Sensitivity variations in insect chemoreceptors: a review. Experientia **42**:13–19.

Blaney, W.M., and Simmonds, M.S.J. 1985. Food selection by locusts: the role of learning in rejection behaviour. Entomol. Exp. Appl. **39**:273–278.

Blaney, W.M., and Simmonds, M.S.J. 1987. Experience: a modifier of neural and behavioral sensitivity. In V. Labeyrie, G. Fabres, and D. Lachaise (eds), Insects–Plants: Proc. 6th Int. Symp. Insect–Plant Relationships. Junk, Dordrecht, pp. 237–241.

Blaney, W.M., Winstanley. C., and Simmonds, M.S.J. 1985. Food selection by locusts: an analysis of rejection behaviour. Entomol. Exp. Appl. **38**:35–40.

Champagne, D.E., and Bernays, E.A. 1991. Phytosterol unsuitability as a factor mediating food aversion learning in the grasshopper *Schistocerca americana*. Physiol. Entomol. **16**:391–400.

Chapman, R.F., Bernays, E.A., and Wyatt, T. 1988. Chemical aspects of host plant specificity in three *Larrea* feeding grasshoppers. J. Chem. Ecol. **14**:561–579.

Cohen, R.W., Waldbauer, G.P., Friedman, S., and Schiff, N.M. 1987. Nutrient self-selection by *Heliothis zea* larvae: a time lapse study. Entomol. Exp. Appl. 44:65–74.

Cottee, P.K., Bernays, E.A., and Mordue, A.J. 1988. Comparisons of deterrency and toxicity of selected secondary plant compounds to an oligophagous and a polyphagous acridid. Entomol. Exp. Appl. 46:241–247.

Culligan, N., and Gelperin, A. 1983. One-trial associative learning by an isolated molluscan CNS: use of different chemoreceptors for training and testing. Brain Res. **266**:319–327.

Daly, M., Rauschenberger, J., and Behrends, P. 1982. Food aversion learning in kangaroo rats: a specialist-generalist comparison. Anim. Learn. Behav. **10**:314–320.

Delaney, K., and Gelperin, A. 1986. Post-ingestive food-aversion learning to amino acid deficient diets by the terrestrial slug *Limax maximus*. J. Comp. Physiol. A. **159**:281–295.

Dethier, V.G. 1980. Food aversion learning in two polyphagous caterpillars, *Diacrisia virginica* and *Estigmene congrua*. Physiol. Entomol. **5**:321–325.

Dethier, V.G. 1988. Induction and aversion-learning in polyphagous caterpillars, arctiid larvae (Lepidoptera) in an ecological setting. Can. Entomol. **120**:125–131.

Dethier, V.G., and Yost, M.T. 1979. Oligophagy and the absence of food-aversion learning in tobacco hornworms, *Manduca sexta*. Physiol. Entomol. **4**:125–130.

Duerr, J.S., and Quinn, W.G. 1982. Three *Drosophila* mutations that block associative learning also affect habituation and sensitization. Proc. Natl. Acad. Sci. USA **79**:3646–3650.

Eisenstein, E.M., and Reep, R.L. 1985. Behavioral and cellular studies of learning and memory in insects. In G.A. Kerkut and L.I. Gilbert (eds.), Comprehensive Insect Physiology, Biochemistry and Pharmacology, Vol 9. Pergamon, New York, pp. 513–547.

Etscorn, F. 1973. Effects of a preferred vs a non-preferred CS in the establishment of a taste aversion. Physiol. Psychol. **1**:5–6.

Frain, J.M., and Newell, P.F. 1982. Meal size and a feeding assay for *Deroceras reticulatum* (Mull.). J. Mollusc Stud. **48**:98–99.

Garcia, J., Kimmeldorf, D.J., and Hunt, E.L. 1961. The use of ionizing radiation as a motivating stimulus. Psychol. Rev. **68**:383–385.

Gelperin, A. 1983. Neuroethological studies of associative learning in feeding control systems. In F. Huber and H. Markl (eds.), Neuroethology and Behavioral Psychology. Springer, New York, pp. 189–205.

Gelperin, A. 1987. Plasticity in control systems for insect feeding. In R.F. Chapman, E.A. Bernays, and J.G. Stoffolano (eds.), Perspectives in Chemoreception and Behavior. Springer-Verlag, New York, pp. 33–46.

Gelperin, A., and Forsythe, D. 1975. Neuroethological studies of learning of mollusks. In J.C. Fentress (ed.), Simpler Networks and Behavior. Sinauer, New York, pp. 239–250.

Gustavson, C.R. 1977. Comparative and field aspects of learned food aversions. In L.M. Barker, M.R. Best, and M. Domjan (eds.), Learning Mechanisms in Food Selection. Baylor University Press, Waco, TX, pp. 23–43.

Harris, L.F., Clay, J., Hargreaves, F., and Ward, A. 1922. Appetite and choice of diet. The ability of the vitamin B deficient rat to discriminate between diets containing and lacking the vitamin. Proc. R. Soc. Lond. B. **113**:161–190.

Jermy, T. 1987. The role of experience in the host selection of phytophagous insects. In R.F. Chapman, E.A. Bernays, and J.G. Stoffolano (eds.), Perspectives in Chemoreception and Behavior. Springer-Verlag, New York, pp. 143–158.

Jermy, T., Bernays, E.A., and Szentesi, A. 1982. The effect of repeated exposure to feeding deterrents on their acceptability to phytophagous insects. In J.H.

Visser and A.K. Minks (eds.), Proc. 5th Int. Symp. Insect–Plant Relationships. Pudoc, Wageningen, pp. 25–32.

Kalat, J.W. 1974. Taste salience depends on novelty not concentration in taste aversion learning in the rat. J. Comp. Physiol. Psychol. **86**:47–50.

Kauffmann, T. 1965. Biological studies on some Bavarian acridoidea (Orthoptera), with special reference to their feeding habits. Ann. Entomol. Soc. Am. **58**:791–801.

Lee, J.C. 1991. Basis of polyphagy in generalist insect herbivores. Ph.D. Thesis, University of California, Berkeley.

Lee, J.C., and Bernays, E.A. 1988. Declining acceptability of a food plant for the polyphagous grasshopper *Schistocerca americana*: the role of food aversion learning. Physiol. Entomol. **13**:291–301.

Lee, J.C., and Bernays, E.A. 1990. Food tastes and toxic effects: associative learning by the polyphagous grasshopper *Schistocerca americana* (Drury) (Orthoptera: Acrididae). Anim. Behav. **39**:163–173.

Logue, A.W. 1985. Conditioned food aversion learning in humans. Ann. NY Acad. Sci. **443**:316–329.

MacFarlane, J.H., and Thorsteinson, A.J. 1980. Development and survival of the twostriped grasshopper, *Melanoplus bivittatus* (Say) (Orthoptera: Acrididae) on various single and mutiple plant diets. Acrida **9**:63–76.

McFarland, D. 1983. Animal Behavior. Benjamin Cummings, Menlo Park, CA, 576 pp.

McGuire, T.R., Tully, T., and Gelperin, A. 1990. Conditioning odor-shock associations in the black blowfly *Phormia regina*. J. Insect. Behav. **3**:49–60.

Papaj, D.R., and Prokopy, R.J. 1989. Ecological and evolutionary aspects of learning in phytophagous insects. Annu. Rev. Entomol. **34**:315–350.

Parker, M.A. 1984. Local food depletion and the foraging behavior of a specialist grasshopper, *Hesperotettix viridis*. Ecology **65**:824–835.

Punzo, F. 1985. Recent advances in behavioral plasticity in insects and decapod crustaceans. Flor. Entomol. **68**:89–104.

Raffa, K.F. 1987. Maintenance of innate preferences by a polyphagous insect despite ingestion of applied deleterious chemicals. Entomol. Exp. Appl. **44**:221–227.

Raubenheimer, D., and Bernays, E.A. 1992. Patterns of feeding in the polyphagous grasshopper *Taeniopoda eques*: a field study. Anim. Behav., in press.

Rozin, P., and Kalat, J.W. 1971. Specific hungers and poison avoidance as adaptive specializations of learning. Psychol. Rev. **78**:459–486.

Schiff, N.M., Waldbauer, G.P., and Friedman, S. 1989. Dietary self-selection by *Heliothis zea* larvae: roles of metabolic feedback and chemosensory stimuli. Entomol. Exp. Appl. **52**:261–270.

Simpson, S.J., Simmonds, M.S.J., and Blaney, W.M. 1988. A comparison of dietary selection behaviour in larval *Locusta migratoria* and *Spodoptera littoralis*. Physiol. Entomol. **13**:225–238.

Simpson, S.J., and Simpson, C.L. 1990. The mechanisms of nutritional compensation by phytophagous insects. In E.A. Bernays (ed.), Insect–Plant Interactions, Vol II. CRC Press, Boca Raton, FL, pp. 111–160.

Stephens, D.W. 1987. On economically tracking a variable environment. Theor. Popul. Biol. **32**:15–25.

Tully, T., and Quinn, W.G. 1985. Classical conditioning and retention in normal and mutant *Drosophila*. J. Comp. Physiol. A **157**:263–277.

Waldbauer, G.P., Cohen, R.W., and Friedman, S. 1984. Self-selection of an optimal nutrient mix from defined diets by larvae of the corn earworm, *Heliothis zea*. Physiol. Zool. **57**:590–597.

Walters, E.T., Carew, T.J., and Kandel, E.R. 1981. Behavioural and neuronal evidence for conditioned fear in *Aplysia*. In J. Salanki (ed.), Neurobiology of Invertebrates: Advances in Physiological Science, Vol. 23, pp. 295–303.

Wang, G.Y. 1990. Dominance in territorial grasshoppers: studies of causation and development. Ph.D. thesis, University of California, Los Angeles, 139 pp.

Wasserman, S.S. 1982. Gypsy moth (*Lymantria dispar*): induced feeding preferences as a bioassay for phenetic similarity among host plants. In J.H. Visser and A.K. Minks (eds.), Proc. 5th Int. Symp. on Insect–Plant Relationships. Pudoc, Wageningen, pp. 261–268.

Zahorik, D.M., and Houpt, K.A. 1981. Species differences in feeding strategies, food hazards, and the ability to learn aversions. In A. Kamil and T. Sargent (eds.), Foraging Behavior: Ecological, Ethological and Psychological Approaches. Garland Press, New York, pp. 289–310.

2

Ethological and Comparative Perspectives on Honey Bee Learning

James L. Gould

The Ethology and Psychology of Learning

Ethologists are concerned with the mechanisms and evolution of behavior. They presuppose that natural selection will have acted as much on behavior as on morphology and physiology (Darwin, 1872). In consequence, some aspects of behavior are likely to be species specific, "tuned" to the contingencies of an animal's niche. Traditional behavioristic psychologists, on the other hand, focus more narrowly on learning and emphasize species-independent behavior in their search for a general-process theory of learning. I have argued that the terminology and results from each perspective are complementary and are useful together in analyzing invertebrate learning. Studies of vertebrates can help illuminate learning in insects, and vice versa (Gould, 1986a).

Two of the best-known concepts of traditional ethology are sign stimuli and fixed-action patterns (reviewed in Gould, 1982). A sign stimulus is a key, innately recognized feature of an object or individual that helps trigger or guide a (sometimes covert) response. The thin, vertically oriented, horizontally moving bill of parent laughing gulls, for instance, is a sign stimulus, as is the red spot at the tip (Tinbergen and Perdeck, 1950; Alessandro et al., 1989). These stimuli trigger and guide both the overt response of pecking by chicks and learning of the parents' distinctive features. Pecking itself is a fixed-action pattern: an innate, coördinated, steretoyped series of muscle movements.

The equivalent terms in behavioristic psychology come from the standard description of classical conditioning: an unconditioned stimulus (US) triggers an unconditioned response (UR); pairing of the US with a conditioning stimulus (CS) leads to an association between the two, and, eventually, the ability of the CS to trigger the UR. This relationship is frequently expressed diagrammatically:

$$US \rightarrow UR; CS + UR \rightarrow UR; \text{then } CS \rightarrow UR$$

These terms can be recast in ethological jargon without much difficulty: "unconditioned," of course, means "innate"; the US is a sign stimulus; the UR is a fixed-action pattern; the CS is what Lorenz (1959) refers to as a learned gestalt—the collection of cues an animal learns to use in identifying a meaningful stimulus. A major philosophical difference between ethology and behavioristic psychology is that ethologists recognized that many (perhaps most) sign stimulus/fixed-action patterns pairs are immune to learning (reviewed in Tinbergen, 1951) whereas behaviorists believed that all were conditionable (Pavlov, 1927; Watson, 1925; for J.B. Watson, in fact, even the circulation of the blood was learned through prenatal conditioning). Behaviorists also believed that all stimuli (CSs) were equally effective for conditioning, but, for example, a gull model lacking a bill will elicit some pecking but cannot be learned by chicks to be a preferred object even when it is exclusively associated with food (Margolis et al., 1986; Alessandro et al., 1989).

The other major form of conditioning recognized by behaviorists is operant conditioning (also called instrumental or trial-and-error learning). Whereas classical conditioning can be thought of as "learning to recognize," operant conditioning is "learning to do": an animal learns and optimizes a novel behavior based on feedback that results from its behavioral experiments (Skinner, 1983). Ethologists refer to the relatively stereotyped behaviors that animals create in this way as learned motor programs. Bird song is a familiar example (reviewed in Marler, 1984); operant learning of how to exploit flower blossoms is clear in insects (Heinrich, 1979; Lewis, 1986; Gould, 1987b).

What Should Animals Learn?

The ethological perspective on learning does not rule out the possible existence of a general-process learning mechanism (indeed, Gould and Marler, 1984, argue that there is good evidence for species-independent patterns and criticize the wholesale abandonment of the general-process perspective advocated by some behaviorists); rather, it supposes that many of the context-specific details of learning will be "customized" to accommodate the necessities and contingencies of a species' natural history. That this point of view has clear predictive power becomes evident when one considers a few common categories of behavior.

Food Acquisition

The ethological perspective leads us to make a distinction between specialist and generalist feeders. A digger wasp that hunts only honey bees, for example, is born able to recognize, capture, and paralyze that species (Tinbergen, 1932, 1935); species with other innate preferences have corresponding behavioral specializations (Fabre, 1921). Food learning is obviously unnecessary for such species; indeed, the complete "hardwiring" of food acquisition allows the adult to begin prompt and efficient harvesting, and learning, it would seem, could lead to delays in rearing offspring (and thus lower reproductive fitness) and even fatal errors in prey choice. In at least some mass-provisioning wasps that provide paralyzed prey, species specificity is important if the larvae are to consume the food source correctly (Fabre, 1921). On the other hand, the narrowness of their dietary niche leaves such species dependent on the abundance, seasonality, and habitat preferences of their prey. The all-too-frequent cool, rainy summers in Scotland, for example, lead to a widely fluctuating honey bee population, unsuitable for a specialist predator; a generalist like the crab spider, on the other hand, can survive by capturing nearly any flower-blossom visitor. It is also likely that there is a limit to what can be innately specified: though it is difficult to find a learned motor task in insects that is obviously more complicated than nest building (which is invariably innate), it is not at all obvious that innate recognition of complex shapes is possible, at least within the limits of the insect brain (Gould, 1982); even among vertebrates, for whom constraints of brain volume are clearly less severe, the evidence for innate shape recognition is equivocable (Gould and Marler, 1984).

Honey bees represent an intermediate case: though bees are specialists in the sense that they harvest flowers, they are generalists in the wide range of blossoms they can exploit. Any innate recognition and harvesting behavior would have to be sufficiently general to allow the use of flowers of nearly every conceivable color, shape, and size. With a relatively wide food niche, learning could obviously be useful in allowing bees to recognize the most rewarding food currently available and to learn how to efficiently handle the blossom so as to maximize the rate of nectar intake. This argument is probably applicable only to highly social bees: solitary bees, for example, must alternate between nest building, searching for new food sources, gathering nectar and water for their own needs, and collecting pollen for brood; with such rapid alternation between tasks and food sources, the optimum degree and organization of learning is probably different. Indeed, since the pollen and nectar of different species offer quantitatively and qualitatively different macro- and micronutrients (Dietz, 1975; Shuel, 1975), high-efficiency learning and flower constancy might constitute a serious disadvantage; honey bees can achieve a relatively "balanced diet" simply by virtue of the large number of independent foragers, and there is no need to switch between source types since recruits automatically favor whatever is in shortest supply (von Frisch, 1967). In short, it may pay solitary insects to forget faster than highly social ones; for semisocial species, the optimum balance should depend on the details of their niche and life history.

For bees, the taste of sugar is the US—the innate releaser of learning. I will review the nature of honey bee food learning in some detail in a later section and present evidence indicating that where possible, innate programming supplies clues that guide searching, and preordained storage systems efficiently filter and organize the learned information—in short, what can be hardwired has been. I will also argue that as a result of having a large forager force, any disadvantages of flower constancy are minimized, and thus the advantages of highly specific flower learning are maximal.

Generalist feeders include omnivous species like humans and rats, for whom an ability to learn and remember what is edible, inedible, and dangerous is essential. For these species we expect the greatest degree of food learning, and the minimum amount of innate guidance. Nevertheless, at least food-avoidance learning is highly programmed (Revusky, 1984) and remains powerful even in our own species (Seawright et al., 1978; Bernstein and Sigmundi, 1980). Examples of food avoidance in invertebrates are known (e.g., Gelperin, 1975; see also Bernays, this volume) but relatively rare. Since the services of bees are critical for pollination, it seems unlikely that toxic nectars could evolve, but cases of flower avoidance based on physical mistreatment by blossoms are known (Reinhardt 1952; Pankiw, 1967). Food-avoidance learning might be a great advantage to generalist hunters like yellow jackets, based on both prey toxicity and ability to fight back.

Some species with generalist potential apparently become specialists by imprinting (learning rapidly during an early experience and becoming committed to the learned stimulus) on a particular food—snapping turtles, for instance, can become imprinted on what they eat during an early meal (Burghardt and Hess, 1966). Such a system combines some of the advantages of hardwired specialists with those of generalists that depend on learning: the species is not limited to a single food type and so enjoys a wider food niche; on the other hand, by concentrating on a single food encountered early in life, individuals can minimize the time devoted to trial-and-error learning and enjoy the efficiency which comes with having a single search image to guide hunting. There is much evidence of this kind of dramatic food learning in insects (reviewed in Papaj and Prokopy, 1989).

Beyond the question of *whether* a species learns about food, there is the issue of *what* is learned. Any particular encounter with food is associated with a limitless set of other stimuli, only some of which are actually pre-dictive—the temperature, wind direction, humidity, pattern of clouds in the sky, and other readily preceivable cues are of little or no importance to flower foragers, for instance. The conventional behavioristic model posits that animals sort out the irrelevant stimuli through repeated experience (reviewed in Schwartz, 1984); ethologists, on the other hand, assume that if there are reliable stimuli, animals will have evolved responses to them in the first place independent of experience and so learn faster and more reliably. Such biases, long obvious in the field, are evident in generalist-food-niche laboratory animals as well. Rats, for instance, will readily associate odor with a noxious food, but not color or sound; birds like pigeons and quail, on the other hand, quickly learn the color but not an associated sound or odor (e.g., Wilcoxon et al., 1971). These sensory preferences do not stem from any simple asymmetries in receptor sensitivity: when danger avoidance is the context, rats readily learn visual cues but not odors, and pigeons associate sounds with danger faster than colors (Forse and Lo-Lordo, 1973; Revusky, 1984). In most cases the ecological logic of these innate biases is fairly clear: the seeds pigeons regularly eat can rarely be identified by the noises they make, whereas the sound of an approaching predator is probably worth learning; similarly, the nocturnal habits of rats make odor a more reliable natural cue than color (and one they continue to prefer even under full illumination), whereas diurnal birds generally have sufficient light available to make visual discriminations.

The same sort of logical biases are evident in the context of motor learning (operant conditioning). Pigeons can easily be taught to peck for food, but resist learning to treadle-hop; when avoiding shock is the reward, hopping is easily learned whereas pecking is nearly impossible to condition (Bolles, 1984). Similarly, rats readily learn to press a lever for food, but nearly refuse to learn to jump; when evading a shock is the goal, jumping is easily conditioned, but lever-pressing is hard to teach (Bolles, 1984). It seems clear that animals may come provided with response predispositions that bias their attempts to create novel motor behavior to solve problems posed by their habitat, and at least in the case of pigeons and rats, these preconceptions are probably adaptive: food is typically handled with the beak/forepaws, whereas danger is most often dealt with by whole-body movement. Just as with associative biases, animals with sensible innate response tendencies are more likely, in their attempts to find a solution through trial and error, to solve the problem quickly and effectively if they try plausible motor experiments first.

Reproduction and Parental Care

In addition to finding food, an animal must be able to locate, identify, court appropriately, and mate with the best reproductively ready member available of the opposite sex of its own species. Often a nest must be constructed in a suitable location. In species that care for their young, the attending animals must recognize the offspring as such, understand when they are hungry, provide the correct amounts of the right food, and offer it in the appropriate way. The young, for their part, must usually know something about how to indicate hunger and obtain food from the parents. In these contexts, all species are specialists, and we would expect, where possible, the error-prone strategy of learning to play as small a role as possible.

The literature overflows with examples of innate species recognition (reviewed in Gould, 1982) and preordained mechanisms for judging mate quality (reviewed in Gould and Gould, 1989). Nests (vertebrate and invertebrate), cocoons, and webs, as far as can be determined, are all built initially according to innate instructions using innately recognized categories of materials (reviewed in Gould, 1982). Offspring care seems equally preprogrammed, both in mass and progressive provisioners (e.g., Fabre, 1921; Baerends, 1941) as does the begging behavior of the young or colony mates. In some cases involving vertebrates, learning seems to play a role, as practice increases the efficiency of nest building in many species, and the quality of offspring care in others (reviewed in Gould and Marler, 1984). Learning is most obvious in certain species that depend in part on learned recognition of species, parents, or offspring. In the latter two cases the need for individual recognition explains the occurrence of the learning, and its timing. For example, in precocial birds the chicks learn to recognize their parents, and the parents their young, in the first 2 days; in ground-nesting gulls, this occurs about 3–4 days, just as the chicks begin to leave the nest; in arboreal-nesting birds, the learning is delayed until fledging; in birds that do not feed their young after fledging, it never happens (reviewed in Gould and Marler, 1984). There is nothing in the nature of classical conditioning that demands such temporal specializations. Similarly, the pattern of egg learning in birds tracks ecological necessity: most species learn nothing about their eggs, but those that nest in dense colonies and many that are parasitized by cuckoos and other brood parasites do memorize their eggs (reviewed in Gould and Marler, 1984).

Species learning—what is often called sexual imprinting—is also well known, particularly considering how rare it is (Immelmann, 1984). Innately recognized cues are used to trigger learning in a confined social context, where errors are unlikely. Such imprinting in some cases serves to distinguish the adults of the species from those of similar species, and it is striking how specific the cues committed to memory can be—the eyes, for example, in a group of sympatric gulls that differ only in eye and eye-ring color (Smith, 1967). Other species seem to use imprinting as a mechanism for incest avoidance, and again the need for learning in this context is obvious

(Bateson, 1982). But these examples of learning should not distract us from the main pattern: learning appears to be rare where innate guidance can suffice, and where learning does occur, its timing and cue sensitivity seem adaptively restricted. In insects we might expect to encounter colony-level recognition, and there is some suggestion of learned, olfactory-based kin (patriline) recognition in honey bees (Breed, 1983; Getz and Smith, 1983), but other aspects of species recognition and parental care seem so sufficiently predictable that they can be prewired and learning excluded, which would seem to be usually the case (e.g., Fabre, 1921).

A prominent exception to this generalization demonstrates the inflexible nature of learning even when it is needed (Baerends, 19841). At least one species of digger wasp provisions several burrows at once, providing only enough food to maintain the developing larva for a day at a time. Because newly hatched larvae require much less food than older larvae, while those about to pupate need none, a hunting female should tailor her provisioning to the needs of her individual offspring. Baerends found that the wasp makes a morning trip to inspect each burrow and then sets off to hunt. Over the course of the day she provisions each burrow appropriately. The inflexibility of this system came to light when Baerends exchanged larvae between burrows after the morning inspection: the hunting wasp ignored the changes, provisioning each burrow according to the needs of the earlier occupant. The learning is more in the nature of calibration, something like filling in the blanks on a preprinted form; it is not easily interpreted in the terms of conventional classical conditioning, in which each encounter with US and CS contributes to learning or extinction.

Defense

Escape and defense behaviors are crucial facets of an animal's repertoire: individuals need to recognize danger and react appropriately to the threat. When the characteristics of the danger are sufficiently predictable and simple to permit innate recognition, natural selection ought to have favored the use of sign stimuli rather than the slow process of associative learning. The limited resolution of the insect compound eye (about $1.5°$, roughly four times the sun's diameter) makes visually based learning of danger relatively useless: the threat would need to be very close before it could be recognized; on the other hand, the fact that crab spiders are cryptic might argue that they have been selected to avoid the notice of prey, though it could as easily serve to conceal them from their own predators. In honey bees, a rapidly vibrating dark object seems to be interpreted innately as threatening, but there is no evidence that the shape of threat can be learned. The need for olfactory-based learning of danger seems an equally unlikely contingency, and I know of no examples.

With regard to sound, things may be more promising. Some butterfly larvae have an innate response to the sound (air-particle movements) of hunting wasps (Tautz and Markl, 1978), and a number of moths react appropriately to the echolocation sounds (pressure waves) of hunting bats. The only case of clear auditory-based danger learning I know of is the ability of honey bees to associate sound with impending shock (Gould and Towne, 1988); since bees are surrounded by constant buzzing, any differential reaction to hunting wasps would probably have to be based on learning; on the other hand, it is not clear that bees are equipped with the necessary sensory apparatus to allow sufficiently fine tone discrimination that would make such learning useful. Perhaps the context—loud sound at a food source (one of the two favorite hunting spots for bee-killing wasps)—makes detailed frequency discrimination unnecessary.

Orientation and Navigation

Animals with a home (either localized like a nest, or more diffuse, as with a home range) may need to learn in order to navigate. But animals rarely if ever learn *to* navigate; instead, learning serves as calibration of an innate orientation program—that is, they acquire the data necessary to implement a prewired behavior (reviewed in Gould, 1982; Gould and Gould, 1988). Birds and monarch butterflies, for instance, generally require no experience in order to select an appropriate direction for migration, or to know when to make major course alterations, or when to stop their journey. What they do need to do is to learn the direction and degree of deviation between the axis of rotation of celestial cues vs. magnetic north. This calibration occurs during an early sensitive period (reviewed in Gould, 1990a). Nocturnal migrants (as most birds are) also learn the pattern of stars overhead.

The role of learning as calibration is particularly clear in the case of honey bees (reviewed in Gould, 1982; Gould and Gould, 1988). Scout bees keep track of each leg of their searching flights relative to the sun's azimuth. Since the sun appears to move from east to west, bees must allow for this shift in their primary orientation cue; this is all the more difficult because the sun moves at a rate that varies with date, latitude, and time of day. Bees appear to learn the direction of sun movement (left to right, or right to left) before they commence foraging, and the memory seems to be irreversible. Learning about the rate of movement seems to depend on two alternative systems: when large, unambiguous landmarks (at least 4° wide and 6° high) are visible from the hive, bees memorize the sun's course relative to these markers; otherwise they measure the sun's rate of movement over the previous 40 minutes and use that value to extrapolate the change in solar azimuth.

Experienced bees learn the terrain within their flight range, and, if the landmarks are large and unambiguous, may use them in preference to celestial cues. There is even some evidence that bees and wasps (and perhaps even some butterflies—reviewed in Papaj and Prokopy, 1989) can use landmarks to return home or locate familiar food sources after displacement. As in the case of food learning, selection appears to have acted to hardwire the predictable (how to use orientation data in order to navigate) and impelled the creature to learn the unpredictable (the direction and rate of sun movement, the nature and position of landmarks, and so on). And once again the learning is focused on cues that are relevant, so that less useful cues are ignored (like, for instance, the sun's size, shape, color, and elevation; each of these characteristics is studiously ignored; Brines and Gould, 1979).

Social Behavior and Communication

All animals must understand the social signals of their species, but other than learning to recognize individuals in a group or subgroup (rarely necessary or even plausible for insects), social signals are usually innate. For instance, the most elaborate example of social communication in invertebrates—the dance language of honey bees—shows no evidence of learning (reviewed in von Frisch, 1967; Gould and Gould, 1988). Not only are bees able to perform and understand dances when the necessity first arises, rearing bees from subspecies with different dialects together has no effect: each group dances according to its own race's convention and misunderstands dances by members of the other race. What learning does occur during the dance (learning the odor, distance, direction, and quality of the food source) serves simply to allow various decision-making and navigation circuits to operate—that is, which of the many advertised sources offers the best return for distance traveled given the current hive needs, and what direction and how far to fly to get there.

The Evolution of Learning

The general pattern evident in this survey of learning is that animals are usually unable to learn in situations in which the information needed to guide behavior is completely predictable, or too complex to be innately specified. This makes sense because learning is less efficient and certain than reliable innate information—that is, it takes time to learn, mistakes are inevitable as irrelevant correlations and overly broad stimulus generalizations are edited, and so on. What is most dramatic is that the ability to learn can vary between species in direct relation to this degree of certainty. So too, the cue preferences (and, as we shall see, the preferences within a cue modality) now so obvious in learning vary in ways that often suit the likely contingencies an animal will encounter. This innate guidance of learning is especially appropriate for what I have called the "predictably unpredictable" (Gould, 1984): if the information needed is predictable, it can be innately coded; if it is unpredictable, but if the situation can be innately recognized, the useful cues to remember anticipated, and the way in which the information is most effectively stored and used predicted, then innate guidance of learning will be more adapative than the sort of blank-slate learning that early behaviorists championed. Contexts that are unpredictably unpredictable, if any exist, may actually require an utterly naïve kind of learning, or may make learning useless (see Stephens, this volume).

Learning therefore exists to solve problems that innate information cannot satisfactorily deal with. This being the case, it is somewhat misleading to talk in terms of "constraints" on learning. The implication is that learning has *become* constrained, that flexibility has been lost as selection has narrowed an animal's sensitivity (e.g., Terrace, 1984). This evolutionary scenario parallels the history of behavioristic psychology: in the first half of the century, animals were thought by behaviorists to be able to learn essentially anything; by 1965 a few biases were evident; now, innate predispositions in learning seem almost universal, up to and including human language acquisition (Gleitman, 1984).

The important evolutionary question of whether learning was born constrained cannot be answered with certainty. It is possible that animals originally learned everything (which must have kept them pretty busy) and subsequently became more limited in their abilities, but it seems to me more likely that specialized learning was the first step away from innate control.

The issue is somewhat clearer when viewed from a more mechanistic perspective. The circuitry of conditioning has been worked out for one behavior in one invertebrate species—gill withdrawal in *Aplysia* (Hawkins et al., 1983). The basic US $\rightarrow$ UR pathway—a touch on the gill or siphon, followed by contraction of the gill—is well understood, as is the wiring for two nonassociative forms of behavioral plasticity: habituation and sensitization. Habituation, which allows the animal to adjust to the background level of stimulation, occurs in synapses on interneurons along the pathway from sensor to muscle (Fig. 2.1). Sensitization—the ability of certain irrelevant stimuli to "alert" the animal and instantly remove habituation—is mediated by inputs from other sensors onto the synapses that can be habituated. Since habituation and sensitization are seen in many behaviors that are otherwise innate, it is possible (even likely, though by no means certain) that these two mechanisms for fine tuning behavioral responsiveness evolved before associative learning. If this is the case, then they could

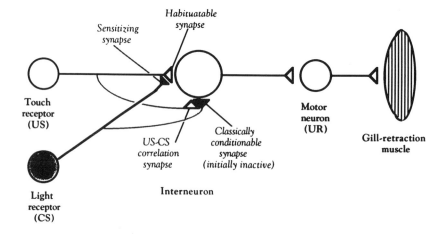

Figure 2.1. The gill-withdrawal circuit in *Aplysia* connects touch-sensitive receptors in the gill and siphon, through interneurons, to the motor neurons that effect withdrawal. Habituation can occur on the synapses onto the interneuron. Inputs from other sensory modalities onto these synapses can sensitize them when certain irrelevant stimuli are encountered. Associative learning occurs on normally inactive synapses from some of the sensitization axons which terminate directly onto the interneuron. These synapses become active when stimulation from a sensitization input (the CS) occurs repeatedly at the same time that the touch-sensitive cells (the US) are being stimulated; branches from the touch-sensor axons terminate on the conditional synapses and cause the conditioning.

have provided the raw material from which associative learning evolved; I make this assumption in the rest of this discussion.

Conditioning *Aplysia* (in this case teaching the animal that a flash of light or a touch on the tail—stimuli that normally do not elicit gill withdrawal— predicts a touch on the gill) occurs on the same interneurons that are involved in habituation and sensitization, using the same inputs. The axons arriving from the sensitization pathway not only synapse on the habituation synapses, they connect directly (but weakly) to the interneurons; inputs from the touch sensors in the siphon and gill not only communicate along the direct line to the muscles, they also send side branches to the direct inputs from the sensitization axons (Fig. 2.1).

When both the US (touch) and CS (light, say) provide signals that arrive essentially simultaneously at the direct inputs from the sensitization pathway, they strengthen that route. Given enough neural associations, the direct pathway from the light receptors becomes strong enough to trigger withdrawal on its own; the CS achieves the status of the US. The number of associations that will be necessary will depend on the number and strength of the preexisting connections, which automatically explains the differential conditioning rates mentioned above. Note also that a major consequence of this arrangement is that an animal using this kind of circuitry can only learn what its wiring allows it to learn—that is, if the conditioning synapses from a potential CS are not already in place, the CS cannot be learned. This fact readily accounts for learning biases. Whether the *Aplysia* system is common to higher invertebrates is not known, but it seems likely since the molecular details, which would seem to depend on the neuroanatomical underpinning, appear to be identical in *Drosophila* (reviewed in Quinn, 1984).

It is possible, therefore, that only after the US → UR pathway evolves, and then habituation, and then selective sensitization, does the potential for evolving associative learning exist, and even then it is restricted to the subset of sensitization cues whose inputs are in place. As a result, learning is, of necessity, initially very selective; a more generalized ability could only evolve later. Of course, as in other evolutionary progressions, it is possible to evolve back to an earlier state: specialized learning could secondarily evolve from generalized learning, just as a few species of bees have secondarily returned to the hunting lifestyle of their wasp ancestors.

Food Foraging and Learning in Honey Bees

Because of their experimental convenience, far more is known about how honey bees learn than about any other species. Most of the evidence comes from the food-foraging cycle and is a result of the honey bee's life-history strategy (reviewed in Gould and Gould, 1988). I will make comparisons with other species (mainly bumble bees and butterflies) where appropriate.

The Honey Bee's Niche

Like most bees, honey bees make their living collecting nectar and pollen from flowers (reviewed in Gould and Gould, 1988). The nectar is used as an energy source, whereas pollen supplies protein. Honey bees are unusual in that they are highly social (living in colonies of 10,000–60,000), perennial (they overwinter as a group), and reproduce by colony fission (swarming) rather than through the production of many individual reproductives. By maintaining their numbers in winter, thermoregulating the hive temperature, and beginning brood rearing in late winter in anticipation of the need for workers in spring, honey bees are able to exploit the highly productive flowers of springtime with a minimum of competition from other species. (By contrast, honey bee colonies often operate at a slight loss—consuming more honey than they make—in the fall, when there is both intensive competition from the growing population of wasps, butterflies, and solitary and semisocial bees, and a dearth of sources of concentrated nectar.) To exploit this strategy, however, honey bees must collect and store enormous quantities of nectar for use in winter heating.

As Aristotle first noted, honey bees are flower constant: they stick to one species on any given foraging trip, bypassing flowers of other species encountered *en route* (Waser, 1986). Darwin (1876) suggested that this strategy might lead to greater efficiency in searching for and exploiting flowers; this is clear in the context of learning how to handle flowers, both for honey bees (as discussed below) and bumble bees (Heinrich, 1979; Laverty, 1980). Flower constancy requires learning at least how to recognize blossoms with great certainty, and it is this aspect of forager behavior that has been studied most.

Bees find flowers either as recruits, directed to food sources by dances of returning foragers, or as scouts, a much smaller group that looks for new sources of food. Each class of bee faces the problem of minimizing its searching time. One way in which searching is made more efficient is through landing on only the objects most likely to be flowers—targets that are colorful, relatively small, contrast with their background, have relatively complex shapes (i.e., shapes with high spatial frequency), and produce a floral odor (reviewed in Gould and Gould, 1988).

Odor Learning

Von Frisch (review, 1967) first demonstrated that odors carried back on the waxy hairs of foragers are learned by potential recruits in the hive when they come into contact with a dancing forager, and their memory of these odors is used to localize the food source in the field. Menzel and his colleagues (Menzel et al. 1974; Menzel and Erber, 1978; Menzel, 1985) have greatly extended our understanding of odor learning, demonstrating that it occurs quickly, that the memory allows highly reliable choice behavior (with accuracies of about 98%), and that it is never forgotten unless another odor is learned at the same time of day. Koltermann (1974) and Lindauer (1976) report that the rate of learning of different odors varies between subspecies in an ecologically sensible way: odors common to the habitat of one race are learned more readily than some that are common to the habitat of another race, and vice versa. Floral odors are learned faster than most others, while putrid odors are learned more slowly, though with equal accuracy (Menzel, 1985). Erber (1981, 1984) has shown that cells in the olfactory pathway that ultimately respond to a pairing between a particular odor (the CS) and food (the US; the UR is proboscis extension) show a slight initial response to both sugar water and the odor; learning serves to amplify this activity greatly, a strong indication that the wiring pattern uncovered in *Aplysia* is at work in honey bees.

Odor learning is evident in a variety of other insects. One of the most illuminating cases involves parasitic wasps (Lewis and Tumlinson, 1988). Though the species studied is a specialist on one kind of caterpillar, and initially locates its victims on the basis of an innately recognized chemical sign stimulus (US) emitted by the host's frass, it associates any stronger odor (CS) in the frass—a volitile chemical from the plant the host is foraging on—with the presence of the caterpillars it is hunting. It uses the learned odor to detect host organisms from a greater distance, and so increases its hunting efficiency. The learning, then, accommodates the variability in host diet rather than that of the parasite. Since most wasps also need to forage for nectar to satisfy their need for carbohydrates, it may be that flower learning is widespread even among these hunters.

Color Learning

Von Frisch (review, 1967) demonstrated color learning at a time when only humans were thought to have color vision. In the course of his tests, he discovered that honey bees can see and learn ultraviolet but are blind to red. Again, Menzel and colleagues (Menzel et al., 1974; Menzel and Erber, 1978; Menzel, 1985) have uncovered much more about this feat. They have shown that the rate of learning is slower than that for odor, and a thoroughly trained bee still selects the correct color with less precision than it displays in choosing the correct odor (Fig. 2.2). Though all visible colors can be learned about as well, during the initial stages of training bees learn violet much faster (and green more slowly) than other colors (Fig. 2.3). They confirmed Opfinger's (1931) report that color learning occurs only as a bee approaches a flower, and showed that the last 2–3 seconds before landing is critical.

Bitterman and his colleages (e.g., Couvillon and Bitterman, 1980; Bitterman, 1988; Bitterman and Couvillon, 1991) have challenged the idea that there is any difference between the rates of color and odor learning (or between these and pattern or time learning, discussed below), or any

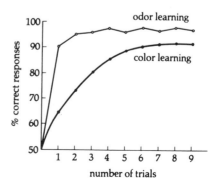

Figure 2.2. In a two-choice test, foragers learn odor faster than color and choose with a higher degree of accuracy. (The stimulus intensities have been balanced to yield a 50:50 choice ratio among naïve bees.)

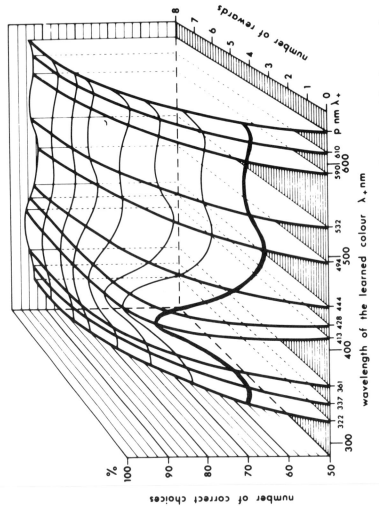

Figure 2.3. Although bees can learn any color they can sense, they learn violet fastest and green more slowly. (The heavy line emphasizes the difference after one visit.)

variation within modalities, or, for topics to be discussed below, any differences in the timing of learning, or any specialization in the organization of memory, or any adaptive hierarchy in cue use, or the existence of any niche-specific specializations. They castigate any "undue preoccupation with relatively minor ecological details" (Bitterman, 1988), referring to the niche differences between, in particular, rats and bees—a difference most biologists would consider to be enormous.

Bitterman's results are so at odds with those of other workers in the field that some comment is necessary. The normal technique for studying honey bee learning is to train the bees from the hive to a convenient location using dilute unscented sucrose on a nondescript feeder. When the experiment is to begin, a new feeder is substituted with the cues to be learned. (The food-reinforced feeder is often referred to as the S^+; any unreinforced feeder present during the training is denoted by S^-; any novel feeder offered during testing is called S^0.) Bitterman instead trains with a feeder offering concentrated sucrose and a mixture of the S^+ and S^- cues (e.g., two colors, a mixture of two odors, or both, for example).

Training with the experimental solution and feeder can profoundly alter forager behavior—indeed, this was the change in conventional techniques that led to the anomolous results that created the dance-language controversy (reviewed in Gould, 1976). The use of dilute food during training inhibits dancing and recruitment; recuits must be captured before testing begins (otherwise they distract the single forager being tested and can greatly alter its behavior), and the capture process itself often disturbs the bee being readied for testing. Training with a minimum of cues, or at least changing the cues at the onset of the experiment, allows the researcher to begin with a bee that is as naïve as possible; this is also the procedure that comes closest to matching the experience of a scout bee finding a new food source, or a recruit bee locating a resource advertised by dancing in the hive. Offering both the right and wrong cues during the extensive training and maintenance period before a learning experiment begins, on the other hand, seems likely to seriously alter and degrade the performance of bees after the experiment begins. For example, Bogdany (1978) showed that *removal* of one cue (one or more of the S^+s in this case) has no effect on memory; this means that a mixture of training cues will be remembered without significant alteration. As a result, a Bitterman experiment pits a subset of the S^+ cues against another subset of S^+ cues, but which are now the S^- cues. The problem here is that honey bees can learn through conditioned inhibition (Gould, 1986b)—that is, they can learn which cues predict the absence of reward. (So, for instance, if we train with an S^+ vs. an S^-, and then offer a choice of the S^- against an S^0—a novel cue or set of cues—the bees choose the S^0; they have learned that the S^- is nonrewarding, but know no evil of the S^0.) The consequence is a procedural quagmire: Bitterman's bees have already learned the S^+ subset whose acquisition he thinks he is studying, but choose randomly at the outset (as though they are naïve) because they are also familiar with the S^-, which they have been taught throughout training is rewarding; the task before them is one of suppressing the latter response—that is, conditioned inhibition. Although the time course of conditioned inhibition and the organization of the associated memory are not known, there is no reason to suppose it is the same—indeed, in vertebrates where this has been studied, the learning, inhibition, and extinction curves are rarely the same. My guess is that the reason Bitterman's results look like they are from a different species of animal, therefore, is not because everyone else in the field is wrong, but because his unique training technique systematically eliminates the very process he seeks to study. Equally important, they have no counterpart in the natural world, and thus are not well conceived for studying a system adapted to the task of flower foraging.

Though honey bees learn color relatively quickly, they are by no means the fastest insect at this task. Lewis and Lipani (1990) found that cabbage butterflies in a two-choice test selected the color they had been fed on with about 82% accuracy after one visit; honey bees, by contrast, learn most colors to about 65% accuracy in one trial (Fig. 2.3). It would be interesting to know what the rest of the learning curve for butterflies look like.

Color learning has also been reported in other species of butterflies, as well as a few other insects (reviewed in Papaj and Prokopy, 1989).

Shape Learning

For many years honey bees were thought incapable of representing learned patterns in memory in the vertebrate ("photographic") manner (reviewed in Gould, 1984); instead, they were supposed to store only a list of defining "parameters" like spatial frequency. There is now clear evidence for photographic storage (Gould, 1985a, 1986b). The experimental test involved training with an S^+ and S^- which, though different, were identical in all the parameters said to be remembered by bees. Trained foragers chose as accurately as when taught patterns of equal complexity but which differed significantly in many parameters.

The resolution of the visual memory is on the order of 8–10° (vs. a real-time visual resolution of 1–2°), and the learning takes place on arrival (Gould, 1985a, 1988a); the relevance of these two facts will become apparent when we look at landmark learning below. That the memory is not eidetic—that is, not of the same resolution as real-time vision—is not unusual; indeed, eidetic memory is unknown beyond a few remarkable human individuals. Perhaps the lower resolution represents a compromise between the increased accuracy of matching, or the increased range at which a match can be made, when the memory is eidetic vs. the increased storage space required to achieve such resolution.

Pattern learning takes place more slowly than either color or odor learning, but reaches a level of accuracy roughly equivalent to that of color memory.

Just as with odor and color learning, there are within-modality biases evident in pattern learning. The most memorable is a comparison of the learning curves for a four-pointed target vs. a 23-pointed figure of equal area (Fig. 2.4): the more complicated figure is learned much faster, though to the same level of accuracy (Schnetter, 1972). The internal biases in all three modalities parallel the realities of flower learning: blossoms are more likely to have floral odors than putrid ones, will more often be violet than green, and are usually complex rather than simple. In short, bees are able to learn common cues faster than uncommon ones; humans, offered the same tests, might well learn other cues faster—nonfloral food odors, simple shapes, and subtle shades of green, for instance.

Shape learning has not been much studied in other insects. Leaf shape is learned as a CS for oviposition-site selection in *Battus* butterflies (Rausher, 1978; Papaj, 1986); innately recognized chemicals act as the US sign stim-

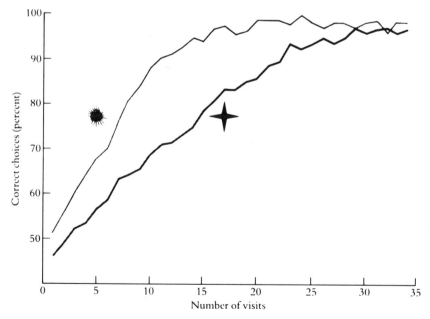

Figure 2.4. Foragers learn a 23-pointed figure faster than a four-pointed one. Note that the curve for shape learning is not as steep as that for odor and color learning (Fig. 2.2).

ulus. The consequence is probably a higher efficiency of searching. Surprisingly, there is little or no evidence for shape learning in the context of food-foraging, though this may reflect more an experimental oversight than an absence of learning.

Animals that do learn shapes do not just commit a picture of what they see to memory; the information is processed before storage, and additional processing must occur when matching is attempted. One of the most common matching transformations is the mirror-image ambiguity: animals are said to confuse left and right, perhaps as some necessary consequence of bilateral nervous systems (e.g., Corballis and Beale, 1970). An alternative view is that left-right ambiguities are adaptive, since virtually any object in nature presents at least one range of views that look like its mirror image when seen from another angle. Honey bees, at least, provide evidence for the more evolutionary hypothesis (Fig. 2.5): trained on one figure, they strongly prefer it over a mirror-image alternative when offered a choice; but when presented with the mirror image and a novel pattern, they opt for the reversal (Gould, 1988b). This suggests that mirror images are recognized on the basis of specific processing, and treated as they deserve.

Some animals are also capable of recognizing rotations of familiar targets (e.g., Hollard and Delius, 1983). Bees appear unable to recognize a vertically oriented target rotated by 90° or more as similar to the training pattern (Gould, 1988b); this is unlikely to create any problem during foraging since most flowers are either rotationally symmetric with a repeat angle of 60° or less, or bilaterally symmetric. For horizontally oriented blossoms, however, the problem is different: a forager unable to recognize a target from a novel angle might have to circle each potential target to determine if it is of the correct species.

I trained bees on a horizontal pattern with four "petals" of different colors, positioned in an apparatus so that they could only view the pattern from the front quadrant; during testing the patterns had been rotated 180°, and the choice had to be made from a perspective unavailable during training (Fig. 2.6). The foragers were able to select the training pattern from the novel viewpoint with high accuracy (Gould and Gould, 1988). This observation suggests a relatively sophisticated matching ability, as do preliminary indications (unpublished data) that bees can recognize familiar patterns from novel distances (a phenomenon known as size constancy).

Hierarchies

Though honey bees can learn about odor, color, and shape, the information from the various modalities is not equally valued. Odor takes precedence over color, and color "overshadows" shape (Lindauer, 1969; Hoefer and Lindauer, 1975, 1976; Kriston, 1973). As a result, a bee trained to a

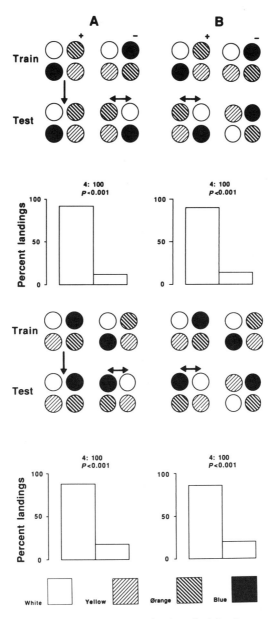

Figure 2.5. Bees were trained to find food on one pattern and then offered a choice between that pattern and its mirror image (left); they selected the original pattern overwhelmingly. When, however, the choice was between the mirror image and a novel pattern, bees chose the reversal.

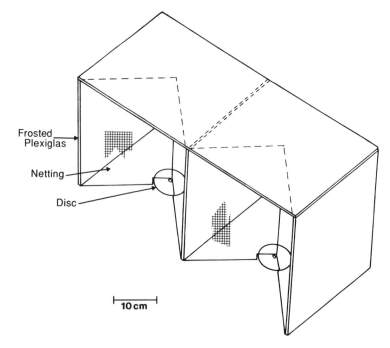

Figure 2.6. Bees were trained to an apparatus constructed out of frosted Plexiglas—that prevented them from seeing the targets from the top, back, and sides—and netting, which denied them access to any but the front quadrant. When presented a choice in the same apparatus between the S$^+$ and S$^-$, each rotated 180°, bees reliably chose the training pattern even though they had never seen it from that perspective before with an accuracy comparable to that observed with unrotated patterns.

blue, triangular, peppermint-scented target will select a circular yellow pattern with peppermint odor over a triangular blue alternative with, say, orange scent (Fig. 2.7). (This seems surprising because our species weights these modalities differently.) If the odor offered by two alternatives is correct, the forager will use color to guide her choice. Only if the odor and color are correct does the bee depend heavily on shape. This hierarchy might correlate with the predictiveness of these cues in nature: odor may be the most reliable guide to floral identity; color perhaps varies more from blossom to blossom and plant to plant, whereas shape probably varies much more depending on the angle of approach.

Landmark Learning

Bees have very poor visual acuity—1–2° in most of the visual field; the sun, by comparison, is 0.4° in diameter. It follows that targets as small as

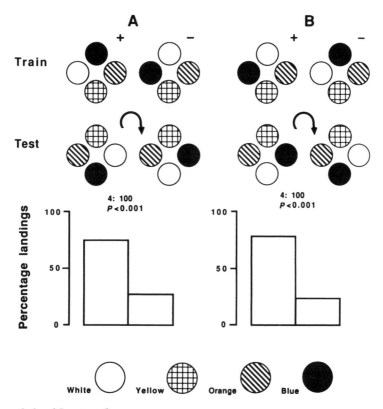

Figure 2.6. (*Continued*)

flowers (not to mention the hive opening) will be hard to see at a distance. Foragers returning to a familiar patch of flowers compensate for this shortcoming by using prominent landmarks for guidance. Landmarks may be used to guide flights, particularly when they lie near the flight path (von Frisch and Lindauer, 1954; Dyer and Gould, 1981); alternatively, they can be used to triangulate an inconspicuous source (Anderson, 1977; Cartwright and Collett, 1982, 1983; Gould 1985b, 1987a). The interesting thing is that, unlike color and shape, which are learned on arrival, landmark learning occurs on departure (Opfinger, 1931; Gould, 1988c) and is stored at a resolution of about 3.5° (Gould, 1987a), as compared with 8–10° for flower shapes. It is as though foragers cannot commit two pictures to memory at once, and so must segregate them temporally. The higher resolution of landmark memory may be a consequence of the constraints reality places on matching in the two contexts: A forager can always compensate for the low resolution of shape memory by flying closer to the flower. Landmark discrimination, on the other hand, has to be made near

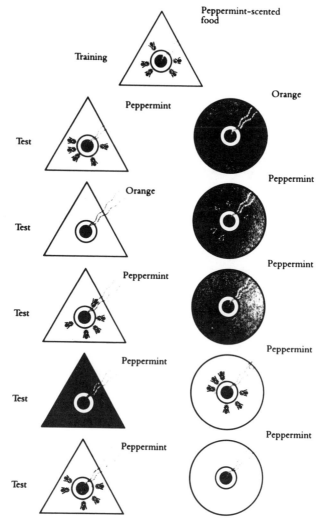

Figure 2.7. The cue hierarchy of bees can be determined by training foragers to one constellation of cues (a blue, triangular, peppermint-scented target in this case) and then offering choices between various dissociations of cues. Bees select a feeder with the correct odor regardless of its shape or color over a feeder with the wrong odor. When both alternatives have the correct odor, foragers choose on the basis of color. Shape comes into play only when both feeders present the training odor and color.

the flower; flying closer to a landmark to see it better moves the bee away from its goal—the location at which the picture was originally taken.

There are some logical biases evident in landmark memory—or, at least, in the use of this memory. Foragers tend to use landmarks far enough away to be useful for triangulation, but close enough so that small errors in location can be recognized (Cartwright and Collett, 1983; Gould, 1987a). Just as Tinbergen (1932, 1935) found with hunting wasps, the spatial arrangement of the markers takes precedence over their shape, though bees do remember the shape and color of landmarks and use them when spatial arrangement is not useful (Gould 1985b, 1987a; Cheng et al., 1986). In addition, bees tend to be selective about which landmarks they use: given a surfeit, they rely on a subset that has the greatest spatial separation (and therefore the most useful information) to the exclusion of intermediate markers (Gould, 1987a).

Bitterman and Couvillon (1991) report that landmark learning occurs on a bee's *approach* to a feeder rather than during departure. However, their experiment involved using a single marker actually touching the feeder, which bees might well consider as part of the blossom. When this experiment is repeated with the same marker at different distances during training (Fig. 2.8), it becomes clear that close markers are learned as part of the flower pattern (on approach) while more distant ones are learned on departure as landmarks (Opfinger 1931; Gould, 1991b).

Lauer and Lindauer (1971) report that the position of landmark use in the cue hierarchy of bees varies from subspecies to another. For *Apis mellifera carnica* (the German honey bee), landmarks take precedence over shape, whereas for *A. m. lingustica* (the Italian honey bee), shape is more important. Such genetic differences also exist between individuals in a single subspecies as well (Brandes 1988), which means that at least some parameters of learning remain open to potential selection.

Time Learning and the Organization of Foraging Memory

Honey bee memory can be remarkably persistent. Menzel and Erber (1978) report that after three visits to a feeder, a forager will remember a food source indefinitely without further training. The measured record must surely be that documented by Lindauer (cited in von Frisch, 1967) of a forager trained in the fall who overwintered and returned 182 days later in the spring to choose correctly.

As von Frisch noted (review, 1967), many species of flowering plants provide nectar only during a restricted part of the day. Doubtless this strategy increases the probability of pollination. It comes as no surprise, therefore, to discover that bees also learn the time of day that a source provides food (reviewed in von Frisch, 1967). More interesting is the dis-

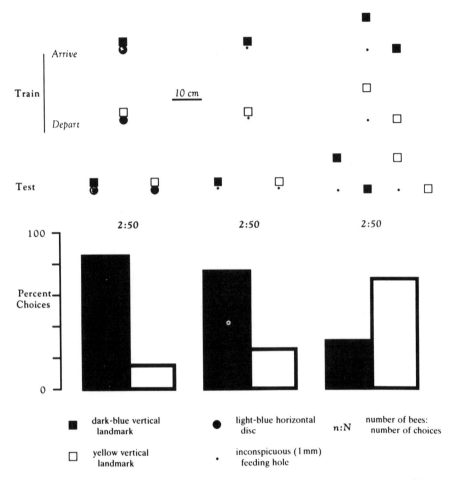

Figure 2.8. When, during training, approaching bees are offered a feeder with a "landmark" of one color immediately adjacent to the food, and then the color of the marker is switched before departure (left and center), the bees prefer the color seen during approach. The marker is treated as part of the blossom. When the experiment is repeated with two landmarks located at least 15 cm away (right), returning bees prefer the color observed on departure.

covery by Bogdany (1978) that bees apparently store information about flowers according to time of day. Hence, though a forager can learn about a second source of food (with different odor, color, shape, and landmark cues) at a different time of day without affecting its memory of the first, if the new source is offered at the same time the first had been presented on previous days, all memory of the first source seems to disappear.

Bogdany also found that if just one cue in a set was altered at a feeder, the forager had to relearn all the cues; on the other hand, if a cue had been missing and was simply added, the bee learned the new cue without altering its memory of the others. It is as though forager memory is organized like an appointment book, with time as the index variable, and blanks available for writing in odor, color, and so on. When one of the cues is changed, the processing logic erases all entries (on the sensible basis that this must be a new species of blossom), whereas when one is added, the empty blank is simply filled in. It is difficult to imagine a more efficient way to organize the memory of a bee facing the predictably unpredictable task of flower learning. This analogy also helps explain the inability of foragers to learn cues they can readily perceive—bees fail to learn flashing, rotating, or polarized targets (Menzel, 1985): as these are cues with no natural value, selection has not created blanks for this information in the mental appointment book; apparently selection has not favored wiring this sort of irrelevant input into the learning circuit for flowers.

Although bee memory appears to be organized like an appointment book, we have no direct information on the number of entries that can be accommodated. Numerous experiments have shown that individual foragers can learn two different sets of cues for feeders offering rewards at two different times of day. Koltermann (1974) demonstrated that foragers could learn that a particular source provided nectar at nine different times over the course of a day, and return with an accuracy of better than 15 minutes. But since Koltermann's test used the same target at each time of day, it does not necessarily follow that bees can learn at least nine different cue sets.

Some butterflies, though they harvest a similar resource, apparently do not have the elaborate memory structure so evident in bees. For example, when cabbage butterflies were trained to land on targets of a particular color in a two-choice test, experience with the S^- on previous days severely hampered learning (Lewis and Lipani, 1990). This suggests an inability to remember more than one set of cues, an interpretation that is consistent with the observation by Lewis (1986) that when cabbage butterflies were trained first on one source and then, later, another, their memory of the first was affected; control butterflies, which were trained on the first target and then held without further experience while the experimental group was trained on the second, still remembered how to handle the first source (see also Lewis, this volume).

Operant Learning

Classical and operant learning are usually treated as wholly distinct, but a little reflection reminds us that, in nature, learning to recognize often exists in order to permit an animal to initiate a learned response. If bees store cues according to time of day, we might wonder if learned flower-handling techniques are filed the same way. In fact, when foragers are taught to land on one petal of a feeder at one time of day, and a different petal of the same feeder at another (Fig. 2.9), they remember this piece of feeding etiquette on subsequent days (Gould, 1987b).

The organization of operant learning in animals is unclear, but evidence from birdsong learning (reviewed in Gould and Marler, 1984) suggests that individuals may possess an innate repertoire of motor "gestures," a context-specific subset of which are used in developing a novel motor behavior by arranging them in an efficient order and modifying certain of them to a limited degree. This model is similar in many ways to the long stimulus-response chains that Watson (1925) imagined being created by classical conditioning to generate complex behavior; a CS was pictured as producing

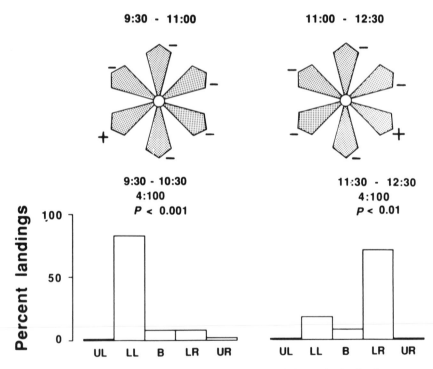

Figure 2.9. Bees were trained to land on the lower left petal of a feeder during the first half of the morning and then the lower right petal later on. (The training was accomplished by flipping the forager off of incorrect petals.) On subsequent days, bees landed on the correct petal during the corresponding period even in the absence of punishment for incorrect choices.

a UR, which served as the CS for the next UR in the chain, and so on. This hypothesis was largely discarded when Skinner (1938) proposed his model of operant conditioning. Nevertheless, Watson's ideas are more nearly consistent with the motor biases discussed earlier, and receive additional support from the only study I know of on the acquisition of operant skills by insects: Laverty (1980) reports that bumble bees used a limited set of motor gestures when exploring a novel blossom, only some of which were modifiable by experience. This would seen a fertile area for future research.

Conclusion

It seems reasonably clear that animal learning in general, and honey bee learning in particular, is specialized, full of omissions and oversights and biases, rather than flexible and open and free from preconceptions. Associative and operant biases are evident, and differ between contexts as well as between species. At least in bees, where the question has been asked in some detail, the same is true for the organization and resolution of memory. The larger issue, however, is whether these differences are, in general, just differences, or are in fact adaptive consequences of natural selection. Nearly all of the known biases have clear adaptive interpretations, but are these correct explanations or simply just-so stories? In some cases it has been possible to make a prediction and test it; the issue of whether operant learning would be stored according to time of day is a case in point. But there are other instances of demonstrated learning in bees for which an adaptionist account is more difficult to imagine—the ability to learn to associate magnetic-field reversals or sounds with an impending shock, for instance (Walker and Bitterman, 1985; Gould and Towne, 1988). Of course, not all learning need be strictly adaptive, any more than all morphology or biochemistry is; there will always be compromises, relics of obsolete characters, and artifacts. On balance, the anomalies are few and the general pattern of adaptive specialization of learning is obvious; the learning programs of honey bees seem clearly a consequence of the ecological pressures and challenges they face. What is perhaps more puzzling is why learning in nectar-feeding butterflies, whose niche overlaps that of bees to a large extent, seems more limited and less specialized.

References

Alessandro, D., Dollinger, J., Gordon, J.D., Mariscal, S.K., and Gould, J.L. 1989. Ontogeny of the pecking response of herring gull chicks. Anim. Behav. **37**:372–382.

Anderson, A.M. 1977. A model for landmark learning in the honey bee. J. Comp. Physiol. **114**:335–355.

Baerends, G.P. 1941. Fortpflanzungsverhalten und Orientierung der Grabwespe. Tijdschr. Entomol. **84**:68–275.

Bateson, P.P.G. 1982. Preferences for cousins in Japanese quail. Nature **295**:236–237.

Bernstein, P.P.G., and Sigmundi, R.A. 1980. Tumor anorexia: a learned food aversion? Science **209**:416–418.

Bitterman, M.E. 1988. Vertebrate-invertebrate comparisons. In H.J. Jerison and I. Jerison (eds.), Intelligence and Evolutionary Biology. Springer-Verlag, Berlin, pp. 251–276.

Bitterman, M.E., and Couvillon, P.A. 1991. Failures to find evidence of adaptive specialization in the learning of honeybees. In L.J. Goodman and R.C. Fisher (eds.), The Behaviour and Physiology of Bees. CAB International, Wallingford, pp. 288–305.

Bogdany, F.J. 1978. Linking of learning signals in honey bee orientation. Behav. Ecol. Sociobiol. **3**:323–336.

Bolles, R.C. 1984. Species-typical response predispositions. In P. Marler and H.S. Terrace (eds.), The Biology of Learning. Springer-Verlag, Berlin, pp. 435–446.

Brandes, C. 1988. Estimation of heritability of learning behavior in honeybees. Behav. Genet. **18**:119–132.

Breed, M.D. 1983. Nestmate recognition in honey bees. Anim. Behav. **31**:86–91.

Brines, M.L. and Gould, J.L. 1979. Bees have rules. Science **206**:571–573.

Burghardt, G.M., and Hess, E.H. 1966. Food imprinting in the snapping turtle. Science **151**:108–109.

Cartwright, B.A., and Collett, T.A. 1982. How honey bees use landmarks to guide their return to a food source. Nature **295**:560–564.

Cartwright, B.A., and Collett, T.A. 1983. Landmark learning in bees. J. Comp. Physiol. **151**:521–543.

Cheng, K., Collett, T.S., and Wehner, R. 1986. Honey bees learn the color of landmarks. J. Comp. Physiol. **159**:69–73.

Corballis, M.C., and Beale, I.L. 1970. On telling left from right. Psychol. Rev. **77**:89–116.

Couvillon, P.A., and Bitterman, M.E. 1980. Some phenomena of associative learning in honey bees. J. Comp. Physiol. Psychol. **94**:878–885.

Darwin, C. 1872. The Expression of the Emotions in Man and Animals. Appleton, London.

Darwin, C. 1876. On the Effects of Cross- and Self-Fertilisation in the Vegetable Kingdom. John Murray, London.

Dietz, A. 1975. Nutrition of the adult honey bee. In Dadant & Sons (eds.), The Hive and the Honey Bee. Dadant & Sons, Hamilton, IL, pp. 125–156.

Dyer, F.C., and Gould, J.L. 1981. Honey bee orientation: a backup system for cloudy days. Science **214**:1041–1042.

Erber, J. 1981. Neural correlates of learning in the honeybee. Trends Neurosci. **4**:270–273.

Erber, J. 1984. Response changes of single neurons during learning in the honey bee. In F. Huber and H. Markl (eds.), Neuroethology and Behavioral Physiology. Springer-Verlag, Berlin, pp. 216–230.

Fabre, J.H. 1921. Hunting Wasps. Dodd Mead, New York.

Forse, D.D., and LoLordo, V.M. 1973. Attention in the pigeon: differential effects of food-getting versus shock-avoidance procedures. J. Comp. Physiol. Psychol. **88**:551–558.

Frisch, K.v. 1967. The Dance Language and Orientation of Bees. Harvard University Press, Cambridge, MA.

Frisch, K.v., and Lindauer, M. 1954. Himmel und Erde in Konkurrenz bei der Orientierung der Bienen. Naturwissenschaften **41**:245–253.

Gelperin, A. 1975. Rapid food aversion learning by a terrestrial mollusc. Science **189**:567–570.

Getz, W.M., and Smith, K.B. 1983. Genetic kin recognition: honey bees discriminate between full and half sisters. Nature **302**:147–148.

Gleitman, L.C. 1984. Biological predispositions to learn language. In P. Marler and H.S. Terrace (eds.), The Biology of Learning, Springer-Verlag, Berlin, pp. 553–584.

Gould, J.L. 1976. The dance-language controversy. Q. Rev. Biol. **51**:211–244.

Gould, J.L. 1982. Ethology: the Mechanisms and Evolution of Behavior. W.W. Norton, New York.

Gould, J.L. 1984. The natural history of honey bee learning. In P. Marler and H.S. Terrace (eds.), The Biology of Learning. Springer-Verlag, Berlin, pp. 149–180.

Gould, J.L. 1985a. How do honey bees remember flower shape? Science **227**:1492–1494.

Gould, J.L. 1985b. Honey bee learning and memory. In G. Lynch, J.L. McGaugh, and N. Weinberger (eds.), The Neurobiology of Learning and Memory. Guildford, New York, pp. 193–210.

Gould, J.L. 1986a. The biology of learning. Annu. Rev. Psychol. **37**:163–192.

Gould, J.L. 1986b. Pattern learning by honey bees. Anim. Behav. **34**:990–997.

Gould, J.L. 1987a. Landmark learning by bees. Anim. Behav. **35**:26–34.

Gould, J.L. 1987b. Honey bees store learned flower-landing behavior according to time of day. Anim. Behav. **35**:1579–1581.

Gould, J.L. 1988a. Resolution of pattern learning by honey bees. J. Insect Behav. **1**:225–233.

Gould, J.L. 1988b. A mirrow-image ambiguity in honey bee visual memory. Anim. Behav. **36**:487–492.

Gould, J.L. 1988c. Timing of landmark learning in honey bees. J. Insect Behav. 1:373–378.

Gould, J.L. 1990. Honey bee cognition. Cognition **37**:83–103.

Gould, J.L. 1991a. Why birds still fly south. Nature **347**:331.

Gould, J.L. 1991b. The ecology of honey bee learning. In L.J. Goodman and R.C. Fisher (eds.), The Behavior and Physiology of Bees. CAB International, Wallingford, pp. 306–322.

Gould, J.L., and Gould, C.G. 1988. The Honey Bee. W.H. Freeman, New York.

Gould, J.L., and Gould, C.G. 1989. Sexual Selection. W.H. Freeman, New York.

Gould, J.L., and Marler, P. 1984. Ethology and the natural history of learning. In P. Marler and H.S. Terrace (eds.), The Biology of Learning. Springer-Verlag, Berlin, pp. 277–288.

Gould, J.L., and Towne, W.F. 1988. Honey bee learning. Adv. Insect Physiol. **20**:55–86.

Hawkins, R.D., Abrams, S.W., Carew, T.J., and Kandel, E.R. 1983. A cellular mechanism of classical conditioning in *Aplysia*. Science **219**:400–404.

Heinrich, B. 1979. Bumblebee Economics. Harvard University Press, Cambridge, MA.

Hoefer, I., and Lindauer, M. 1975. Das Lernverhalten zweier Bienenrassen unter veränderten Orientierungsbedingungen. J. Comp. Physiol. **99**:119–138.

Hoefer, I., and Lindauer, M. 1976. Der Einflub einer Vordressur auf das Lernverhalten der Honigbiene. J. Comp. Physiol. **109**:249–264.

Hollard, V.D., and Delius, J.D. 1983. Rotational invariance in visual pattern recognition by pigeons and humans. Science **218**:804–806.

Immelmann, K. 1984. The natural history of bird learning. In P. Marler and H.S. Terrace (eds.), The Biology of Learning. Springer-Verlag, Berlin, pp. 289–310.

Koltermann, R. 1973. Rassen-bzw. artspezifische Duftbewertung bei der Honigbiene und ökologische Adaption. J. Comp. Physiol. **85**:327–360.

Koltermann, R. 1974. Periodicity in the activity and learning performance of the honey bee. In L.B. Brown (ed.), Experimental Analysis of Insect Behaviour. Springer-Verlag, Berlin, pp. 218–227.

Kriston, I. 1973. Die Bewertung von Duft- und Farbsignalen als Orientierungshilfen an der Futterquelle durch *Apis mellifera*. J. Comp. Physiol. **84**:77–94.

Lauer, J., and Lindauer, M. 1971. Genetische fixierte Lerndisposition bei der Honigbiene. In Informationsaufnahme und Informationsverarbeitung im Lebenden Organismus l. Franz Steiner Verlag, Wiesbaden, pp. 1–87.

Laverty, T.M. 1980. The flower-visiting behavior of bumblebees: floral complexity and learning. Can. J. Zool. **58**:1324–1335.

Lewis, A.C. 1986. Memory constraints and flower choice in *Pieris rapae*. Science **232**:863–865.

Lewis, A.C., and Lipani, G. 1990. Learning and flower use in butterflies: hypotheses from honey bees. In E.A. Bernays (ed.), Insect–Plant Interactions, 2. CRC Press, Boca Raton, FL, pp. 95–110.

Lewis, W.J., and Tumlinson, J.H., 1988. Host detection by chemically mediated associative learning in a parasitic wasp. Nature 331:257–259.

Lindauer, M. 1969. Lernen und Vergessen bei der Honigbiene. Proc. of the VI Congress IUSSI, Bern.

Lindauer, M. 1976. Recent advances in the orientation and learning of honey bees. In Proc. of the XVth Int. Congress on Entomology. International Entomological Association, Washington, DC, pp. 450–460.

Lorenz, K.Z. 1959. Gestaltwahrnehmung als Quelle wissenschaftlicher Erkenntnis. Z. Exp. Angew. Psychol. 4:49–67.

Margolis, R., Mariscal, S.K., Gordon, J.D., Dollinger, J., and Gould, J.L. 1986. Ontogeny of the pecking response of laughing gull chicks. Anim. Behav. 35:191–202.

Marler, P. 1984. Song learning: innate species differences in the learning process. In P. Marler and H.S. Terrace (eds.), The Biology of Learning. Springer-Verlag, Berlin, pp. 289–310.

Menzel, R. 1985. Learning in honey bees in an ecological and behavioral context. In B. Hölldobler and M. Lindauer (eds.), Experimental Behavioral Ecology and Sociobiology. Fischer Verlag, Stuttgart, pp. 55–74.

Menzel, R., and Erber, J. 1978. Learning and memory in bees. Sci. Am. 239(1):102–110.

Menzel, R., Erber, J., and Mashur, J. 1974. Learning and memory in the honey bee. In L.B. Browne (ed.), Experimental Analysis of Insect Behavior. Springer-Verlag, Berlin, pp. 195–217.

Opfinger, E. 1931. Uber die Orientierung der Biene an der Futterqueulle. Z. Vergleich. Physiol. 15:431–487.

Pankiw, P. 1967. Studies of honey bees on alfalfa flowers. J. Apic. Res. 6:105–112.

Papaj, D.R. 1986. Conditioning of leaf-shape discrimination by chemical cues in the butterfly *Battus philenor*. Anim. Behav. 34:1281–1288.

Papaj, D.R., and Prokopy, R.J. 1989. Ecological and evolutionary aspects of learning in phytophagous insects. Annu. Rev. Entomol. 34:315–350.

Pavlov, I. 1927. The Conditioning Reflex. W.W. Norton, New York.

Quinn, W.G. 1984. Work in invertebrates on the mechanisms underlying learning. In P. Marler and H. Terrace (eds.), The Biology of Learning. Springer-Verlag, Berlin, pp. 197–246.

Rausher, M.D. 1978. Search image for leaf shape in a butterfly. Science 200:1071–1073.

Reinhardt, J.F. 1952. Responses of honey bees to alfalfa flowers. Am. Nat. 86:275–275.

Revusky, S. 1984. Associative predispositions. In P. Marler and H.S. Terrace (eds.), The Biology of Learning. Springer-Verlag, Berlin, pp. 447–460.

Schnetter, B. 1972. Experiments on pattern discrimination in honey bees. In R. Wehner (ed.), Information Processing in the Visual System of Arthropods. Springer-Verlag, Berlin, pp. 195–200.

Schwartz, B. 1984. Psychology of Learning and Behavior. W.W. Norton, New York.

Seawright, J.E., Kaiser, P.E., Dame, D.A., and Lofgren, C.S. 1978. Learned taste aversion in children receiving chemotherapy. Science 200:1302–1304.

Shuel, R.W. 1975. The production of nectar. In Dadant & Sons (eds.), The Hive and the Honey Bee. Dadant & Sons, Hamilton, IL, pp. 265–282.

Skinner, B.F. 1938. The Behavior of Organisms. Appleton-Century-Crofts, New York.

Smith, N.G. 1967. Visual isolation in gulls. Sci. Amer. 217(4):94–102.

Tautz, J., and Markl, H. 1978. Caterpillars detect flying wasps by hairs sensitive to airborne vibrations. Behav. Ecol. Sociobiol. 4:101–110.

Terrace, H.S. 1984. Animal learning, ethology, and biological constraints. In P. Marler and H.S. Terrace (eds.), The Biology of Learning. Springer-Verlag, Berlin, pp. 15–46.

Tinbergen, N. 1932. Über die Orientierung des Bienenwolfes. Z. Vergleich. Physiol. 16:305–334.

Tinbergen, N. 1935. Über die Orientierung des Bienenwolfes. II. Die Bienenjagd. Z. Vergleich. Physiol. 21:699–716.

Tinbergen, N. 1951. The Study of Instinct. Oxford University Press, Oxford.

Tinbergen, N., and Perdeck, A.C. 1950. On the stimulus situation releasing the begging response in the newly hatched herring gull chick. Behaviour 3:1–38.

Walker, M.W., and Bitterman, M.E. 1985. Conditional responding to magnetic fields by honeybees. J. Comp. Physiol. 157:67–71.

Waser, N.M. 1986. Flower constancy: definition, cause, and measurement. Am. Nat. 127:593–603.

Watson, J.B. 1925. Behaviorism. W.W. Norton, New York.

Wilcoxon, H.C., Dragoin, W.B., and Kral, P.A. 1971. Illness-induced aversions in rat and quail: relative salience of visual and gustatory cues. Science 171:826–828.

3

Learning of Host-Finding Cues by Hymenopterous Parasitoids

Ted C. J. Turlings, Felix L. Wäckers,
Louise E. M. Vet, W. Joseph Lewis, and
James H. Tumlinson

Introduction

Interactions between insect parasitoids and their arthropod hosts characteristically result in the premature death of the hosts, and are obligatory for the development of the parasitic insects. This obviously places strong pressure on the hosts to avoid detection by parasitoids, and on the parasitoids themselves to improve encounter rates with suitable hosts. To confront the challenge of finding the often-inconspicuous, well-hidden hosts, parasitoids have developed various sophisticated searching strategies that depend on a vast array of environmental cues.

A wealth of information generated over the last few decades has enabled several experts to categorize the different elements of the searching behavior of parasitoids (Doutt, 1959; Vinson, 1975, 1976, 1981, 1984; Lewis et al., 1976; Weseloh, 1981; van Alphen and Vet, 1987). Although a great variety of specialized lifestyles have been described, the ability of parasitoids to modify their responses to foraging cues based on experience seems to be characteristic of many species. The ability to learn profitable cues has now been demonstrated for almost 20 different species. This chapter first presents a selective review of learning in parasitoid foraging behavior and then discusses some intriguing ecological aspects of this phenomenon.

Several studies suggest that learning may take place during the immature stage as well as during the adult stage. We will argue that adult learning generally contributes more to the foraging success of the insects then pre-adult learning. Recent evidence indicates that the adult parasitoids' responses are mainly modified through the process of associative learning: the wasps innately recognize host-derived stimuli (unconditioned stimuli) upon contact, and they associate these stimuli with surrounding stimuli (conditioned stimuli) to which they originally show no or limited responsiveness. Subsequently, the wasps become responsive to the newly learned stimuli and use them in their search for hosts.

While work has focused on the olfactory sense, parasitoids use other sensory modalities (visual and mechanosensory) as well to find hosts. Several studies now show that all three modalities can be affected by learning. We will point out that parasitoids appear to be particularly effective at using and learning a combination of olfactory and visual cues.

Next we will argue that associative learning is most evident in responses to cues from the environment of the host, especially its food, while responses to cues that are directly host-derived tend to be more congenitally fixed. Flexible responses to cues from the host's food and environment should be favored over fixed responses because of the enormous variability in time and space of such cues and their relatively low reliability (over evolutionary time) in indicating host presence and suitability. Several behavioral studies will be reviewed that show how wasps deal with this variability.

We will also discuss the observation that learning generally requires only very brief contact with specific innately recognized stimuli. However, learned responses are strongest and wane the least if such unconditioned stimuli are contacted in conjunction with an actual oviposition in a suitable host. Additional positive experiences reinforce and strengthen the learned responses, while lack of reinforcement (no encounters with suitable hosts) or perhaps even negative experiences (encounters with unsuitable hosts) may depress responses to certain learned cues. We will argue that these observations agree with our concept of optimal foraging.

In making our arguments, we will focus mainly on our own work involving the solitary larval endoparasitoids *Cotesia marginiventris* (Cresson), which attacks a wide range of lepidopterous species (Turlings, 1990), and *Microplitis croceipes* (Cresson), which specializes on *Heliothis* spp. (Eller, 1990). We will summarize by speculating on how learning may assist a single parasitoid female to overcome foraging problems with which she will be confronted in nature.

Onset of Parasitoid Learning

Parasitoids' searching behavior can be affected by experience at various stages of their life cycle. Sometimes cues appear to be learned by the immature insect (i.e., preimaginal conditioning), which are subsequently manifested in their responses as adults. Generally, however, experiences during the adult stage have a greater effect on the insects' responses. These experiences, as we will see, involve contacts with specific host-derived stimuli that the wasps recognize innately. In addition to this general increase in responsiveness, which we refer to as "priming," the wasps may also learn to respond to previously unrecognized stimuli by linking these new stimuli to the contact stimuli. Such associative learning can significantly alter the parasitoids' preferences for specific stimuli.

Learning by Immatures

Thorpe and Jones (1937) conducted one of the earliest detailed studies on parasitoid learning. They reared the ichneumonid parasitoid *Venturia (Nemeritis) canescens* (Grav.) on its regular host, the meal moth *Ephestia kühniella* (Zell.), but also on *Meliphora grisella* (F.), the small wax moth, which the wasp does not attack under natural conditions. Wasps reared on *E. kühniella* showed a strong preference for the odor of meal moth larvae. However, attraction to the odor of wax moth larvae was induced by rearing the wasps on the unusual host. Thorpe and Jones termed this "pre-imaginal conditioning," implying that the female parasitoids acquired a preference for a specific host species (or associated cues) during the immature stage. A similar idea had been put forth by Hopkins (1917; Craighead, 1921), who observed that herbivorous beetles selected oviposition sites on plant species similar to the ones on which they had been reared. This explanation has been used to explain host preference or selection in various insect groups (Thorpe, 1939; Hershberger and Smith, 1967; Jermy et al., 1968; Jaenike, 1982, 1983), and is often referred to as Hopkins' host-selection principle. Pre-imaginal learning has also been implicated in olfactory kin recognition in social Hymenoptera (Isingrini et al., 1985).

Various theories have been brought forward to explain how experiences of immatures may cause internal changes that increase sensitivity to certain stimuli as adults. Hopkins' host-selection principle was the starting point for other studies (Jermy et al., 1968; Jaenike, 1982, 1983). In 1985, Corbet put forth the chemical legacy hypothesis, which states an alternative explanation for the above phenomena. It suggests that actual traces of chemical cues inside or outside the immature parasitoid are carried over into the adult stage, where they directly affect the sensitivity of the insect to these chemicals.

Hérard et al. (1988) shed some light on a mechanism by which chemical cues from the host in which the immatures developed may be carried over to condition the adult parasitoid. *Microplitis demolitor* were reared from *Helicoverpa zea* (Boddie) larvae fed either artificial diet or a diet of cowpea seedlings. In flight-tunnel tests, wasps reared from plant-fed hosts were readily attracted to semiochemicals emitted from host larvae feeding on cowpea. Similar responses could only be obtained from the wasps reared from hosts fed on artificial diet if the adult wasps first were given a contact experience with the cowpea-host complex. Hérard et al. (1988) went on to demonstrate that experience with the cocoon increased the wasps' responsiveness to the semiochemicals. When plant-reared wasps were excised from their cocoons shortly before their emergence, the adult wasps responded poorly in flight tunnel bioassays. When the excised wasps were allowed to contact the cocoons prior to introduction into the flight tunnel, however, their responsiveness increased significantly. Apparently, chemicals emanating from the cocoons provide emerging wasps with information that they use in host-searching as adults. This seems to be also the case for the leafminer parasitoid *Opius dissitus* Muersebeck (Petitt et al., unpublished data). When reared on leafminers in lima bean, adults were more attracted to the odors of leafminer-infested lima bean than to those of leafminer-infested eggplant. This preference is not apparent if the parasitoids are excised from their cocoons before eclosion (Petitt et al., unpublished data). The environment in which the immature insect grows up can often have an apparent effect on the responses of the adult wasp. However, no study has shown that learning actually takes place in the immature. As Corbet (1985) suggested and Hérard et al. (1988) and Petitt et al. (unpublished data) demonstrate, cues carried by the immature may only affect the insect's responsiveness after adult emergence.

All of the above examples may be the result of early adult learning rather than learning by the immatures. Early adult learning was demonstrated by Kester and Barbosa (1991) who found that the gregarious parasitoid *Cotesia congregata* is only sensitive to learning plant odors for a few hours after emergence. They suggest that postemergence learning not only retains wasps in the habitat of potential hosts but keeps them in the vicinity of potential mates as well.

For some parasitoids, rearing environment seems to have no effect on their subsequent responses. McAuslane et al. (1990a), for example, reared *Campoletis sonorensis* (Cameron) on *Heliothis virescens* F. larvae feeding on either artificial diet or cotton foliage or sesame foliage. Insects of all three treatments responded equally well to host-damaged cotton or sesame plants. Likewise, Mueller (1983) found that the plants on which hosts were fed had no effect on preference for plants exhibited by *Microplitis croceipes* that emerged from these hosts. Adult experience, on the other hand, has a significant effect on responsiveness to semiochemicals by both *C. sonorensis* (McAuslane et al., 1990b, 1991) and *M. croceipes* (e.g., Eller et al., 1992, see below). Although the effects of both immature and adult learning are seldom studied, it appears that preadult experience generally has only a minor effect on adult host-searching behavior, compared to adult experience (e.g., Vet, 1983; Drost et al., 1988; Mandeville and Mullens, 1990; Petitt et al., 1992).

Learning by Adults

In their studies on *V. canescens*, Thorpe and Jones (1937) also presented some of the earliest evidence for learning by adults. They showed that responses to odors from wax moth larvae (which the wasps normally do not attack) could be induced by allowing adults to contact such larvae for a period of time upon emergence.

Likewise, Vet (1983) demonstrated that the responses by the parasitoid *Leptopilina clavipes* (Hartig) are affected by both rearing environment and adult experience. This wasp normally attacks fungivorous Drosophilidae and is attracted to the odor of decaying mushrooms (a potential habitat of its hosts). Wasps emerging from hosts that were reared on a yeast medium were significantly more attracted to yeast odors than mushroom-reared wasps, but they still preferred the odor of decaying mushrooms. Obviously, the rearing environment had only a limited effect. By contrast, an adult oviposition experience in hosts feeding on the yeast medium altered their preference significantly in favor of yeast odors.

Numerous studies have now shown that adult learning can strongly modify the responses to host-related cues in many parasitoid species. Experiences can influence responses in two different ways. It can cause a general increase in the responsiveness of a female (i.e., priming), but it can also alter a female's preference for specific cues. In both cases the same or a similar mechanism may be at work, but the effects are distinct enough to discuss them separately.

Priming vs. Preference Learning

Definitions to describe the various effects that experiences may have on insect responses continue to be the cause of confusion (e.g., McGuire, 1984; Tully, 1984; Papaj and Prokopy, 1989). Without trying to add to this confusion we would like to emphasize a distinction between what we regard as two separate phenomena, priming and preference learning. The former refers to the observation that certain experiences merely make the parasitoids more responsive to foraging cues, while the latter includes those cases where the increase in responsiveness is specific for the cues that the insects encounter during the experience.

McAuslane et al. (1991b) found that the ichneumonid *Campoletis sonorensis* (Cameron) is more responsive to plant odors after contacting host larvae in the absence of plants. Hence, wasps did not need to experience plant odors in order to become more responsive to these odors. A general increase in responsiveness to odors of host feces was observed by Eller et al. (1992) for *Microplitis croceipes* after these wasps were allowed to oviposit in larvae on different plants. Even contact with hosts on artificial diet increased responsiveness to feces from plant-fed hosts. We think these are examples of priming whereby the insects come in contact with an innately recognized stimulus (unconditioned stimulus = US) and become more receptive to other cues (not necessarily present during the experience) to which they already show some degree of responsiveness. Turlings et al. (1989) and McAuslane et al. (1991a) termed this sensitization. However, the term sensitization has been used by others (McGuire, 1984; Tully, 1984) in a very different way (see Smith, this volume). The term "priming" may be more suitable since it has been used to describe similar interactions (Birch, 1974).

Cotesia marginiventris was also considerably more responsive to odors associated with its hosts following contact with a host or its feces in the host's microhabitat (Turlings et al., 1989). Wasps became most receptive to the odor that they encountered during an experience, but the experience also caused a less dramatic increase in responsiveness to the odor of an alternative host microhabitat. The odors may have had something in common, but they were distinct enough for the wasps to differentiate between them. That the experience did not merely cause a general increase in responsiveness was also shown in olfactometer choice tests, where *C. marginiventris* exhibited minor shifts in preference in favor of the odor of the plant-host combination it had experienced (Turlings et al., 1990a).

The preference shifts, however, were not very strong. Eller et al. (1992) suggested that experience may affect preferences more strongly. They offered *M. croceipes* females choices between the odors of feces of hosts that were fed on different diets. The wasps showed no changes in preference after one experience with a complete plant-host complex (including oviposition). However a single experience did increase their overall responsiveness drastically. When the experience was repeated two or four times, the wasps exhibited a significant preference for the odor of the feces they had experienced, even if they strongly preferred the alternative odor before experience or after only one experience (Eller et al., 1992).

In summary, these studies suggest a twofold effect of experience on the wasps. First, experience makes the insects immediately more alert and responsive to odor (priming) and, second, the insects learn to respond to the specific odors that they encounter during the experience (preference learning). We will discuss now how preference, and perhaps aversion, for specific cues can be brought about through associative learning.

Associative Learning

Arthur (1971) was perhaps the first to demonstrate learning of novel odors by a parasitoid. He demonstrated that the ichneumonid *V. canescens* could be conditioned to search for hosts in a medium impregnated with geraniol by presenting hosts to the wasps in the presence of this odorous chemical. The wasps apparently associated this novel odor with the presence of hosts.

Similar associations were found by Vinson et al. (1977) and Wardle and Borden (1989). Vinson et al. (1977) found that the parasitoid *Bracon mellitor* Say exhibited ovipositor probing in response to an antimicrobial additive (methyl parahydroxy-benzoate) in artificial diet used to rear their hosts, the boll weevil (*Anthonomus grandis* Boheman). They demonstrated that *B. mellitor* had learned this novel chemical cue and suggested that this was a "classical" form of associative learning (*sensu* Pavlov, 1941). Wardle and Borden (1989) found that the polyphagous ectoparasitoid *Exeristes roborator* (F.) learned apple odors after the wasps had experienced the odor during encounters with hosts.

We found that contacting host feces was sufficient experience for *C. marginiventris* wasps to increase dramatically their responses to odors experienced during contact (Turlings et al., 1989, 1990a). It was suggested that, upon contact, wasps recognized specific semiochemicals in the feces of suitable hosts. The parasitoids associated surrounding odors with the possible presence of hosts and subsequently used these odors as cues in host searching. That parasitoids indeed are capable of learning specific cues through such an association was demonstrated by Lewis and Tumlinson (1988) and Vet and Groenewold (1990).

Lewis and Tumlinson (1988) found that a water-soluble nonvolatile contact kairomone in the frass of host larvae served as the key stimulus in learning by *M. croceipes*. After contact with this kairomone the wasp was found to be attracted to odors that were present during the contact experience, even if these odors are not normally associated with hosts. Vet and Groenewold (1990) found that a similar mechanism triggered learning in *Leptopilina heterotoma* (Thomson), a parasitoid of *Drosophila* species. A kairomone was extracted from yeast media in which *Drosophila* larvae had been crawling. Wasps were found to be attracted to a synthetic odor (Z)-3-hexen-1-ol after contacting yeast containing kairomone in presence of that odor. Recently, de Jong and Kaiser (1991) demonstrated that a related specialist parasitoid, *L. boulardi*, is also capable of learning a novel odor (perfume) in association with a successful oviposition in a host larva.

With these experiments Lewis and Tumlinson (1988) and Vet and Groenewold (1990) presented direct evidence for unconditioned stimuli (US) in host by-products, which the parasitoids recognize upon contact. During a contact, the wasps associate the US with surrounding conditioned stimuli (CS). As a result of the association the wasps will be responsive to the CS and use them as cues in subsequent host-seeking efforts. It has been suggested that, after this association, the CS may serve as US during subsequent experiences, a phenomenon termed "second-order learning" (Vet et al., 1990a; see also Menzel et al., this volume).

The physiological processes behind associative learning have not yet been elucidated. Evidence indicates that experience actually causes sensitivity changes in the olfactory receptors on the insects' antennae (Vet et al., 1990b), but, unlike behavioral studies, research on physiological aspects of learning in parasitoids has only just begun. Many similarities with other Hymenoptera (e.g., Gould, this volume; Menzel et al., this volume; Smith, this volume) can be expected. Associative learning is not limited to olfactory stimuli; parasitoids are also able to link visual (next section) and mechanosensory stimuli (Monteith, 1963) with an US that indicates the presence of hosts.

Visual Learning

A Neglected Aspect of Learning in Parasitoids

As indicated above, research into sensory orientation in insect parasitoids has long focused on the capacity of parasitoids to detect and learn chemical information. This emphasis on olfaction has sometimes overshadowed the role of other sensory modalities in parasitoid foraging. The use of visual stimuli by parasitoids, for example, has received only limited attention (Wäckers and Lewis, 1992, and references within). This one-sided approach to sensory orientation in parasitoids is remarkable, considering that most of our knowledge of insect visual ecology is based on work done with Hymenoptera. Besides the prominent work on vision in honey bees (for an overview see Gould and Towne, 1988), aspects of visual orientation have been studied extensively in digger wasps (Tinbergen and Kruyt, 1938; van Iersel, 1975; Rosenheim, 1987) and ants (Hölldobler and Wilson, 1990, and references within).

Two factors may explain the limited attention paid to visual orientation in parasitoids as compared to other Hymenoptera. First, bees, ants, and digger wasps, being central-place foragers, have to commute between a home base and foraging sites. This immediately raised in the minds of investigators the question of how these insects are able to find their way back to the nest and to profitable foraging locations. Subsequent research has revealed such intriguing visual mechanisms as landmark learning, and orientation to the sun, moon, and polarized light (Gould and Towne, 1988; Hölldobler, 1976; van Iersel, 1975; Gould, this volume). Parasitoids, on the other hand, are expected to abandon host sites when prolonged searching no longer contributes to fitness optimization (MacArthur and Pianka, 1966). Landmark learning or navigation by a sun-compass could enable parasitoids to search more systematically at the habitat level, allowing them to avoid previously exploited areas. At specific locations, visual recognition of previously visited sites could enable parasitoids to avoid repeated parasitization of the same host (visual discrimination). Evidence for such visual discrimination has recently been reported by van Giessen et al. (unpublished data) and Sheehan et al. (1992).

The second aspect in which parasitoids differ from pollinators is in their interaction with resources. Interest in visual learning by flower pollinators such as honey bees was aroused by the striking visual display of insect-pollinated flowers. Darwin (1876) proposed that this visual display was a consequence of coevolution between plants and pollinators (see Lewis, this volume). Von Frisch (1915) showed long ago that bees indeed use visual information to locate nectar sites, and that associative learning of visual stimuli enables them to specialize on the most rewarding nectar sites.

Parasitoids and predators, on the other hand, might well put selection pressure on their host resource to minimize chances of being detected. Therefore, in contrast to the mutualistic interaction between plants and their insect pollinators, parasitoids and their hosts are involved in an evolutionary game of hide-and-seek. This game, in combination with the small size of hosts should restrict the role of host-derived stimuli in parasitoid foraging including visual stimuli. Parasitoids, however, appear to have adopted a different strategy to employ visual information while foraging for hidden hosts. Responses to visual stimuli have been found to be modifiable by experience analogous to olfactory learning. Visual cues from the hosts are limited, but associative learning of visual stimuli from the hosts' environment (e.g., the plants they feed on) enables parasitoids to exploit visual information during their search.

What Visual Cues Do Parasitoids Employ?

The ability of hymenopteran parasitoids to learn visual stimuli has been known since Arthur (1966, 1967) showed that the ichneumonid parasitoid *Itoplectis conquisitor* (Say) could learn to discriminate visually between rewarded and unrewarded microhabitats. Although learning of colors was the likely basis of this conditioned preference, his experiments did not exclude the possibility that the parasitoids were distinguishing microhabitats on the basis of their brightness (i.e., difference in light intensity). More recently, Wardle (1990) provided direct evidence for color learning. She demonstrated unambiguously that the parasitoid *Exeristes roborator* (F.) could be conditioned to the color of rewarding microhabitats, using differently colored microhabitats of equal intensity within the range of insect visible wavelengths. In subsequent experiments, Wardle and Borden (1990) showed that the parasitoid was also able to learn to distinguish microhabitats on the basis of their form. Form and pattern learning have also been demonstrated in work with *Microplitis croceipes* (Wäckers and Lewis, 1992).

Visual stimuli can be used by parasitoids for various purposes. For instance, *Campoletis sonorensis* females navigate by visual orientation to plants in search of hosts while visual plant stimuli also enhance mate location by males (McAuslane et al., 1990a). Wäckers and Lewis (unpublished data) reported that visual stimuli are involved in several stages of the host location sequence in *M. croceipes*. During target-oriented flight, parasitoids distinguish among targets on the basis of their visual characteristics. Conspicuous visual targets improve the accuracy of the landing. After alighting, parasitoids respond to moving objects and surface vibrations by assuming an "attack posture" often leading to an oviposition attempt in the moving subject.

Visual vs. Olfactory Cues

The ease with which an insect can visually detect an item is a function of the item's dimensions, pattern, and contrast against background, as well as the distance between the insect and the item, and the intensity of illumination (Prokopy and Owens, 1983). Visual detection is independent of air currents, and detection of visual stimuli is not altered by small changes in distance to the source (Miller and Strickler, 1984). On the other hand, detection of olfactory stimuli is influenced by the rate of emission of odor molecules, the release area, the distance between insect and odor source, wind speed, turbulence, and contrast against background odors. Since odors are transmitted as meandering plumes in moving air, an insect downwind from an odor source will encounter odor stimuli in bursts, which makes olfactory detection variable in time and in distance from the source.

In short, visual signals supply more reliable information on the direction of and the distance to the source, independent of wind direction (Prokopy, 1986). However, since physical barriers obstruct visual signals more than odor plumes, olfactory orientation would be more useful in situations where vision is hindered, such as in dense plant canopy.

The results obtained with *M. croceipes* demonstrate how parasitoids may enhance search efficiency by learning both olfactory and visual cues. In this wasp, host-finding success could be increased when information from the two sensory modalities was combined. When *M. croceipes* females experienced odor cues in association with specific visual information during encounters with hosts, they exhibited a stronger conditioned response than when they had experience with only one of the two stimuli (Wäckers and Lewis, unpublished data).

How May Learning Increase Parasitoid Foraging Success?

Sources and Reliability of Cues

The current consensus is that many wasps should learn because cues that may guide wasps to their hosts at a certain time in a certain place are unpredictable (Vet et al., 1990a; Lewis et al., 1990; Vet and Dicke, 1992). Over evolutionary time, associations between hosts and cues should be minimized by selection and most cues should be unreliable indicators of host presence and host suitability over multiple generations (Vet et al., 1991; Wäckers and Lewis, unpublished data). Yet within the life span of an individual wasp, a few cues could be highly reliable, and learning of these cues through experiences would allow many parasitoids to more effectively exploit the diversity of potential cues that may lead them to hosts (Tumlinson et al., 1992). Learning will enable wasps to adjust their responses to changes in host quality and abundance. Vet et al. (1990a) argue that responses to cues that are directly host-derived (such as kairomones) should be congenitally fixed, as such cues are intimately and reliably linked with the material presence of the host. Responses to such cues are expected to be conservative to change in both an ontogenetic and an evolutionary sense. Although host cues may be highly reliable, they will not be readily available to foraging parasitoids. After all, selection will place constant pressure on hosts to minimize the production and/or release of signals that may give away their presence (Tumlinson et al., 1991; Vet et al., 1991). Many parasitoids therefore rely on cues that are furnished not by the hosts, but by the hosts' environment. Cues from the hosts' environment such as host plant volatiles, however, are presumably less reliable because each host species can generally feed on more than one plant species and different plant structures. Until recently, evidence of variability in cues has been lacking. We can now substantiate the variability of plant and host cues with some of our own work on *Cotesia marginiventris* and *Microplitis croceipes*.

Variability in Volatile Cues

Cotesia marginiventris is a generalist that attacks the larvae of many Lepidoptera (Turlings, 1990). *Microplitis croceipes* is more specialized in that it can only successfully develop in *Heliothis* and *Helicoverpa* larvae (Eller, 1990). Despite its specialization this wasp, like *C. marginiventris*, finds its hosts in many different habitats. To locate hosts in these habitats, both *C. marginiventris* (Turlings et al., 1990b, 1991a,b) and *M. croceipes* (Drost et al., 1986, 1988; Elzen et al., 1987; Eller et al., 1988a,b; Zanen and Cardé, 1991; McCall and Turlings, unpublished data) rely principally on volatile cues emitted by plants that have been damaged by the hosts. The chromatograms in Figure 3.1 illustrate the enormous variety in volatile blends that several of these plants release when damaged by hosts. Each of the depicted volatile blends was the result of feeding-damage inflicted by one particular host, the beet armyworm, *Spodoptera exigua*. *C. marginiventris* is capable of learning to distinguish between odors released by the same host feeding on different plants (Turlings et al., 1990a). The differences in the volatile blends released by different plants also explain why a specialist parasitoid like *M. croceipes* is capable of odor learning

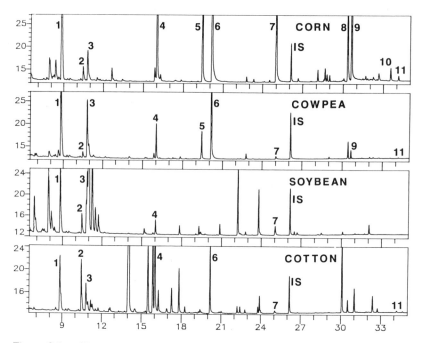

Figure 3.1. Chromatographic profiles that illustrate the differences and similarities in volatiles released by various plants damaged by beet armyworm caterpillars. The volatiles released by corn, cowpea, soybean, and cotton seedlings were collected for 2 hours while the plants were fed upon by 15 caterpillars. The volatiles were trapped on Super Q adsorbent and analyzed by means of gas chromatography. For details on collection and analysis techniques see Turlings et al. (1991b). An internal standard (IS = nonyl-acetate) was added to the samples for reference. The numbered compounds are: **1**, (*Z*)-3-hexenal; **2**, (*E*)-2-hexenal; **3**, (*Z*)-3-hexen-1-ol; **4**, (*Z*)-3-hexen-1-yl acetate; **5**, linalool; **6**, (3*E*)-4,8-dimethyl-1,3,7-nonatriene; **7**, indole; **8**, α-*trans*-bergamotene; **9**, (*E*)-(β)-farnesene; **10**, (*E*)-nerolidol; **11**, (3*E*,7*E*)-4,8,12-trimethyl-1-3-7-11-tri-decatetraene. Volatiles were collected and analyzed by P.J. McCall and T.C.J. Turlings.

(Drost et al., 1986, 1988; Eller et al., 1988b; Lewis and Tumlinson, 1988; Kaas et al., 1990; Zanen and Cardé, 1991; Wäckers and Lewis, unpublished data).

Odors released by different larvae feeding on the same plant species can be quite different as well. When we collected the volatiles released by several lepidopterous pests feeding on corn seedlings we observed consistent differences in ratios in several of the released compounds (Turlings and Tumlinson, unpublished data). *C. marginiventris* is able to distinguish between the odors released by the two closely related host species *Spodoptera frugiperda* J.E. Smith (fall armyworm = FAW) and *S. exigua* Hübner (beet armyworm = BAW). When given a choice in a flight tunnel, female wasps fly more often to corn seedlings with hosts that they had contacted previously than to corn seedlings with the other host (Fig. 3.2). *Microplitis croceipes* also can distinguish between the different odors that

CHOICE TESTS IN FLIGHT TUNNEL
(% LANDINGS ON ODOR SOURCE)

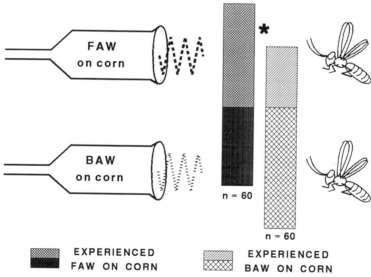

Figure 3.2. Effect of learning on preferences for odor cues exhibited by *C. marginiventris* in a two-choice flight-tunnel bioassay. Females received a 30-second contact experience with either fall armyworm (FAW) or beet armyworm (BAW) on corn. They were then offered a choice between odor sources with the two host species feeding on corn placed next to each other in a flight tunnel. The asterisk indicates a significant shift in odor preference (chi-square = 5.69, $p < 0.02$).

are released when hosts and nonhosts are feeding on the same species of plant (Zanen and Cardé, 1991).

When caterpillars feed on different structures of the same plant the odors emitted vary also. Figure 3.3 shows the odor blends that can be obtained when *H. zea* feeds on different parts of cotton plants. Wäckers and Lewis (unpublished data) were able to demonstrate that *M. croceipes* can actually learn to distinguish odors emitted by feces from hosts that feed on these different plant structures.

From these examples it is clear that, at least in those cases where the hosts are larval herbivores, the plants are essential contributors of host location cues. In fact, as we will see next, plants may actively provide the wasps with reliable chemical information, thereby significantly adding to the maze of information with which the wasps will have to deal.

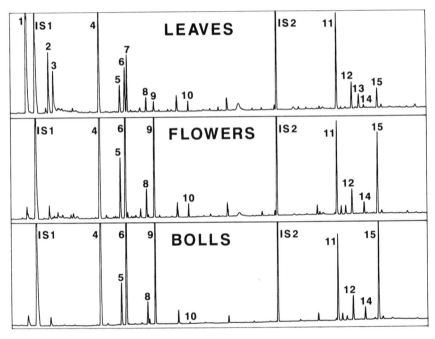

Figure 3.3. Volatiles released by various parts of cotton plants when fed upon by corn earworm caterpillars. For more details on procedure see legend with Figure 1 and Turlings et al. (1991b). Tentative analysis of the volatiles by mass spectrometry indicate the following identities: **1**, (*Z*)-3-hexenal; **2**, (*E*)-2-hexenal; **3**, (*Z*)-3-hexen-1-ol; **4**, α-pinene; **5**, β-pinene; **6**, myrcene, **7**, (*Z*)-3-hexen-1-yl acetate; **8**, limonene; **9**, ocimene; **10**, (3*E*)-4,8-dimethyl-1-3,7-nonatriene; **11**, caryophyllene; **12–15**, various sesqueterpenes. IS1 and IS2 are internal standards (*n*-octane and nonyl-acetate). Volatiles were collected and analyzed by P.J. McCall and T.C.J. Turlings.

Active Role of Plants

Many parasitoids are attracted to the odors from plants on which their hosts feed (Vinson, 1975, 1981; Vinson et al., 1987; Nordlund et al., 1988; Whitman, 1988; Whitman and Eller, 1990; Williams et al., 1988). Recent research shows that such plants are actively involved in attracting natural enemies of their herbivores. For example, when spider mites feed on lima bean, leaves of this plant release a blend of volatiles that attracts predatory mites (Dicke and Sabelis, 1988; Dicke et al., 1990a). The release of this specific blend of volatiles cannot be induced by artifically damaging the leaves (Dicke and Sabelis, 1988).

Similarly, corn leaves initiated the release of relatively large amounts of terpenoids in response to damage inflicted upon them by caterpillars (Turlings et al., 1990b). Several observations suggest that this is an active response by the plants: (1) Only a minor response can be induced by artificial damage. However, a strong volatile release can be induced when artifically damaged sites are treated with regurgitate of the caterpillars (Turlings et al., 1990b). (2) The plant's response is not instantaneous. Terpenoid release reaches significant amounts only several hours after damage (Turlings et al., 1990b). (3) Undamaged leaves of herbivore-damaged corn plants will release terpenoids in unusually large amounts as well (Turlings and Tumlinson, 1992).

Cotesia marginiventris females are strongly attracted to herbivore-damaged corn leaves (Turlings et al., 1990b, 1991a). Again, prior experience of the wasps was a very important factor in their responses. When the wasps were experienced on freshly damaged seedlings (which did not release significant amounts of terpenoids) they would fly to freshly damaged leaves just as readily as to leaves with 15-hour-old damage (which released large amounts of terpenoids) (Fig. 3-4). On the other hand, wasps that had experienced seedlings with old damage showed a strong preference for the terpenoid-releasing plants over plants with fresh damage. Apparently, the wasps were able to learn the odors that the plants with old damage were emitting.

In the earlier discussion on visual learning, we mentioned the mutually beneficial interactions between pollinators and their nectar sources (i.e., plants) and pointed out that selection would work in a different direction for interactions between parasitoids and their resources (i.e., hosts). Yet here is a case in which plants and parasitoids too may have evolved an interaction from which both profit (earlier suggested by Price, 1981; Price et al., 1980, 1986; Vinson, 1975; Vinson et al., 1987; Dicke and Sabelis, 1989; Dicke et al., 1990b; Turlings et al., 1990b; Turlings and Tumlinson, 1991). It is premature to conclude that plants are purposely signaling the presence of herbivores to natural enemies, since other functions could form

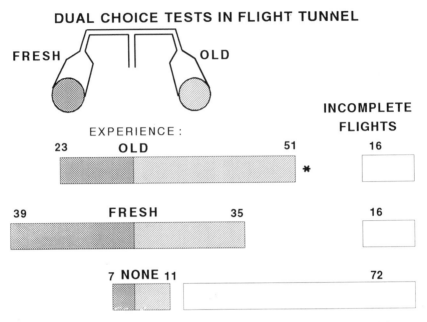

Figure 3.4. Effect of experience on preferences for odor cues exhibited by *C. marginiventris* in a two-choice flight-tunnel bioassay. Females received a 30-second contact experience with beet armyworm on either corn that had been damaged overnight (OLD) or on corn with fresh beet armyworm damage (FRESH). A third group of wasps received no experience (NONE). They were then offered a choice between two odor sources placed next to each other in a flight tunnel. The sources consisted of beet armyworm larvae feeding on corn with fresh or old damage. The asterisk indicates a significant preference for odors from corn with old damage (Chi-square test; $p < 0.05$).

the basis for the observed volatile release (Turlings and Tumlinson, 1991). It is clear, however, that the induced responses in plants add a new dimension to variation in host-location cues. Odors will vary depending upon the degree and age of the inflicted damage, and induced odor emissions will again differ for different parts of the plants. It is even possible, as speculated by D.R. Papaj (personal communication), that plants not only emit chemical "signals," but change visually (color and/or shape) as well, thereby providing the wasps with even more cues.

Increasing Search Efficiency Through Learning

The above examples illustrate the variability in cues that may guide wasps to hosts and the way in which learning assists parasitic wasps in coping with this variability. Two recent field studies demonstrate that learning aids the wasps in their foraging efforts. Papaj and Vet (1990) released *Leptopilina heterotoma* (Thomson) in a forest with systematically arranged apple-yeast and mushroom baits, both infested with hosts (*Drosophila melanogaster* or *D. phalerata*). They were able to show that a 2-hour experience on the artificial apple-yeast or mushroom microhabitats had increased the wasps' foraging success threefold. First, experienced wasps found baits more often than naive ones. Second, experienced wasps found baits faster than naive ones. Third, females were more likely to find the habitat that they had experienced than the alternative habitat (Papaj and Vet, 1990).

Lewis and Martin (1990) demonstrated the importance of host density and associated cues, as well as experience, for the foraging efficiency of *M. croceipes*. They showed that in a soybean plot artificially infested with variable host and host feces densities, both a more complete experience and high host density increased foraging efficiency (Lewis and Martin, 1990; Martin and Lewis, unpublished data).

These are encouraging results for the planned application of prerelease conditioning of beneficial insects for biological control purposes (Prokopy and Lewis, this volume). These field studies present convincing evidence that odor learning increases parasitoid foraging success in nature.

Retention of Learned Responses Depends on Reward

Learning can occur very rapidly (Vet, 1988; Vet and Groenewold, 1990; Turlings et al., 1989, 1990a; Martin and Lewis, 1992; Poolman-Simons et al., 1992). Effects of experience, however, may be short lived (Martin and Lewis, 1992) and may have only limited consequences for the long-term odor preferences exhibited by insects (Eller et al., 1992). Repeated experience will strengthen the effects and influence the wasps' foraging strategies in a more permanent way.

Cotesia marginiventris' response was significantly altered after a brief (20-second) contact experience with a plant-host complex (Turlings et al., 1989). In fact, contact with the host was not even necessary; simply contacting host frass in the presence of host-damaged leaves was sufficient to increase the responses to odors from the plant-host complex dramatically (Turlings et al., 1989). These single brief experiences may have an immediate strong effect on the responsiveness, but do not necessarily influence a female's preference significantly. In experiments with *M. croceipes* females, wasps required more than one experience before the odor of the feces they had experienced was preferred over an alternative odor (Eller et al., 1992). Poolman-Simons et al. (1992) showed that, in parasitoids of *Drosophila* larvae, brief experiences have an immediate effect on preference, but not on acceptance of host microhabitats. A single oviposition experience was sufficient to alter the insects' preferences; repeated experiences increased the time that wasps would spend in the habitat they had experienced.

Repeated experience should increase responsiveness and acceptance of certain cues and the wasps then should focus more strongly on a particular resource after repeated successful encounters with it. One-time experiences with either frass or an oviposition may not provide the wasps with adequate information about what is available within their foraging area. Repeated experience should allow them to determine where and which host species are available and adjust their foraging strategy accordingly.

The most positive experience for a female parasitoid would seem to be successful oviposition in a suitable host. If that were the only measure used to assess host presence, encounters with host feces without an actual contact with the host would not alter a female's responsiveness. In fact, contact with feces or plant damage without oviposition actually reduces the responsiveness of *C. sonorensis* (McAuslane et al., 1990b). On the other hand, the chances that a female contacting feces will actually find a host might sometimes be high. Therefore, it is not surprising that, in some instances mere contact with host products increases responses significantly (Lewis and Tumlinson, 1988; Vet and Groenewold, 1990; Eller, 1990; Turlings et al., 1989; van Giessen et al., unpublished data). Several studies show that actual oviposition does increase responsiveness, and particularly preference more than contact with host products. *Leptopilina heterotoma* females are strongly attracted to (Z)-3-hexen-1-ol after smelling this volatile when contacting a *Drosophila*-produced kairomone, but responsiveness was stronger if, in addition to contacting the kairomone, they were allowed to oviposit in hosts (Vet and Groenewold, 1990). Likewise, Martin and Lewis (unpublished data) found that an oviposition plus contact with host feces had a stronger effect on the responsiveness of *M. croceipes* than just contact with feces. Those wasps that contacted only feces showed an increase in responsiveness that decreased within an hour, while responsiveness for those that had experienced feces with an oviposition remained high for more than 48 hours (Martin and Lewis, unpublished data). Drost et al. (1986) also showed a relatively long-lasting effect of a complete experience for *M. croceipes*.

In the absence of continued experience, parasitoids tend to "forget" what they learned within a few days (e.g., Sheehan and Shelton, 1989; Papaj and Vet, 1990; McAuslane et al., 1991a). Intuitively, this makes sense. Cues that are associated with hosts should be learned by parasitoids but, when the wasps' subsequent efforts to locate hosts by tracking the same or similar cues are in vain, they may profit by orienting to other cues that are more reliable indicators of host presence.

Van Giessen et al. (unpublished data) showed that wasps will not give up easily. In fact, if *M. croceipes* encountered only frass at an odor source to which they had flown in a flight tunnel, they would readily fly to the same source a second time. Yet if wasps oviposited in a host after landing at the odor source, they appeared to avoid that particular source during additional flights. Van Giessen et al. (unpublished data) went on to show that, through visual learning, wasps were able to recognize sites at which they successfully parasitized a host which they would avoid subsequently. This avoidance through learning is likely to reduce self-superparasitism.

It seems that the wasps are capable of exploiting the most profitable cues. They will use those cues that repeatedly led them to successful encounters with suitable hosts. The wasps' tendency to "forget" may enable them to resort to different cues when the cues that they have learned are no longer rewarded. This will only be true if the drop in response to the learned cues increases the chances of encountering new hosts or host sites that are not associated with those cues. Thus far, no evidence on this point is available.

To Summarize: a Short Story

The following paragraphs present a conjectural scenario that illustrates how a female parasitoid might use her learning abilities to most effectively exploit various chemical and visual cues in the field. It is a step-by-step description of a parasitoid's host-searching efforts and emphasizes the complexity of cues that a parasitoid may encounter during her lifetime. The story may be removed from biological reality, but it touches on the essential aspects of how learning may assist parasitoids in locating hosts.

A female *C. marginiventris* parasitoid emerges from a fall armyworm (*S. frugiperda*) larva in a cotton field. The newly emerged wasp may initially not be ready to search for hosts; she may first require food and a mate to develop and fertilize her eggs. Mated females respond more intensely to host contact kairomones than unmated females (Loke and Ashley, 1984). To locate food and perhaps also mates, the wasp will use various cues which will be quite different from those that she will eventually use to locate hosts. The wasps associate different odor cues with different types of resources (food or hosts). The intensity of a wasp's response to these cues depends on her physiological state: a hungry wasp will respond strongly to the odor that she previously encountered during feeding, while a food-satiated wasp will respond to odors that she perceived while contacting host kairomones (Lewis and Takasu, 1990).

A satiated and mated mature female will subsequently spend most of her time searching for hosts. Although her mother successfully located the larva feeding on the leaves of a young cotton plant only 12 days earlier, suitable (early) stages of this lepidopterous host are no longer available.

However, earlier-instar *H. zea* larvae are quite abundant, as are cabbage loopers (*Trichoplusia ni*), which are far less suitable as hosts (Turlings et al., 1990a; M.R. Strand, personal communication). The *H. zea* larvae feed predominantly on the bolls of the cotton plants, while cabbage loopers feed mainly on the leaves. The chromatograms in Figure 3.3 indicate that odor blends vary when a herbivore feeds on different parts of the cotton plants. Most obvious is the absence of the typical low-molecular-weight green leaf volatiles in collections from cotton bolls and flowers. The wasp may initially follow odor cues, the traces of which she perceived on the cocoon from which she emerged. By tracking odor blends that are roughly similar to the one that led her mother to her host, she should eventually encounter corn earworm and cabbage looper larvae in the field. Contact kairomones will allow her to assess the suitability of the alternative hosts. During each of her encounters with the hosts, she will learn more and more to distinguish one host species from the other by both smell and vision. She will find the corn earworms mainly on bolls and thus learn to visually search for bolls. Furthermore, she will respond to the specific odor emanating from infested cotton bolls. Eventually she will search primarily for the most abundant suitable host present, i.e., corn earworm.

We can make the story more complicated by introducing a new generation of highly suitable fall armyworm larvae into the field a week or so after our female wasp emerges. By that time, many corn earworm larvae have been parasitized not only by *C. marginiventris* but also by the competing specialist parasitoid *M. croceipes*. The remaining corn earworm larvae have become too large for parasitization. At this time it is much more profitable for the wasp to track odors released by fall armyworm on cotton. Negative reinforcement may facilitate the transition. The wasp will encounter more and more unsuitable corn earworm larvae (already parasitized or too large) and she will therefore learn to avoid the odor blend that is associated with corn earworm and become more responsive to new blends. After several successful encounters she will search primarily for fall armyworm larvae and respond mainly to odor and visual stimuli associated with that host.

We probably exaggerated the dynamics of host populations in a cotton field but, even if such changes only occur over longer periods of time, it is clear that fixed responses to specific volatiles will not allow generalist wasps to deal with these changes. We limited the story to one plant species. Obviously, adding plant species to the searching range of the wasps would increase the complexity of the host environment and perhaps add to the benefits of associative learning.

Conclusions

Behavioral studies on parasitoid foraging behavior clearly indicate that learning is the rule rather than the exception for many species. Our theories (*sensu* Vet et al., 1990a; and Lewis et al., 1990) on how odor learning will modify the wasps' responses require that females innately respond to a diversity of chemicals commonly associated with hosts and/or their habitat. In the immobile immature insects the sensitivity to some of these chemicals may be elevated due to their presence in the insect's local chemical millieu, but more likely chemicals carried over from the immature affect the adult directly. This heightened sensitivity is expressed in the increased responsiveness to these odors by the adult. As an adult, a female wasp will continue to adjust her responses to specific stimuli in accordance with her experience. Each time a chemical or another type of cue is encountered in conjunction with innately recognized stimuli, the wasp will increase her response levels to this cue and/or lower her responses to others.

The process of associative learning is not limited to olfaction. Several examples now show that visual learning plays a key role as well. The different sensory modalities complement each other, each adding to the efficiency with which wasps are able to locate the hosts in specific habitats.

Although responsiveness can be dramatically increased in parasitoids by simply allowing them to contact host-related kairomones, the strength and tenacity of their responsiveness may strongly depend upon the reward (oviposition) that was associated with this contact. Repeated experiences may be required before wasps lock onto specific cues. This may allow them to assess the host situation in an area and adjust their foraging strategies accordingly. Effects of experiences are not lasting; in the absence of continued experience, they will wane within days or even hours. This may enable the wasps to switch more easily to the use of more reliable and more profitable cues when the cues that they had previously learned become less lucrative.

As suggested by many investigators (e.g., Nordlund et al., 1981; Wardle and Borden, 1985; Vet and Groenewold, 1990; Lewis et al., 1990; Vet and Dicke, 1991), the phenomenon of associative learning in parasitoids may be exploited for purposes of biological control (Prokopy and Lewis, this volume). It might be possible to condition mass-reared parasitoids prior to their release in a target area. When done properly, this may increase strongly the searching efficiency of the released insect such that control will be more effective. Some field data that support the potential of this procedure have recently been published (Lewis and Martin, 1990; Papaj and Vet, 1990).

Learning in parasitoids is not only of interest for future exploitation for biological control purposes. Parasitoids are ideal for comparative ecological studies because of the enormous range of different lifestyles that are found among the numerous parasitoid species. Detailed studies on parasitoid learning may answer many questions in insect behavior. For instance, how do instincts evolve (Papaj, this volume)? At what stages of their lives are insects sensitive to learning? Which different sensory modalities are affected by learning? Also questions regarding the link between dietary specialization of insects and their ability to learn can be answered by studying parasitoids. As several chapters in this volume indicate, parasitoid ecology is particularly useful for modeling the advantages and consequences of learning by insects.

Acknowledgments

We thank Dan Papaj, Barbara Dueben, Heather McAuslane, Aileen Wardle, Fred Petitt, and Peter Landolt for their useful comments on earlier versions of this chapter. We also thank Philip McCall for his advice and comments and particularly for providing us with some of the chemical information presented here.

REFERENCES

Alphen, J.J.M. van, and Vet, L.E.M. 1987. An evolutionary approach to host finding and selection. In J.K. Waage and D.J. Greathead (eds.), Insect Parasitoids, Academic Press, London, pp. 23–61.

Arthur, A.P. 1966. Associative learning in *Itoplectis conquisitor* (Say) (Hymenoptera: Ichneumonidae). Can. Entomol. **98**:213–223.

Arthur, A.P. 1967. Influence of position and size of host shelter on host-searching by *Itoplectis conquisitor* (Hymenoptera: Ichneumonidae). Can Entomol. **99**:877–886.

Arthur, A.P. 1971. Associative learning by *Nemeritis canescens* (Hymenoptera: Ichneumonidae). Can Entomol. **103**:1137–1141.

Birch, M. 1974. Introduction. In A. Neuberger and E.L. Tatum (eds.), Frontiers of Biology, Vol. 32: Pheromones. Amsterdam, North-Holland, pp. 1–7.

Corbet, S.A. 1985. Insect chemosensory response: A chemical legacy hypothesis. Ecol. Entomol. **10**:143–153.

Craighead, F.C. 1921. Hopkins host-selection principle as related to certain cerambycid beetles. J. Agric. Res. **64**:189–220.

Darwin, C. 1876. The Effects of Cross- and Self-Fertilization in the Animal Kingdom. (Murray, London).

De Jong, R., and Kaiser, L. 1991. Odor learning by *Leptopilina boulardi*, a specialist parasitoid (Hymenoptera: Eucoilidae). J. Insect Behav. **4**:743–750.

Dicke, M., van Beek, T.A., Posthumus, M.A., Ben Dom, N., van Bokhoven, H., and de Groot, A.E. 1990a. Isolation and identification of a volatile kairomone that affects acarine predator-prey interactions. Involvement of host plant in its production. J. Chem. Ecol. **16**:381–396.

Dicke, M., and Sabelis, M.W., 1988. How plants obtain predatory mites as bodyguards. Neth. J. Zool. **38**:148–165.

Dicke, M., and Sabelis, M.W., 1989. Does it pay plants to advertize for body-guards? Towards a cost-benefit analysis of induced synomone production. In H. Lambers (ed.), Causes and Consequences of Variation in Growth Rate and Productivity of Higher Plants. Academic, The Hague, pp. 341–358.

Dicke, M., Sabelis, M.W., Takabayashi, J., Bruin, J., and Posthumus, M.A. 1990b. Plant strategies of manipulating predator-prey interactions through allelochemicals: Prospects for application in pest control. J. Chem. Ecol. **16**:3091–3118.

Doutt, R.L. 1959. The biology of parasitic Hymenoptera. Annu. Rev. Entomol. **4**:161–182.

Drost, Y.C., Lewis, W.J., and Tumlinson, J.H. 1988. Beneficial arthropod behavior mediated by airborne semiochemicals. V. Influence of rearing method, host-plant, and adult experience on host-searching behavior of *Microplitis croceipes* (Cresson), a larval parasitoid of *Heliothis*. J. Chem. Ecol. **14**:1607–1616.

Drost, Y.C., Lewis, W.J., Zanen, P.O., and Keller, M.A. 1986. Beneficial-insect behavior mediated by semiochemicals. I. Flight behavior and influence of pre-flight handling of *Microplitis croceipes* (Cresson). J. Chem. Ecol. **12**:1247–1262.

Eller, F.J. 1990. Foraging behavior of *Microplitis croceipes*, a parasitoid of *Heliothis* species. Ph. D., dissertation, University of Florida, Gainesville, FL, 221 pp.

Eller, F.J., Tumlinson, J.H., and Lewis, W.J. 1988a. Beneficial arthropod behavior mediated by airborne semiochemicals. II. Olfactometric studies of the host-location by the parasitoid *Microplitis croceipes* (Cresson) (Hymenoptera: Braconidae). J. Chem. Ecol. **14**:425–434.

Eller, F.J., Tumlinson, J.H., and Lewis, W.J. 1988b. Beneficial arthropod behavior mediated by airborne semiochemicals: Source of volatiles mediating the host-location flight behavior of *Microplitis croceipes* (Cresson) (Hymenoptera: Braconidae), a parasitoid of *Heliothis zea* (Boddie) (Lepidoptera: Noctuidae). Environ. Entomol. **17**:745–753.

Eller, F.J., Tumlinson, J.H., and Lewis, W.J. 1992. Effect of host diet and preflight experience on the flight response of *Microplitis croceipes* (Cresson). Physiol. Entomol. **17**:in press.

Elzen, G.W., Williams, H.J., Vinson, S.B., and Powell, J.E. 1987. Comparative flight behavior of parasitoids *Campoletis sonorensis* and *Microplitis croceipes*. Entomol. Exp. Appl. **45**:175–180.

Frisch, K. v. 1915. Der farbensinn und formensinn der Biene. Zool. Jahrb. Abteil. Zool. Physiol. **35**:1–182.

Gould, J.L. and Towne, W.F. 1988. Honey bee learning. Adv. Insect Physiol. **20**:55–86.

Hérard, F., Keller, M.A., Lewis, W.J., and Tumlinson, J.H. 1988. Beneficial arthropod behavior mediated by airborne semiochemicals. IV. Influence of host-diet on host-oriented flight chamber responses of *Microplitis demolitor* Wilkinson. J. Chem. Ecol. **14**:1597–1606.

Hershberger, W.A., and Smith, M.P. 1967. Conditioning in *Drosophila melanogaster*. Anim. Behav. **15**:259–262.

Hölldobler, B. 1976. Recruitment behavior, home range orientation and territoriality in harvester ants, *Pogonomyrmex*. Behav. Ecol. Sociobiol. **1**:3–44.

Hölldobler, B., and Wilson, E.O. 1990. The Ants. Springer-Verlag, Berlin.

Hopkins, A.D. 1917. [Contribution to Discussion]. J. Econ. Entomol. **10**:92–93.

Iersel, J.J.A. van 1975. The extension of the orientation system of *Bembix rostrata* as used in the vicinity of its nest. In G. Baerends, C. Beer, and A. Manning (eds.), Function and Evolution in Behavior. Clarendon, Oxford, pp. 143–157.

Isingrini, M., Lenoir, A., and Jaisson, P. 1985. Preimaginal learning as a basis of colony-brood recognition in the ant *Cataglyphis cursor*. Proc. Natl. Acad. Sci. USA **82**:8545–8547.

Jaenike, J. 1982. Environmental modification of oviposition behavior in *Drosophila*. Am. Nat. **119**:784–802.

Jaenike, J. 1983. Induction of host preference in *Drosophila melanogaster*. Oecologia (Berlin) **58**:320–325.

Jermy, T., Hanson, F.E., and Dethier, V.G. 1968. Induction of specific food preference in lepidopterous larvae. Entomol. Exp. Appl. **11**:211–230.

Kaas, J.P., Elzen, G.W., and Ramaswamy, S.B. 1990. Learning in *Microplitis croceipes* Cresson (Hym., Braconidae). J. Appl. Entmol. **109**:268–273.

Kester, K.M., and Barbosa, P. 1991. Postemergence learning in the insect parasitoid, *Cotesia congregata* (Say) (Hymenoptera: Braconidae) J. Insect Behav. **4**:727–742.

Lewis, W.J., Jones, R.L., Gross, H.R., Jr., and Nordlund, D.A. 1976. The role of kairomones and other behavioral chemicals in host finding by parasitic insects. Behav. Biol. **16**:267–289.

Lewis, W.J., and Martin, W.R. 1990. Semiochemicals for use with parasitoids: Status and future. J. Chem. Ecol. **16**:3067–3089.

Lewis, W.J., and Takasu, K. 1990. Use of learned odours by a parasitic wasp in accordance with host and food needs. Nature **348**:635–636.

Lewis, W.J., and Tumlinson, J.H. 1988. Host detection by chemically mediated associative learning in a parasitic wasp. Nature **331**:257–259.

Lewis, W.J., Vet, L.E.M., Tumlinson, J.H., Lenteren, J.C. van, and Papaj, D.R. 1990. Variations in parasitoid foraging behavior: Essential element of a sound biological control theory. Environ. Entomol. **19**:1183–1193.

Loke, W.H., and Ashley, T.R. 1984. Behavioral and biological responses of *Cotesia marginiventris* to kairomones of the fall armyworm, *Spodotera frugiperda*. J. Chem. Ecol. **10**:521–529.

MacArthur, R.H. and Pianka, E.R. 1966. On optimal use of a patchy environment. Am. Nat. **916**:603–609.

Mandeville, J.D., and Mullens, B.A. 1990. Host preference and learning in *Muscidifurax zaraptor* (Hymenoptera: Pteromalidae). Ann. Entomol. Soc. Am. **83**:1203–1209.

McAuslane, H.J., Vinson, S.B., and Williams, H.J. 1990a. Influence of host plant on mate location by the parasitoid *Campoletis sonorensis* (Hymenoptera: Ichneumonidae). Environ. Entomol. **19**:26–31.

McAuslane, H.J., Vinson, S.B., and Williams, H.J. 1990b. Effect of host diet on flight behavior of the parasitoid *Campoletis sonorensis* (Hymenoptera: Ichneumonidae). J. Entomol. Sci. **25**:562–570.

McAuslane, H.J., Vinson, S.B., and Williams, H.J. 1991a. Influence of adult experience on host micro-habitat location by the generalist parasitoid, *Campoletis sonorensis*. J. Insect Behav. **4**:101–113.

McAuslane, H.J., Vinson, S.B., and Williams, H.J. 1991b. Stimuli influencing host microhabitat location in the parasitoid *Campoletis sonorensis*. Entomol. Exp. Appl. **58**:267–277.

McGuire, T.R. 1984. Learning in three species of Diptera: The blow fly *Phormia regina*, the fruit fly *Drosophila melanogaster* and the house fly *Musca domestica*. Behav. Genet. **14**:479–526.

Miller, J.R., and Strickler, K.L. 1984. Finding and accepting host plants. In W.J. Bell and R.T. Cardé (eds.), Chemical Ecology of Insects. Sinauer, Sunderland, MA, pp. 128–157.

Monteith, L.G. 1963. Habituation and associative learning in *Drino bohemica* Men. (Diptera: Tachinidae). Can. Entomol. **95**:418–426.

Mueller, T.F. 1983. The effect of plants on the host relations of a specialist parasitoid of *Heliothis* larvae. Entomol. Exp. Appl. **34**:78–84.

Nordlund, D.A., Jones, R.L., and Lewis, W.J. (eds.). 1981. Semiochemicals, Their Role in Pest Control. Plenum, New York, 306 pp.

Nordlund, D.A., Lewis, W.J., and Altieri, M.A. 1988. Influences of plant produced allelochemicals on the host and prey selection behavior of entomophagous insects. In P. Barbosa and D.K. Letourneau (eds.), Novel Aspects of Insect-Plant Interactions. John Wiley, New York, pp. 65–90.

Papaj, D.R., and Prokopy, R.J. 1989. Ecological and evolutionary aspects of learning in phytophagous insects. Annu. Rev. Entomol. **34**:315–350.

Papaj, D.R., and Vet, L.E.M. 1990. Odor learning and foraging success in the parasitoid, *Leptopilina heterotoma*. J. Chem. Ecol. 3137–3150.

Pavlov, I.P. 1941. Lectures on Conditioned Reflexes, 2 vols. International Publishers, New York.

Petitt, F.L., Turlings, T.C.J., and Wolf, S.P. 1992. Adult experience modifies attraction of the leafminer parasitoid *Opius dissitus* Muesebeck to volatile semiochemicals. J. Insect Behav., in press.

Poolman-Simons, M.T.T., Suverkropp, B.P., Vet, L.E.M., and de Moed, G. 1992. Comparison of learning in related generalist and specialist eucoilid parasitoids. Entomol. Exp. Appl., in press.

Price, P.W. 1981. Semiochemicals in evolutionary time. In D.A. Nordlund, R.L. Jones, and W.J. Lewis (eds.), Semiochemicals: Their Role in Pest Control. John Wiley, New York, pp. 251–279.

Price, P.W., Westoby, M., Rice, B., Atsatt, P.R., Fritz, R.S., Thompson, J.N., and Mobley, K. 1986. Parasite mediation in ecological interactions. Annu. Rev. Ecol. Syst. **17**:487–505.

Price, P.W., Bonton, C.E., Gross, P., McPheron, B.A., Thompson, J.N., and Weis, A.A.E. 1980. Interactions among three trophic levels: Influence of plant interactions between insect herbivores and natural enemies. Annu. Rev. Ecol. Syst. **11**:41–65.

Prokopy, R.J. 1986. Visual and olfactory stimulus interaction in resource finding by insects In T.L. Payne, M.C. Birch, and C.E.J. Kennedy (eds.), Mechanisms in Insect Olfaction. Oxford University Press, Oxford, pp. 81–89.

Prokopy, R.J., and Owens, E.D. 1983. Visual detection of plants by herbivorous insects. Annu. Rev. Entomol. **28**:337–364.

Rosenheim, J.A. 1987. Host location by the cleptoparasitic wasp *Argochrysis armilla*: The role of learning (Hymenoptera: Chrysididae). Behav. Ecol. Sociobiol. **21**:401–406.

Sheehan, W., and Shelton, A.M. 1989. The role of experience in plant foraging by the aphid parasitoid *Diaretiella rapae* (Hymenoptera: Aphidiidae). J. Insect Behav. **2**:743–759.

Sheehan, W., Wäckers, F.L., and Lewis, W.J. 1992. Discrimination of previously searched host sites by microplitis croceipes. J. Insect Behav., in press.

Thorpe, W.H. 1939. Further studies on pre-imaginal olfactory conditioning in insects. Proc. R. Soc. Lond. B **127**:424–433.

Thorpe, W.H., and Jones, F.G.W. 1937. Olfactory conditioning and its relation to the problem of host selection. Proc. R. Soc. Lond. B **124**:56–81.

Tinbergen, N., and Kruyt, W. 1938. Über die Orientierung des Bienenwolfes (*Philanthus triangulum* Fabr.) III. Die Bevorzügung bestimmter Wegmarken. Zs. Vergl. Physiol. **25**:292–334.

Tully, T. 1984. *Drosophila* learning: Behavior and biochemistry. Behav. Genet. **14**:527–557.

Tumlinson, J.H., Turlings, T.C.J., and Lewis, W.J. 1992. The semiochemical complexes that mediate insect parasitoid foraging. Agric. Zool. Rev., in press.

Turlings, T.C.J. 1990. Semiochemically mediated host searching behavior of the endoparasitic wasp *Cotesia marginiventris* (Cresson) (Hymenoptera: Braconidae). Ph.D. dissertation, University of Florida, Gainesville, FL, 178 pp.

Turlings, T.C.J., Scheepmaker, J.W.A., Vet, L.E.M., Tumlinson, J.H., and Lewis, W.J. 1990a. How contact foraging experiences affect the preferences for host-related odors in the larval parasitoid *Cotesia marginiventris* (Cresson) (Hymenoptera: Braconidae). J. Chem. Ecol. **16**:1577–1589.

Turlings, T.C.J., and Tumlinson, J.H. 1991. Do parasitoids use herbivore-induced plant chemical defenses to locate hosts? Fl. Entomol. **74**:42–50.

Turlings, T.C.J., and Tumlinson, J.H. 1992. Systemic release of chemical signals by herbivore-injured corn. Proc. Natl. Acad. Sci. USA, in press.

Turlings, T.C.J., Tumlinson, J.H., Eller, F.J., and Lewis, W.J. 1991a. Larval-damaged plants: Source of volatile attractants that guide the parasitoid *Cotesia marginiventris* to the micro-habitat of its hosts. Entomol. Exp. Appl. **58**:75–82.

Turlings, T.C.J., Tumlinson, J.H., Heath, R.R., Proveaux, A.T., and Doolittle, R.E. 1991b. Isolation and identification of allelochemicals that attract the larval parasitoid *Cotesia marginiventris* (Cresson) to the microhabitat of its hosts. J. Chem. Ecol. **17**:2235–2251.

Turlings, T.C.J., Tumlinson, J.H., and Lewis, W.J. 1990b. Exploitation of herbivore-induced plant odors by host-seeking parasitic wasps. Science **250**:1251–1253.

Turlings, T.C.J., Tumlinson, J.H., Lewis, W.J., and Vet, L.E.M. 1989. Beneficial arthropod behavior mediated by airborne semiochemicals. VII. Learning of host-related odors induced by a brief contact experience with host by-products in *Cotesia marginiventris* (Cresson), a generalist larval parasitoid. J. Insect Behav. **2**:217–225.

Vet, L.E.M. 1983. Host-habitat location through olfactory cues by *Leptopilina clavipes* (Hartig) (Hym.: Eucoilidae), a parasitoid of fungivorous *Drosophilia*: The influence of conditioning. Neth. J. Zool. **33**:225–248.

Vet, L.E.M. 1988. The influence of learning on habitat location and acceptance by parasitoids. Proceedings of the Third European Workshop on Insect Para-sitoids, Les Colloques de l'INRA. **48**:29–34.

Vet, L.E.M., and Dicke, M. 1992. Ecology of infochemical use by natural enemies in a tritrophic context. Annu. Rev. Entomol. **37**:141–172.

Vet, L.E.M., and Groenewold, A.W. 1990. Semiochemicals and learning in par-asitoids. J. Chem. Ecol. **16**:3119–3135.

Vet, L.E.M., De Jong, R., Giessen, W.A. van, and Visser, J.H. 1990b. A learning-related variation in electroantennogram responses of a parasitic wasp. Physiol. Entomol. **15**:243–247.

Vet, L.E.M., Lewis, W.J., Papaj, D.R., and Lenteren, J.C. van, 1990a. A variable-response model for parasitoid foraging behaviour. J. Insect Behav. **3**:471–490.

Vet, L.E.M., Wäckers, F.L., and Dicke, M. 1991. How to hunt for hiding hosts: The reliability-detectability problem in foraging parasitoids. Neth. J. Zool. **41**:202–213.

Vinson, S.B. 1975. Biochemical coevolution between parasitoids and their hosts. In P.W. Price (ed.), Evolutionary Strategies of Parasitic Insects and Mites. Plenum, New York, pp. 14–48.

Vinson, S.B. 1976. Host selection by insect parasitoids. Annu. Rev. Entomol. **21**:109–133.

Vinson, S.B. 1981. Habitat location. In D.A. Nordlund, R.L. Jones and W.J. Lewis (eds.), Semiochemicals—Their Role in Pest Control. John Wiley, New York, pp. 51–77.

Vinson, S.B. 1984. Parasitoid-host relationship. In W.J. Bell, R.T. Cardé (eds.), Chemical Ecology of Insects. Sinauer Associates Inc., Sunderland, MA, pp. 111–124.

Vinson, S.B., Barfield, C.S., and Henson, R.D. 1977. Oviposition behaviour of *Bracon mellitor*, a parasitoid of the boll weevil (*Anthonomus grandis*). II. Associative learning. Physiol. Entomol. **2**:157–164.

Vinson, S.B., Elzen, G.W., and Williams, H.J. 1987. The influence of volatile plant allelochemics on the third trophic level (parasitoids) and their herbivorous hosts. In V. Labeyrie, G. Fabres, and D. Lachaise (eds.), Insects-Plants. W. Junk Publishers, Dordrecht, 109–114.

Wardle, A.R. 1990. Learning of host microhabitat colour by *Exeristes roborator* (F.) (Hymenoptera: Ichneumonidae). Anim. Behav. **39**:914–923.

Wardle, A.R., and Borden, J.H. 1985. Age-dependent associative learning by *Exeristes roborator* (F.) (Hymenoptera: Ichneumonidae). Can. Entomol. **117**:605–616.

Wardle, A.R., and Borden, J.H. 1989. Learning of an olfactory stimulus associated with a host-microhabitat by *Exeristes roborator*. Entomol. Exp. Appl. **52**:271–279.

Wardle, A.R., and Borden, J.H. 1990. Learning of host microhabitat form by *Exeristes roborator* (F.) (Hymenoptera: Ichneumonidae). J. Insect Behav. **3**:251–263.

Weseloh, R.M. 1981. Host location by parasitoids. In D.A. Nordlund, R.L. Jones, and W.J. Lewis (eds.), Semiochemicals: Their Role in Pest Control. John Wiley, New York, pp. 79–95.

Whitman, D.C. 1988. Plant natural products as parasitoid cuing agents. In H.G. Cutler (ed.), Biologically Active Natural Products. American Chemical Society, Washington, DC, pp. 386–396.

Whitman, D.W., and Eller, F.J. 1990. Parasitic wasps orient to green leaf volatiles. Chemoecology **1**:69–75.

Williams, H.J., Elzen, G.W., and Vinson, S.B. 1988. Parasitoid host plant interactions, emphasizing cotton (*Gossypium*). In P. Barbosa, and D.K. Letourneau (eds.), Novel Aspects of Insect Plant Interactions. New York, pp. 171–200.

Zanen, P.O., and Cardé, R.T. 1991. Learning and the role of host-specific volatiles during in-flight host finding in the specialist parasitoid *Microplitis croceipes*. Physiol. Entomol. **16**:381–389.

4

Functional Organization of Appetitive Learning and Memory in a Generalist Pollinator, the Honey Bee

Randolf Menzel, Uwe Greggers, and Martin Hammer

Individual experience with environmental stimuli leaves multiple traces of neuronal plasticities in the nervous system. Receptors adapt to prolonged stimulation; neural circuits habituate to repeated stimuli and dishabituate or sensitize to arousing stimuli; and new functional connections are formed or existing ones abolished by associative and latent learning. What are the rules of neural plasticity and how do they relate to the biological constraints under which they have evolved? The neuroethological approach taken in the study of honey bee learning and memory tries to understand the neuronal mechanisms of the multiple memory traces as adaptations to the particular demands of foraging by a generalist pollinating insect. The study of the functional dynamics of memory thus serves two goals: to unravel the informational sources which guide the sequences and time dependencies of the animal's choice behavior, and to better understand the neural correlates of the various forms of memory.

In the case of a social insect like the honey bee, motivational states, activation of alternative behavioral sets (like resting, food collecting, searching, nest keeping, feeding, etc.), and decisions within any one set depend on innate and acquired experiences, both of the individual animal and of the whole society. Individual behavior and social phenomena as sources of information are equally important (Frisch, 1967; Seeley, 1985; Lindauer, 1959, 1955). The memories from recent and remote experiences, from multiple or single learning events, from mere exposure to stimuli or from contingent pairing of stimuli with other significant stimuli differ in their informational content, in the balance between innate and acquired behavioral routines, in their time dependencies, and in their sensitivities ιo new experience. This paper will focus on the implications that multiple memory traces have for choice behavior of individual honey bees in the context of appetitive learning. We will argue that, at any moment, time- and event-dependent processes inherent in the memory traces provide the animal with expectations about the consequences of its responses to stimuli. Most importantly, these expectations are heavily dependent on intrinsic, automatic memory processes and not just on exposure to external events. This perspective differs fundamentally from those developed in the framework of behaviorism (Hull, 1943; Bitterman, 1988) and traditional Pavlovian reflexology (see Pavlov, 1967). Both of the latter approaches focus on stimulus-response properties and neglect the autonomous contributions of intrinsic functional elements of the nervous system, which reflect both constraints on the cellular machinery and evolutionary adaptations designed to satisfy an animal's needs. Our own approach embraces concepts in both ethology and cognitive psychology. Cognitive terms such as expectation, prediction, attention, decision, orientation in time and space, and communication between members of a society will be applied here to learning and memory in honey bees. However, notions about consciousness or mental operations, even at a rudimentary level, will not be entertained here. Aspects of this fascinating and controversial issue are discussed by Griffin (1984) and Menzel (1990).

Functional Properties of Unconditioned and Conditioned Stimuli

Most stimuli inform the animal and are thus meaningful even to naïve, unexperienced animals. Sign stimuli release whole sets of innate behaviors while arousing stimuli alter the status and direction of attention. Few (if any) stimuli are strictly neutral in the sense that their appearance is irrelevant to the animal. Learning is a property of the nervous system in which informational status of stimuli is changed as a consequence of being passively or actively exposed to stimuli and their combinations. Repeated exposure to a stimulus without any relevant consequences for the animal leads to habituation of the response initially evoked. The sudden, unexpected appearance of a strong and meaningful stimulus arouses the animal, dishabituates habituated responses, and may transfer its informational capacity to less effective and more neutral stimuli which are temporally closely related to its appearance. These strong and meaningful stimuli are usually called unconditioned stimuli (US), while stimuli with more neutral and less obvious significance to a naive animal are called conditioned stimuli (CS). Although unsatisfactory in many respects, these terms (introduced by Pavlov, 1967) have become accepted as technical abbreviations for two extremes in the informational content of stimuli, and will be used here. Insects appear to differ from other animals with rich behavioral repertoires (such as mammals) by exhibiting more instinctive behavior, assigning more innate meaning to stimuli, and being more "prepared to learn" particular stimuli.

Insects should therefore be ideal for learning studies because changes in the informational content of stimuli should be particularly dramatic and selective (Gould and Marler, 1984). Greater preparedness for selective associations among insects may in turn be reflected in lower complexities in the neural substrate of learning and memory, more automatic intrinsic properties for the formation of the memory trace, and a closer relationship between the plasticities of the neural network and the behavioral consequences.

Training freely flying honey bees to a colored or scented target is a very useful and powerful method with which to study visual or olfactory abilities and numerous other aspects of honey bee behavior (Frisch, 1967). The natural context for these training experiments is the fidelity of individual bees to a particular flower species as a food source, a fact well known for more than 100 years (Fabre, 1879; Forel, 1910). An example of this fidelity is illustrated by the behavior of three honey bees observed in a patch offering flowers of four different species (Fig. 4.1). Each honey bee is perfectly tuned to one of the plant species and does not collect nectar or pollen from any other species. The learning underlying this choice behavior is very fast and establishes a long-lasting memory (review Menzel, 1990). Examples of the acquisition functions of two colors are given in Figure 4.2. In a dual-choice situation a violet target is preferred at a very high level after a single learning trial, while a similar preference for a green target is acquired only after several trials. Odorants are learned even faster than color (Koltermann, 1969). A single learning trial on a floral odor results in nearly 100% preference for that odor.

The memory for the CS depends on the number of learning trials. As Figure 4.3 shows, honey bees remember a blue-color target for several days, even after a single learning trial if they are prevented from learning new signals by enclosing them in the colony until the test (Menzel, 1968). If honey bees are rewarded three times on the blue target, they do not forget it for their lifetime. Interestingly, learned performance improves during an initial period, indicating that the memory controlling choice behavior may be strengthened with time since the last experience. This aspect of the memory was examined further in experiments in which shorter time intervals were chosen (Fig. 4.4). It was found that immediately after the first trial, the probability of choosing the rewarded color target was very high, then decreased to a minimum ca. 3 minutes after the learning trial, and then increased again. Such a dual-phase time course indicates that the memory trace may consist of different functional components and that an early form may be consolidated into a later form (Menzel, 1987). Below, we will use this observation as a basis for analysis of the underlying mechanisms.

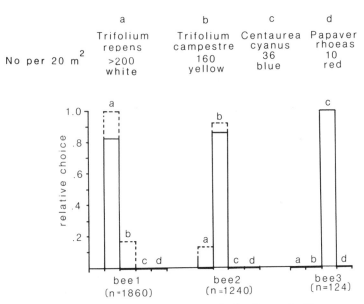

Figure 4.1. Choice behavior of three individually marked bees (Nos. 1, 2, 3, abscissa) in an area of 20 m² with four species of simultaneously blooming plants (a: *Trifolium repens*, > 200 flowers, white; b: *Trifolium campestre*, 160 flowers, yellow; c: *Centaurea cyanus*, 36 flowers, blue; d: *Papaver rhoeas*, 10 flowers, red (= bee UV-violet). Nectar is provided by the two *Trifolium* species, pollen by the poppy, and pollen and nectar by the cornflower. The choices are counted during several foraging bouts of each bee (n gives the number of choices). The dotted line gives the relative frequency of approaches without landing and the bars with the solid line, the relative frequency of landings. The cornflower was also visited, but only by unmarked bees.

The search for functional mechanisms requires more manageable experimental conditions. A very suitable paradigm is olfactory conditioning of the proboscis extension reflex (PER), a paradigm which was developed by Kuwabara (1957) and used successfully for odor discrimination tests by Vareschi (1972) (Fig. 4.5). Honey bees learn quickly to associate an odor with a sucrose reward. The sucrose stimulus delivered to contact chemoreceptors at the antennae and subsequently to the extended proboscis is a strong appetitive, unconditioned stimulus. Sucrose functions to (1) release reflexes; (2) modulate ongoing activities, enhancing the probability or the strength of responses to other stimuli; and (3) reinforce a conditioned stimulus. These properties (termed "releaser," "modulator," and "reinforcer" properties, respectively) are discussed in turn below (Fig. 4.6):

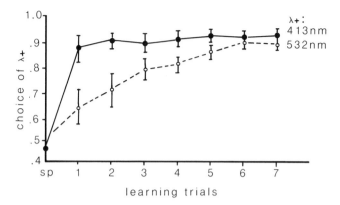

Figure 4.2. Average acquisition functions for two different spectral lights (413 nm = violet, 532 nm = green) in a dual-choice test with two color targets as alternatives. The spontaneous-choice test (sp) was performed after three initial rewarded trials on unilluminated ground glass. After the spontaneous-choice test, the single bee was either rewarded on the violet (upper curve) or the green target (lower curve). The test bee flew back to the colony after each reward and was tested for its choice behavior after it came back to the experimental setup. The tests lasted for 4 minutes, and direct approaches in flight were counted as choices. (Number of test bees: 14 for 413 nm, 34 for 532 nm; number of choices: 1532 for 413 nm, 3,872 for 532 nm, from Menzel, 1967, redrawn).

1. *Releaser function.* Several responses are released by sucrose stimulation: movement of both antennae toward the sucrose solution, extension of the proboscis with sideward and upward searching movements, rhythmic licking movements of the glossa even without contact to the solution. Several additional reflexes are released (e.g., increase in body temperature, ventilation movements of the abdomen, and leg movements), but we will focus on proboscis extension.

2. *Modulator function.* Other responses such as the weak response to an odor are enhanced and ongoing behavior is modulated. For example, the probability that the animal will respond with an extension of the proboscis to pure water or to mechanical stimuli at the antennae is facilitated by sucrose stimulation.

3. *Reinforcer function.* The effects of sucrose on behavior can be transferred to stimuli whose own presentation is contingent upon presentation of sucrose (e.g., during classical conditioning).

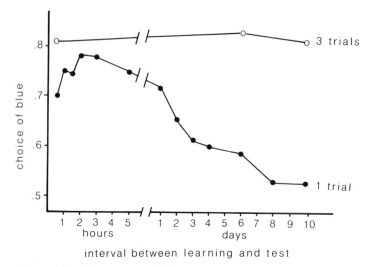

Figure 4.3. Time course of the memory for a blue target after a single learning trial and after three learning trials; from Menzel, 1968, redrawn). The freely flying bees were introduced into the dual-choice test as described for Figure 4.2 and rewarded for 20 seconds, either once or three times on the blue target. The alternative target was always yellow. The spontaneous choice between the two colors was balanced (close to 50% each). The bees were caged within the hive for various time periods and released shortly before the test (abscissa). Each bee was tested only once for about 4 minutes. Similar memory functions were found for a yellow target.

Both the modulatory and reinforcing function of the US will be discussed at length below. All three functional properties—reflex releaser, arousing modulator, appetitive reinforcer (Fig. 4.6)—act together in classical conditioning, and it is important to analyze the different memories initiated by each of these US properties.

A certain low proportion of honey bees respond to an odor stimulus with the PER even before conditioning. We usually term this a "spontaneous response," although it is not clear whether learning prior to the experiment induces the response or whether there is a weak response tendency even without prior learning. The distinction is of general importance, and it is unfortunate that the critical experiments (i.e., experiments which would integrate odor deprivation during larval and pupal development with adult conditioning) have yet to be performed in a satisfactory way.

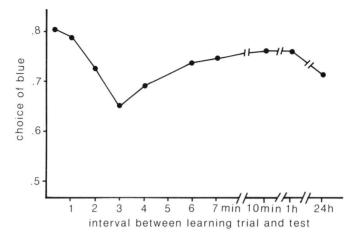

Figure 4.4.   Time course of the memory after a single
learning trial on a blue-color target (freely flying bees, same
experimental arrangement as in Figure 4.2 and 4.3; redrawn
from Menzel, 1968; and Erber, 1972, 1975a,b). Each
experimental bee was tested only once for four minutes after
the time interval indicated at the abscissa. Bees tested up to
an interval of 10 minutes were kept flying freely; those for
tests after longer intervals (1 hour to 24 hours) were caged in
the colony (see Figure 4.3). The minimum around 3 minutes is
highly significantly different from the initial and later high
levels of choice behavior (χ^2 test, $p \leq 0.01$).

Some definitions of learning require that associative learning lead to a
new behavioral response, the β-response, although in many learning paradigms the result of a CS/US pairing is just an increase in the probability
or strength of a preexisting response to the CS (α-response). Hull's (1934)
distinction between α- and β-response is taken by certain authors (e.g.,
Schreurs, 1989; Gormezano, 1984) to imply two classes of associative learning, whereas others see a continuum from pairing-specific sensitization,
through protection from habituation of a weak α-response to the emergence
of a new conditioned response (β-response) (Carew et al., 1984; Hawkins
et al. 1989; Colwill and Rescorla, 1988).

The discussion may appear at first glance to be an academic quarrel over
semantics. However, the discussion takes on a greater significance when
it is considered in light of evolutionary arguments that lower animals with
their greater development of instinctive behavior may be capable only of
α-conditioning and that the development of a totally new behavior as a
result of CS/US pairing is a property of more highly evolved nervous

Figure 4.5. Experimental arrangement for olfactory conditioning of the proboscis extension reflex (PER). Bees are harnessed in metal tubes by a stripe of sticky tape in the neck region. Sucrose stimulation of the antennae releases the PER and arouses the animal. The US used in conditioning experiments is a compound of sucrose stimulation first of the antennae and then of the proboscis (see text).

systems. Indeed, conditioning of an α-response, pairing-specific sensitization, and protection from habituation refer to mechanisms of neural plasticity which are restricted to specific stimuli and their combinations and lead to a behavioral change in a few or even a single pairing trial. Such prepared associations have been defined as a unique form of learning, "instinctive learning," (Gould and Marler, 1984). The concept is a most useful one, stressing as it does species-specific adaptations in learning abilities. However, the concept also tends to downplay the innovative power of associative learning. In the case of olfactory conditioning in the honey bee, for example, a whole range of chemosensory CS can be conditioned. Some stimuli (e.g., pheromones like citral, geraniol, floral odorants) release an α-response, while others (e.g., propionic acid) appear neutral or slightly aversive, and still others (e.g., fatty acids or the sting pheromone isoamylacetate) are strongly repellent. Honey bees can easily be trained to any of these odorants both in instrumental conditioning paradigms using freely flying bees and in classical PER conditioning paradigms using tethered ones (Menzel, 1990). The acquisition function of these various odors is often the same, though odors may differ in the number of extinction

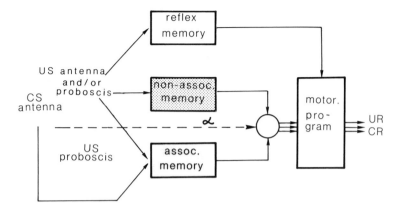

Figure 4.6. Functional organization of the PER-conditioning paradigm. The US (sucrose stimulation of the antenna and/or proboscis) has three properties: reflex releaser, response modulator, and CS reinforcer in CS/US pairing experiments (see text). The CS (olfactory stimulus) is not a completely neutral stimulus but causes a PER in a small proportion of the test animals before conditioning. This response is called α-response according to the nomenclature in animal learning studies (see text).

trials necessary to reach a conditioned response (CR) criterion and in their potential to act as a conditioned US (a phenomenon known as "second-order conditioning"). Other stimuli such as mechanical stimulation of the antennae, which are obviously aversive, can still become a CS after a few CS/US pairings. These observations illustrate two important concepts: (1) an α-response to a CS may facilitate associative learning of the CS, but does not appear to be a prerequisite for it, and (2) the honey bee is fully able to adopt completely new responses to a CS as a result of conditioning. We thus favor the view that α- and β-conditioning represent two extremes of a continuum.

Requirements for Optimal CS/US Pairing

Contiguity between CS and US is the most important factor in the conditioning process (Rescorla, 1967, 1988; Rescorla and Holland, 1982). In the case of PER-conditioning with a single conditioning trial, the optimal CS/US interval is +5 to 0 seconds with the CS preceding the US (so-called "trace or forward conditioning") (Fig. 4.7). Multiple conditioning trials

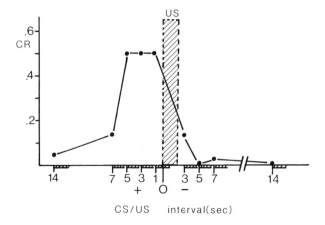

Figure 4.7. Optimal CS/US interval in single-trial olfactory PER-conditioning. The CS (carnation) was presented for 2 seconds at one out of nine different times before or after the US. After this single pairing, the response probability of CR (conditioned response) was tested 20 minutes later. Each point gives the CR probability of a group of animals (10–18 animals in each group). The US lasting for 2 seconds is marked with a striped bar, the CS pulses (abscissa, striped lines) occurring before the US appear at the left side (+), and those after the US appear at the right side (−) of the US. The CR is successfully established if the CS precedes the US by up to +5 seconds. Because the CS lasts only for 2 seconds, a CS trace lasting over 3 seconds can be associated with the US. Backward conditioning is ineffective.

indicate that the animal learns to respond to the CS also at CS-US intervals of +10 seconds, but not at intervals of +30 seconds (Fig. 4.8). The CS must be stored in a kind of sensory memory, which outlasts stimulation by several seconds.

To characterize sensory memory further, an experiment was performed in which two different odors were presented in succession, and the US was applied immediately after the second CS (Fig. 4.9). If the two CSs (denoted O_1 and O_2) are separated by 30 seconds, only O_2 is associated with the US. No effect on O_1 or from O_1 onto O_2 is found, because the response probabilities to O_1 and O_2 alone are the same as in tests when both odors were presented in succession (either in the order presented during con-

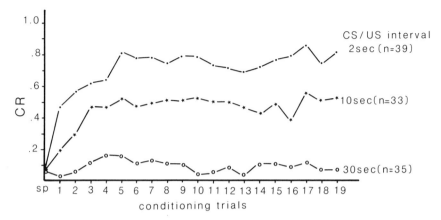

Figure 4.8. Three acquisition functions for different CS/US intervals in olfactory PER-conditioning. The CS (carnation, 2 second) precedes the US (40% sucrose solution, 2 seconds) by either 2 seconds (upper curve), 10 seconds (middle curve) or 30 seconds (lower curve). Multiple-trial conditioning reveals that a CS trace of up to 8 seconds is still successfully associated with the US, but much longer intervals are ineffective. Sp gives the response probability to the CS before conditioning (spontaneous response). Number of animals: n = 39 for 2 seconds CS/US interval, n = 33 for 10 seconds, n = 35 for 30 seconds.

ditioning or in the reverse order). If the two CSs follow each other quickly (1-second interval), however, the animals associate both odors with the US, although O_2 somewhat more than O_1. Interestingly, the animals appear to learn the sequential order of the CSs, because O_2 elicits significantly higher response if the sequence during the test is the same as during the learning trials (O_1, O_2) than if the order is reversed (O_2, O_1).

These kinds of experiments were undertaken to determine whether CS/US contiguity is a fixed, stereotyped property or is flexible and under the control of sensory events or behavioral conditions. In our earlier work with variable CS/US intervals in single conditioning trials, we were impressed to find similar optimal CS/US intervals in instrumental odor learning and olfactory PER-conditioning, and concluded that the optimal CS/US interval is a fixed property of associative learning in honey bees (Menzel, 1990). However, we already know from Grossmann's (1971) experiments with freely flying honey bees that the CS/US interval can be extended considerably by so-called cued delay procedures. Now we find in the multiple-trial conditioning experiment (Fig. 4.8) that the CS/US interval can be extended also in multiple-trial PER-conditioning. Furthermore, we see in

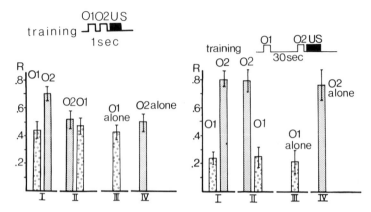

Figure 4.9. The CS trace in olfactory PER-conditioning is characterized by a double CS/single US experiment with two different intervals between the two CSs (1 second and 30 seconds). The two CSs were carnation (C) and propionic acid (P) presented each for 2 seconds, and the US is the usual sucrose solution (40 %, 2 seconds). C and P are two odorants which bees learn to distinguish within a few trials if differentially conditioned. The acquisition functions are similar with C acquired somewhat faster than P. Both series of experiments were run in a balanced fashion, with C being the first CS for some animals and P being the first for the other animals.

The sequence of stimulus presentation and pairing was the following (O_1 and O_2 correspond to either P or C depending on the experimental group; O_1O_2 indicates that the two odors are presented in sequence at the interval of 1 second or 30 seconds depending on the experimental group; + marks US reinforcement; the semicolon indicates an interval of 15–30 minutes): O_1; O_2; O_1O_2+; O_1O_2+; O_1O_2+; O_1O_2+; O_1O_2+; O_1O_2+; O_1; O_1O_2+; O_2O_1; O_1; (*)O_1O_2+; O_2; O_1; O_1O_2+; O_2O_1; O_1O_2+; O_2; O_2O_1; O_1O_2+; O_1; O_2O_1; O_1O_2+; O_2; O_1O_2+; O_2; O_1; O_2O_1; O_1O_2+; O_2; O_1; O_1O_2+; O_1; O_1O_2+; O_2; O_2O_1. The bars in the figure give the response probability to O_1 or O_2 for the different test conditions as cumulative responses after the 12th stimulation (13th–34th from *) during the respective olfactory (O_1, O_2) stimulation.

The four different test conditions are represented by the four groups I to IV. **I**: Sequence of odors during the tests as during conditioning: O_1O_2. **II**: Sequence of odors during the tests reversed to that during conditioning: O_2O_1. **III**: Response to O_1 alone. **IV**: Response to O_2 alone. Two bars are given for the test conditions I and II, because the PER could occur after O_1 and/or O_2. The number of tests for each test condition is: **I**: $n=925$, **II**: $n=523$, **III**: $n=595$, **IV**: $n=514$. Since each animal was tested several times in each test condition, the relative response rate R could be calculated for each animal, and thus the standard deviation of the average could be calculated. Number of test animals; upper graph: $N=97$, lower graph: $N=100$.

the dual CS/single US experiment (Fig. 4.7) that two CSs with different time relationships to the US and their temporal sequence can also be conditioned. We tentatively conclude that contiguity is a flexible component of honey bee learning.

The Intrinsic Dynamics of Memory as Determined by the Single Learning Paradigm

The drastic change in behavior as the consequence of a single trial indicates a strong innate preparedness to learn odors. Aversive learning has often been found to be extremely fast in vertebrates [e.g., food avoidance conditioning (Kamin, 1969; Garcia and Koelling, 1966)], but one-trial appetitive learning as in the honey bee is unusual and most convenient for learning studies which aim to analyze the intrinsic components of memory processing. As described above, free-flying honey bees need only one reward on a scented target (Koltermann, 1969) or on a target with violet color (Menzel, 1967) to choose the target afterward with very high probability. Since the learning trial in olfactory PER-conditioning lasts only a few seconds and the amount of reward can be as little as fractions of a microliter, time-dependent processes following the learning trial can be isolated effectively and distinguished readily from event- or experience-dependent processes. Such intrinsic time-dependent memory processes are evidence of a neural machinery which establishes a preordained memory trace.

The Non-associative and Associative Memory Trace

The modulatory action of a single US exposure leads to an increase in the olfactory α-response immediately after the US and a fast decay within the following 3 minutes (Fig. 4.10). By comparison, an associative learning trial with the optimal CS/US interval for PER-conditioning produces a biphasic time course of the conditioned response which is very much like the time course of conditioned choice behavior in freely flying, color-trained honey bees (Fig. 4.4). The first phase is characterized by fast decay of the CR probability which, although beginning at a higher level, parallels the time course observed after a single sensitization trial. The CR probability in the second phase, 3–10 minutes after the single CS/US pairing, rises slowly over time. One can conclude from these patterns that non-associative memory initiated by the sensitization trial contributes considerably to the high response level within the first minutes and that a specific associative memory component develops slowly over several minutes. During this process of consolidation, responses become con-

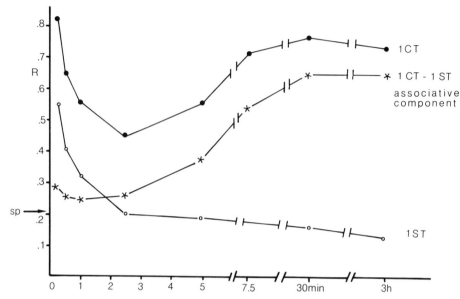

Figure 4.10. Time course of the proboscis extension response to an odor stimulus either as a sensitized α-response after a single US to the antennae and proboscis (lower trace, 1 ST) or as a CR after a single pairing (upper trace, 1 CT). The tests were performed at various intervals (abscissa). Each point represents the response probability of an independent group of animals. The number of animals tested in five series of 1 ST experiments is $n = 2,057$ and in three series of 1 CT experiments is $n = 921$. The middle line is the difference between the two functions and is interpreted to represent the associative component of the memory after a single conditioning trial. The arrow marked with "sp" gives the spontaneous response rate for all eight series of experiments.

trolled more and more by associative memory alone. Associative memory lasts longer than 24 hours (even after a single learning trial) in both PER-conditioned harnessed honey bees as well as in color-trained free-flying ones (Fig. 4.3).

Assuming that the response to the CS at any time after a single CS/US pairing is a simple joined function of the US sensitization effect and the CS/US pairing effect, the biphasic time course can easily be understood, because the nonassociative memory fades faster than the associative memory resulting from the consolidation process strengthens. An important implication of this interpretation is that, immediately after the learning trial, a weak associative memory already exists which is strengthened over time (see Fig. 4.10).

Stability of the Memory Trace

Consolidation of the memory trace after a single learning trial leads to changes in memory not only with respect to its control over choice or response behavior but also with respect to its resistance to change by new information. Consolidation thus influences the extinction of memory (due, for example, to exposure to unreinforced CS presentations), the outcome of reversal learning (i.e., the outcome of reinforcement with a new CS), the resistance of memory to certain experimental treatments which cause retrograde amnestic effects, and the tendency for the learned stimulus to be distinguished from more-or-less similar stimuli (cf. Smith, this volume, for a discussion of stimulus generalization). These phenomena have been well known in vertebrate learning ever since Ebbinghaus's (1885) famous experiments on human memory (Müller and Pilzecker, 1900; Weiskrantz, 1970; Squire and Cohen, 1982). They emphasize the time- and event-dependent character of a memory processor in the nervous system which, after its initiation by a learning trial, proceeds through phases. The properties of these phases are likely to reflect species-specific adaptations to biologically relevant learning processes under natural conditions. In the case of an insect collecting nectar and pollen (which provide only minute amounts of food and exist in a large number of competing plant species), we should expect to find a functional match between properties of these phases and the temporal dynamics of foraging. We shall return to this point after a detailed characterization of the memory processing in the honey bee.

Unreinforced experience with a formerly rewarded CS (i.e., extinction) has little effect on response level in olfactory PER-conditioning and on color-training experiments (Menzel, 1967, 1968, 1990). The conditioned response decreases only slowly after many repetitions of the CS without US. (If, however, the CS is paired with an aversive US such as water under conditions where sucrose solution is expected, the CR decreases much faster.) However, reward duration affects sensitivity to extinction even after a single trial. Figure 4.11 shows the resistance to repeated extinction trials in instrumentally color-trained animals. Resistance is weak after a short reward (2–5 seconds) and strong after a long reward (7–15 seconds). Furthermore, if the time interval between a single PER-conditioning trial and an extinction trial is varied, it becomes clear that the memory trace is more sensitive to extinction during the first 3 minutes after conditioning than later (Fig. 4.12).

The stability of the early memory trace after a single learning trial can also be tested in a dual-reversal learning experiment where the intertrial interval (ITI) between the initial learning trial and the reversal trial is varied between several seconds and 10 minutes (Fig. 4.13). Again, two

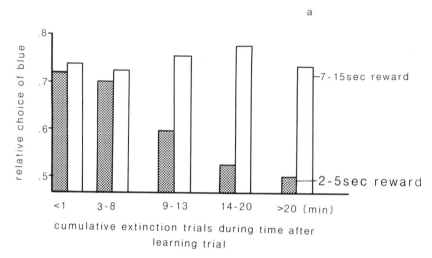

Figure 4.11. Resistance to extinction in freely flying, color-trained bees after a single short (2–5 seconds, dotted bars) or a longer (7–15 seconds, open bars) reward. The instrumental learning trial was given at time zero on a blue target. The alternative color yellow is chosen equally strongly in the dual-choice test before conditioning. Bees are chased from the target after a certain time ranging between 2 and 5 or 7 and 15 seconds. When they return to the experimental setup, they are presented with the two colored alternatives without any reward continuously during the next 30 minutes. Extinction trials accumulate for both groups of animals at about equal frequency. Number of animals tested: 74, number of choices: 2,072 (redrawn from Menzel, 1968).

groups of freely flying honey bees were examined which differed in the strengths of the US (5 seconds vs. 15 seconds during the initial trial). The honey bees were tested for their response to the two color targets in a forced dual simultaneous choice at a time ≥ 10 minutes after the reversal trial. Honey bees are prepared to reverse their learning to the new color target at ca. 3–4 minutes ITI, i.e., when the conditioned response after a single learning trial is minimal (compare with Fig. 4.4). This result suggests that new information is acquired best at the time when the joined action of early (nonassociative) memory and that of consolidated (associative) memory is weakest (Menzel, 1979)

A similar effect was found for olfactory PER-conditioning (Fig. 4.14). The odor conditioned first (O_1) establishes a stronger memory than the odor conditioned next (O_2) if the ITI is either very short (30 seconds) or long (10 minutes), but memories of the two odors O_1 and O_2 are equally strong if the ITI is 3 minutes. The two odors used in this experiment

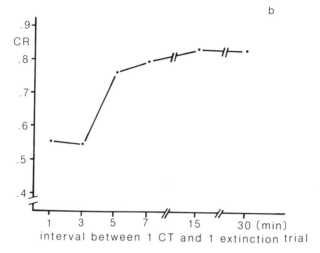

Figure 4.12. Sensitivity to extinction after a single PER-conditioning trial. The time interval between the conditioning trial (1 CT) and the extinction trial was varied between 1 and 30 minutes (abscissa). The test trial was given 1 hour after the CT. Six independent groups of animals were tested at the intervals 1, 3, 5, 7, 15, and 30 minutes respectively (20–30 animals in each group). The difference between the two first and the four later groups is highly significant ($p \leq 0.01$, χ^2 test).

(geraniol, propionic acid) were selected because honey bees are unlikely to generalize between them. This precaution presumably reduced any tendency for the second conditioning trial to reinforce partially the first conditioned odor. Geraniol is acquired faster (CS: 0.57) than propionic acid (CS: 0.34, see Fig. 4.14 control groups). The α-response is considerable for geraniol (spontaneous response probability = 0.18) and nil for propionic acid. Furthermore, proprionic acid releases aversive responses (backward movements of the antennae), while geraniol does not. Nevertheless, the response probability to the first conditioned odor is always higher than in a control group which was conditioned only to that odor. The CR to the second conditioned odor for an ITI of 30 seconds is equal to the CR after conditioning only that particular odor, whereas it is significantly higher for an ITI of 3 minutes. Since a repetition of the US alone at an ITI of 3 minutes does not enhance the CR (see control groups in Fig. 4.14), the second associative learning trial strengthens memory of the first conditioned odor even in the absence of generalization between the

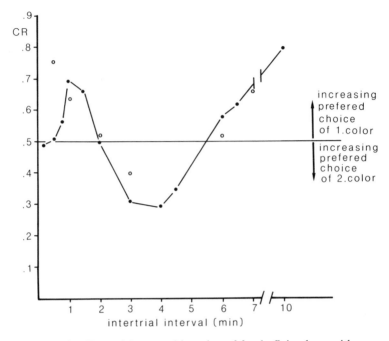

Figure 4.13. Two-trial reversal learning of freely flying bees with varying intertrial intervals (ITI, abscissa) between the initial trial on blue (1. color) and the second trial on yellow (2. color). The bees were tested for the choice between the two colors in a dual forced choice more than 10 minutes after the second learning trial (redrawn from Menzel, 1979). Two groups of bees were distinguished—those rewarded for 5 seconds (●) and those rewarded for 15 sec (○). The ordinate gives increasing choice proportions for the color learned first upward and for the color learned second downwards.

CSs. This strengthening of the first established memory trace appears to be independent of the ITI. However, the second memory trace is enhanced only for the ITI of 3 minutes. We conclude from these results that pro- and retroactive facilitatory processes are strongest at the time when the joined action of the nonassociative and the associative memory on the CR is weakest (ITI = 3 minutes), whereas only retroactive processes are effective when the nonassociative memory is strongest (ITI = 30 seconds).

Results for the two US strengths used in the experiment with the freely flying honey bees differ in the first minute (Fig. 4.13). A long US during the initial trial induces stiffer resistance to reversal learning than does a short US. This result corroborates those of resistance to extinction reported

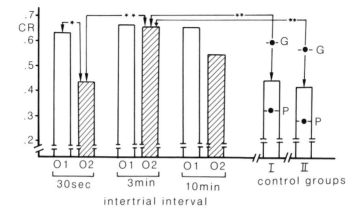

Figure 4.14. Two-trial-reversal conditioning of the olfactory PER. The two odors used were geraniol (G) and propionic acid (P). In one group of animals, geraniol was the first conditioned odor (O_1) and propionic acid the second (O_2); in a second group, propionic acid was O_1 and geraniol O_2. The results are pooled and the probability of the CR to either O_1 or O_2 is given for three intertrial intervals (30 seconds, 3 minutes, 10 minutes). CR was tested 30–60 minutes after the second trial; 141 animals were tested. Control group I gives the CR after one conditioning trial with propionic acid (P) or geraniol (G) alone (the bar corresponds to the average of both); animals of control group II received also only one conditioning trial with geraniol (G) or propionic acid (P) (bar: average), but received an additional US 3 minutes after the conditioning trial.

in Fig. 4.11, if we assume that the balance between the non-associative and the associative memory immediately after the first trial depends on US strength. In particular, a longer and stronger US enhances short-lasting non-associative memory more than it facilitates the consolidation process of associative memory.

The temporal dynamics of the memory trace after a single learning trial are also revealed by retrograde amnestic procedures. Experimental treatments such as narcosis, cooling and weak electric brain stimulation induce amnesia if applied within the first few minutes after the learning trial, but cause no effect if applied at intervals longer than 5 min (Menzel et al. 1974; Erber, 1975a,b; Erber et al. 1980) As Figure 4.15 shows, the amnestic gradient is quite independent of both learning conditions and sensory system (e.g., it is the same for classical olfactory PER-conditioning of har-

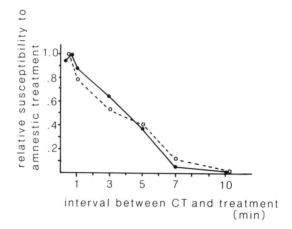

Figure 4.15. Time course of retrograde amnesia induced by weak electric brain stimulation (EBS) after a single instrumental odor-training trial (○; after Erber, 1975a,b, redrawn) or after a single olfactory PER-conditioning trial (●; after Erber et al. 1980, redrawn). The abscissa gives the time interval between the conditioning trial and the EBS. The ordinate expresses the amount of retrograde amnesia as the relative change in the conditioned performance (dual choice or CR probability) as tested more than 30 minutes after the EBS treatment.

nessed honey bees and instrumental color conditioning of freely flying ones). To some extent, the gradient depends on the experimental procedure. For example, cooling and narcosis appear less effective than electric brain stimulation (EBS), perhaps because EBS acts immediately to interfere with ongoing neural activity, whereas cooling to a few degrees above $0°C$ or narcosis with N_2 or CO_2 require up to 1 minute to become effective (Menzel, 1984, 1987, 1990). Significantly less amnesia is observed if the animal has been trained or conditioned for more than two trials and if the EBS is applied immediately after the last trial. The extent of amnesia corresponds to the contribution to memory of the last learning trial (Erber, 1975a,b; Menzel, 1984).

It is concluded that multiple learning trials help to establish a stable memory much faster than just a single trial. We next examined whether the additional learning trial promotes transfer of the susceptible memory trace (resulting from the first learning trial) into a stable memory or whether a new memory is immediately transferred into an unsusceptible form (as, for example, if short-term memory is occupied as a consequence of prior learning trials). If transfer is accelerated by an additional trial, we might ask which of the components of the second trial (CS or US alone or CS/US pairing) is responsible for this effect (Menzel and Sugawa, 1986). It appears (Fig. 4.16) that transfer is promoted by the second trial, because the content of the second learning trial is erased by EBS but not that of the first trial.

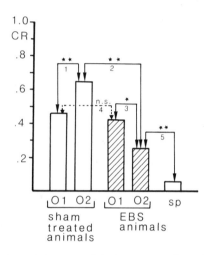

Figure 4.16. Facilitation of the consolidation process in a dual-olfactory PER-conditioning experiment. Groups of bees were conditioned to two odors (geraniol, propionic acid) in quick succession (within 30 seconds). For one group geraniol was the first odor (O_1) and propionic acid the second (O_2); for the other group the order was reversed. The data from both groups are pooled. EBS was applied immediately after the second conditioning trial, thus well within the time period which causes full retrograde amnesia after a single learning trial (see Fig. 4.15). CR tests were performed 1 and 2 hours later. The result of both CR tests are pooled; 100 animals were tested in each of the four groups (temporal order of geraniol and propionic acid, sham treated and EBS). Sp gives the spontaneous response level pooled for both odors and for all four groups.

The results from the sham-treated animals (open bars, left) indicate that response to the second conditioned odor (O_2) is higher than that to the first conditioned odor (O_1) (χ^2 test, $p \leq 0.01$, line 1). EBS has no effect on the response to O_1 (dotted line 4), but induces retrograde amnesia to the second learning trial (line 2). The CR to O_2 is also lower than to O_1 in the EBS group (line 3, $p \leq 0.05$), further indicating that the EBS is selectively acting on the memory for O_2. However, memory for O_2 is not completely erased, because the spontaneous response to either odor before conditioning is less than the CR to O_2 (open bar sp; line 5) (from Menzel and Sugawa, 1986, redrawn).

Together with the additional control experiments in Menzel and Sugawa (1986), these results prove that only a second associative learning trial is able to induce the faster transfer to an unsusceptible memory and not an additional exposure to the CS or US alone.

Although each of the experiments presented in Figs. 4.14 and 4.16 contains the control group necessary to reach the conclusions presented above, there is still an unresolved contradiction between the results for the 30-second interval groups. In one case (Fig. 4.14) the odor conditioned first (O_1) provokes a stronger response than the odor conditioned second (O_2). In the other case (Fig. 4.16) the effect is reversed. We know from the results with freely flying bees (Fig. 4.13) that short-term reversal learning (ITI < 1 minute) is highly sensitive to the strength of the US. A longer lasting US strengthens the stimulus learned first more than a stimulus learned second (*open circles* in Fig. 4.13). It is unknown whether this dependence might explain the discrepancy between the results in Figs. 4.14 and 4.16, and how associative and non-associative memories interact in olfactory conditioning as a function of US-strength and temporal dynamics. These dependencies appear as the most effective components in the control of the choice processes of freely flying bees (Figs. 4.21 and 4.26).

The consolidation process, which establishes a stable memory after a single olfactory learning trial, can be roughly localized in the brain by reversibly blocking neural activity in selected brain areas. This is done through the use of thin, cooled needles inserted into the brain at successive time intervals after the conditioning trial (Menzel et al., 1974; Erber et al., 1980) (Fig. 4.17). Cooling the antennal lobes induces retrograde amnesia only when treatment is applied immediately (i.e., < 2 minutes after a learning trial. By contrast, the mushroom bodies are prone to cold-induced amnesia for a longer time after a learning trial. The output regions of the mushroom bodies (termed α-lobes) are associated with a faster time course than the input regions (termed calyces).

Dynamics of the Memory Content

Consolidation of memory is most likely an active internal process that incorporates the new memory into existing memories. It is possible, for example, that a recent memory may change in content during consolidation as a consequence of interactions between old and new memories. We attempted to evaluate this possibility by looking for stimulus generalization to odorants, either immediately after a conditioning trial or at an interval of 15 minutes (Smith and Menzel, unpublished data) (Fig. 4.18). Responses to four odors (hexanol, citral, geraniol, 2-hexanol) were assayed in a PER-conditioning paradigm which employed a single conditioning trial. These

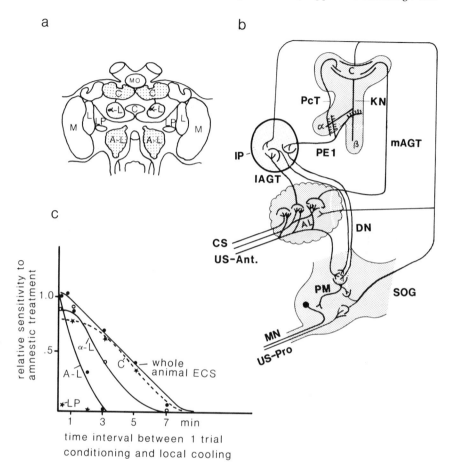

Figure 4.17. *a.* A diagram of the bee brain as it is relevant for the neural circuits and neuropils underlying the olfactory PER-conditioning. The major neuropils in the supraesophageal ganglion (= brain) are M (medulla) and L (lobula), the two inner visual ganglia; A-L (antennal lobe), the primary olfactory neuropils; C (calyx) and α-L (α-lobe) belong to the mushroom bodies; LP (lateral protocerebrum) is an unstructured neuropil ventrolateral to the mushroom bodies; C (central body); Mo (median ocellus). *b.* Major pathways that are involved in PER-conditioning. SN = antennal sensory nerve; MN = motor neurons controlling the mouthparts; mAGT = median antenno-glomerularis tract, a major relay pathway to the chemosensory input region of

odors were selected because honey bees discriminated readily between any pair of them (Smith and Menzel, 1989; Vareschi, 1971). Results of two experimental series are illustrated in Figure 4.18. In one series (trained odor = citral), the response profile to the four odors is quite similar at both test periods (Fig. 4.18). However, the response to the conditioned odor becomes relatively stronger over time, indicating that the degree to which bees will generalize from a conditioned odor to novel odors has been reduced during consolidation. Memory content has thus changed over time so as to favor even more the conditioned odor. This result was also found for hexanol and 2-hexanol. For geraniol, however, the response profile changes drastically during consolidation (Fig. 4.18). Response to the trained odor is highest in the short term, but response to an untrained odor (i.e., citral) is highest in the long term. Analysis of the time course of CR after the conditioning trial with citral or geraniol reveals that the CR to citral follows the usual biphasic time course with a rise at intervals longer than 3 minutes, whereas the time course of the CR to geraniol lacks such a prominent rise.

These results illustrate a very important property of the intrinsic memory processor—namely, its dependence on the nature of the stimulus. Certain CSs evoke stronger changes in response during consolidation than others. For salient stimuli, consolidation does indeed change the content of memory such that an animal is able to discriminate more precisely between conditioned and novel stimuli. For other stimuli, content changes not quantitatively but qualitatively; the bee remembers something other than the actual stimulus/reward pairing. We may summarize by remarking that a stimulus may often be recorded in memory as something different from what was actually experienced. This obviously reflects the influence on

---

Figure 4.17. (Continued) the mushroom bodies; KN = Kenyon cells, the intrinsic neurons of the mushroom body; α and β are the two output lobes of the mushroom body; Pct = proto-cerebro-calycal tract, a feedback tract between the α-lobe and calyx; SOG = subesophageal ganglion; PM, DN = premotor and interneurons that relay the descending commands to the motorneurons. *c.* Time-courses of local cooling which leads to amnestic effects in olfactory PER-conditioning. In all cases, both antennae were exposed to the CS, and the animals were conditioned by one trial. The indicated paired structures were cooled to 1°C for 10 seconds (for experimental details see Erber et al. 1980). The abscissa gives the time interval between the conditioning trial and the onset of cooling. The ordinate gives the sensivity to the amnestic treatment. The latter is the inverse of the proportion of animals responding to the CS 20 minutes after the amnestic treatment, at a time when the animals

memory formation of preexisting information stored in the nervous system, information which may be innate or itself acquired through experience.

Time Course of Choice Behavior Under Natural Conditions

A foraging bout of a honey bee is characterized by two obvious temporal periods: (1) the interval between successive foraging bouts during which the honey bee returns to the colony, unloads the collected nectar and/or pollen, and returns to the food source, and (2) the interval between successive landings within the patch of distributed food sources. A few examples of frequency distributions associated with these periods are given in Figures 4.19 and 4.20. Obviously, the frequency distribution for successive flower choices will depend very much on the spatial distribution of flowers, the availability and quantity of nectar and pollen, and on the proportion of flowers containing a reward. The data in Figure 4.19 were recorded for bees foraging in patches in which flowers were relatively dense and evenly distributed. Thus far, interval distributions have not been examined in which flowers are separated by longer distances and distributed in subdivided patches, e.g., where plants bearing several to many blossoms occur together as discrete units. The frequency distribution of bout intervals (Fig. 4.20) was calculated for training experiments using artificial feeding stations, because no data are available for natural food sources.

Where many flowers are visited during a bout, these two temporal periods are easily distinguished, being in the range of seconds for intervals between successive approaches or landings within a patch, and in the range of minutes to hours for intervals between foraging bouts. Each landing on a flower corresponds to a single learning experience even for very short rewards (Menzel, 1968; Menzel and Erber, 1972). Moreover, unrewarded landings contribute an inhibitory learning component (i.e., contribute to extinction), but the effect on the behavior is much less than that of a rewarded experience (see above discussion on resistance to extinction). The quick succession of visits within the patch ensures that immediate or short-term memory following one choice controls the next choice to a significant degree (the precise rules underlying choice behavior will be discussed below). Only in the next foraging bout are choices controlled by long-term memory. It is this memory which is almost exclusively tested in training experiments, even though sensitization effects of the US and short-term memory clearly contribute to flower choice as well.

In many plant species, a large number of flowers are borne on a single plant and represent a subpatch of food items. Bees are known to apply particular search strategies when exploiting such a subpatch, e.g., they often start collecting from the lower flowers and gradually work upward (review Heinrich, 1984). Consequently, the flight paths between visits within

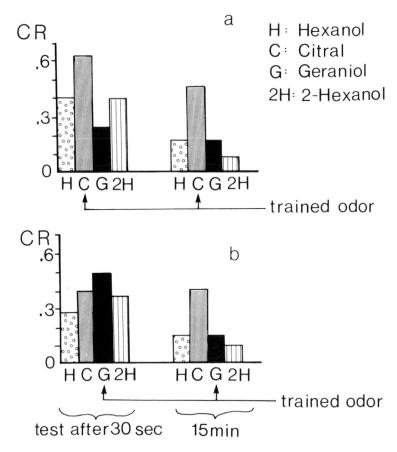

Figure 4.18. The content of short-term and intermediate-term memory. The PER was conditioned to citral (in a) or geraniol (in b) in a single trial. Each group was divided into two subgroups, one of which was tested 30 seconds after conditioning (short-term memory) and the other 15 minutes after conditioning (intermediate-term memory). Each subgroup was again divided into four groups with respect to tests using one of four odorants. If citral is trained (a) consolidation of short-term into intermediate-term memory sharpens the generalization profile in favor of the trained stimulus, whereas after training to geraniol, the generalization profile changes drastically, indicating a reevaluation of the memory during consolidation. Statistics: the χ^2 test reveals that differences in CR exceeding 18% are significant with $p \leq 0.01$ (540 animals in all 16 test groups).

a

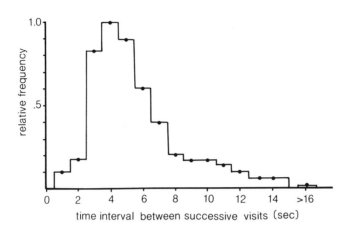

b

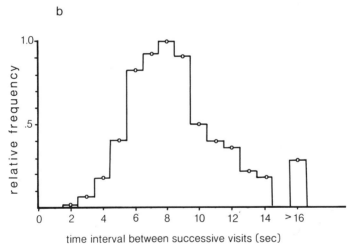

Figure 4.19. Frequency distribution of the time interval between successive landings of individually recognized honey bees on flowers of four different plant species. The time interval (abscissa) includes the handling time on the respective flower. The ordinate gives the frequency for the 1-second bin width in relative proportions of the maximum. a: Pollen-collecting bees on *Doronicum* sp. (*n* = 514). b: Nectar-collecting bees on citrus flowers (*n* = 1,180). c: Nectar-collecting bees on *Corydalis carva*, an early-spring-blooming Papaveraceae (*n* = 182). d: Pollen- and nectar-collecting bees on *Salix* (*n* = 161).

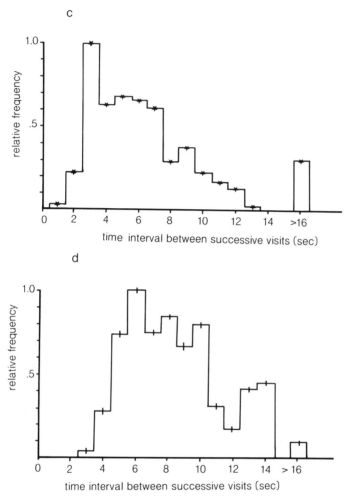

Figure 4.19. (Continued)

a subpatch are more directed and shorter than those between subpatches. Although flower distributions on a single plant and on neighboring plants of the same species differ greatly among species, it still must usually be the case that flights between flowers of the same plant species (subpatch) are shorter than flights between flowers of different plant species. These conditions have an important consequence for flower choice. Where successive choices follow each other quickly, the honey bee is probably landing on flowers of the same plant species. Where more time elapses between landings, the honey bee may well be landing on flowers of different species.

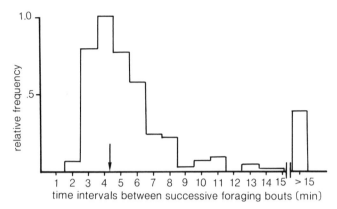

Figure 4.20. Frequency distribution of time intervals between successive bouts. The data were collected during training sessions on an artificial feeding place 100 m away from the hive. Ambient temperature was between 20 and 26°C, and the individually trained bees were fed on a 1.5-m sucrose solution without interruption. The time interval (abscissa) includes the time for unloading the crop within the colony.

It will be shown next that such temporal patterns have a significant impact on choice behavior in an experimental arrangement which resembles quite closely the natural conditions but which permits the behavior of a single honey bee to be monitored continuously.

Consider an experimental setup in which a honey bee collects sucrose solution from four computer-controlled feeders (Greggers, 1989; Menzel and Greggers, 1982; Greggers and Menzel, in press). The four feeders are arranged at distances of between 30 cm and 2 m depending on experimental conditions, are marked with the same or different colors, and offer sucrose solution at a constant flow rate which differs among the four feeders (e.g., 0.062 μl/minute, 0.125 μl/minute, 0.25 μl/minute, and 0.5 μl/minute). Since the feeders differ in reward quantity in a 1:2:4:8 ratio, we shall label the four feeders 1, 2, 4, and 8 respectively. In such a patch the honey bee forages for 20–50 minutes after arrival from the hive, until it has filled its crop and flies back to the hive. During this time each feeder is visited an average of 30 visits during one bout.

The choice frequency on each of the feeders partially matches the reward proportions. Let us first focus our attention on the time dependence of the choice frequency after a visit on any of the four feeders (Fig. 4.21). After leaving a feeder, the bee makes a choice either to return to the same feeder (termed a "return visit," RV) or to visit any of the three other feeders

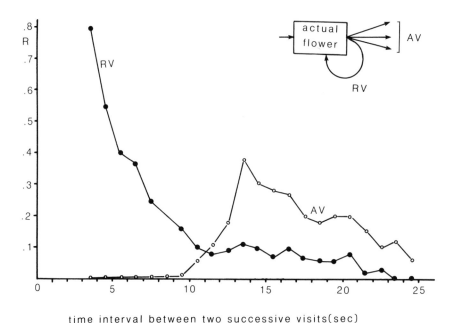

Figure 4.21. Temporal dynamics of the choice behavior of bees collecting nectar from four feeders (see text). Two categories of choices are distinguished; return visits (RV, the bee returns to the same feeder), alternate visits (AV, the bee choses one out of the three alternative feeders). The time interval between the moment when the bee stops sucking on the actual feeder (time zero) and the next arrival is plotted on the abscissa. The ordinate gives the relative frequency. The amount of sucrose solution provided by all four feeders together is 0.83 μl/minute (Greggers, 1989). Number of evaluated visits: $n = 579$.

(termed an "alternate visit," AV). As noted below, the probability of an RV or an AV depends on several parameters (e.g., the amount of reward experienced during the last visit). With respect to the time dependence of RV and AV flights, it is obvious from Figure 4.21 that the probability of an RV is very high immediately after the last visit. With time, the RV probability falls steeply and the AV probability increases. The high probability of RVs at short time intervals is not a consequence of a shorter distance to the just-visited feeder, because a honey bee flying with a speed of approx. 2 m/second can easily reach any feeder within less than 3 seconds.

The temporal dynamics observed for the RV and AV choices indicate that honey bees tend to return to the subpatch if they decide quickly but shift to alternative subpatches if more time elapses. In terms of memory processes, fast RV choices should be dominated strongly by the immediate, short-term memory phase with its strong nonassociative component (i.e., by sensitization). This interpretation is supported by the observation that, immediately after a large reward, RV flights are much more frequent than AV flights. A strong US (i.e., one exceeding the average reward from all feeders) causes strong sensitization and thus induces a strong immediate memory. Our experiments show that such a strong US arouses the animal, which leads to faster movements within the patch, more intensive probing for reward inside the feeder, and higher flight speed after leaving the feeder. Besides these more general effects, the stronger US initiates also a tendency to return to the same food source (RV flights). This indicates clearly that the immediate memory includes CS-specific, associative components.

The paths of bees were reported to be more tortuous after larger rewards than after small or even null ones (Pyke, 1978a; Heinrich, 1979; Schmid-Hempel, 1984, 1985a,b). Our analysis indicates that such movement is much more specific, highly dynamic, and goal-directed than previously thought. The temporal dynamics of the behavior is related to memory processes, particularly the sequence of a strongly US-dependent immediate memory and a later consolidated memory.

Continuous Updating of the Memories in Multiple Learning Conditions

Learning is a continuous process and thus usually involves a large number of successive experiences. Multiple experiences continuously update the memory trace, shape its content, and make it less susceptible to random variation in the environment. Multiple-trial learning in honey bees has been described in several reviews (Bitterman, 1988; Menzel, 1985, 1990; see also Gould, this volume). Here we select examples of our functional approach to honey bee learning which emphasize the plasticity of the memory trace and the intrinsic components of the updating process.

Reversal Learning

Under natural conditions, a generalist pollinator is exposed to changes in food availability over its lifetime and must be able to switch to new food sources when appropriate. Indeed, reversal learning has often been observed in the honey bee (von Frisch, 1967; Menzel, 1969, 1990; Seeley, 1985). The process of reversal learning has been studied in dual-choice experiments with freely flying, color- or odor-trained bees and in olfactory PER-conditioning. These studies reveal a few general features which are also known from the vertebrate-learning literature. For example, Meineke

(1978) performed an experiment in which he trained a bee over 18 days in a multiple blue/yellow reversal task. Each reversal session was continued, until the animal chose the new color at the same high level (i.e., a "criterion level"). He found a reduction in the number of reversal-learning trials necessary to reach criterion after a few reversals when retrained to blue, but variable results and a large number of reversal cycles when retrained to yellow (Fig. 4.22). During the first 2 days, reversals to blue were relatively slow to take place, but performance improved later and reached a low and stable level. By contrast, reversals to yellow were very variable over most of the training time and showed improvement only after 13 days of training with more than 40 reversals.

A higher preparedness to learn blue as a food signal is known from several experiments (Menzel, 1990) and also appears in another reversal-learning experiment in which the initial learning trials were varied (Fig. 4.23). Again, blue and yellow (presented as spectral lights of wavelengths 444 nm and 590 nm, respectively) were trained in a dual-choice experiment and the number of initial trials on either color differed for different animals (Menzel, 1969). Initially reversal to the new color was retarded by increasing the number of learning trials on the first color, and reversal from blue to yellow was slower than from yellow to blue. However, after more than ten initial learning trials, reversal became easier for both colors and the animals switched to the new color as easily as after weak initial training. It thus appears that after an extended experience with a food source, honey

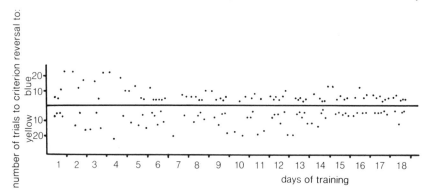

Figure 4.22. Multiple-reversal experiment with a freely flying bee trained in a dual forced choice experiment with the color targets blue and yellow. The number of learning trials on either of the two color targets needed to reach a criterion of correct choices after the reversal is plotted at the ordinate, upward for blue, downward for yellow. The animals were trained over 18 days with a total of 75 reversals to yellow and 76 to blue. The points indicate the number of learning trials needed to reach the criterion at the particular color (after Meineke, 1978, redrawn).

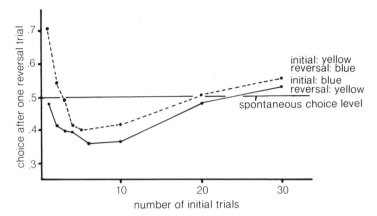

Figure 4.23. The overlearning-reversal effect in freely flying honey bees trained to yellow and blue targets in a dual forced choice situation. Before training the bees chose the yellow (spectral light 590 nm) and the blue (spectral light 444 nm) equally frequently (spontaneous choice level = 0.5). The choice performance was tested after a single reversal trial which followed a variable number of initial learning trials (abscissa) on the other color. The ordinate gives the ratio of choices for the reversal color divided by those for the initially trained color (after Menzel, 1969, redrawn).

bees are prepared to switch to a new one, even though neither the strength of the US nor any other physical parameter has changed. In the vertebrate-learning literature such a phenomenon is known as the overlearning-reversal effect. The overlearning-reversal effect might indicate that a US loses its power as a reinforcer after extended training (Rescorla, 1967, 1988). Proper control experiments (e.g., partial-reinforcement schedules followed by reversals) have yet to be performed for honey bees.

Memory-Based Choice Allocation in Patches With Variable Food Sources

Under natural conditions, the food sources of pollinating insects (i.e., the flowers) compete for pollinators and, in an evolutionary sense, tend to optimize the ratio of investment (=nectar and/or pollen) to profit (=fertilization) in part by minimizing the amount of nectar and/or pollen offered to the pollinator. Small amounts of food force the pollinator to visit many flowers, but the plant runs the risk that it may lose in the competition with other plants because the pollinator seeks the highest net profit. The actual amount of food discovered by the pollinator depends on many factors (e.g., physiological conditions of the plant and the flower, number and species of pollinators working simultaneously in the field, and weather conditions) and thus is highly variable over time and space. Optimization criteria both in the proximate mechanistic context and the ultimate evolutionary context control the choice behavior of the pollinator. For the pollinator, one of several goals is to gain as much food with as little investment and risk as possible by choosing food sources accordingly (MacArthur and Pianka, 1966; Heinrich, 1983; Waddington et al., 1981; Pyke, 1978b). Optimal-decision theories (Maynard Smith, 1978; Krebs et al., 1978; Pyke, 1984) have been applied with limited success to explain the choice behavior of a foraging bee (Waddington and Holden, 1979; Waddington, 1985; Pleasants, 1981; Heinrich 1983). The models developed thus far are inadequate in their focus on the energy budget of the foraging animals and also ignore informational components and mechanisms of memory formation and retrieval. Experimental design thus far has also been unsatisfactory, because only two alternative feeding places are generally used and the animal's behavior is not resolved continuously over time.

The general experimental design underlying our efforts to address these issues was described above. A single bee works on four feeders (33% sucrose solution; relative flow rate in the four feeders in a ratio of 1:2:4:8, total flow rate in all four feeders = 0.94 μl/minute). The bee imbibes all of the sucrose solution available during each visit. Since its potential rate of uptake (ca. 1 μl/second) exceeds the flow rate of any feeder, the bee visits all four feeders at an average frequency of about one visit per minute.

Under these conditions, the bee partially matches its choice behavior (and other parameters such as licking time, flight speed toward the feeder, and the inverse of handling time before licking) with the average amount of reward. The expected amount of reward must be stored in a kind of long-term or reference memory, because this partial matching is found not only during the course of continuous foraging within the patch during a given bout but also during the first approaches after returning from the hive to begin the next bout. The same results were found for conditions in which flowers were set to zero flow rate, a classical test situation in learning experiments.

Under natural conditions, the animal controls the amount of reward in each flower by coordinating its own action with the productivity of the flower. The animal experiences a certain reward during each visit, and its next choice might depend on both an expectancy as a consequence of its immediate experience and long-term memory. The line in Figure 4.24 (all choices) shows the overall-choice frequencies under the conditions of our experiment, conditions which quite closely resemble natural conditions. Compared to a perfect matching between choice proportions and the flow rate of the reward, it is obvious that the low-reward feeders (Nos. 1 and 2) are more frequently visited, and the highest reward feeder (No. 8) is

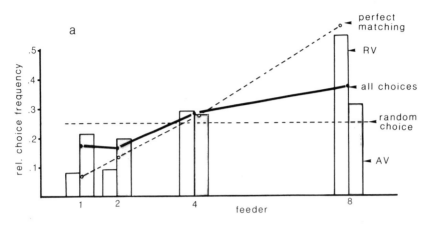

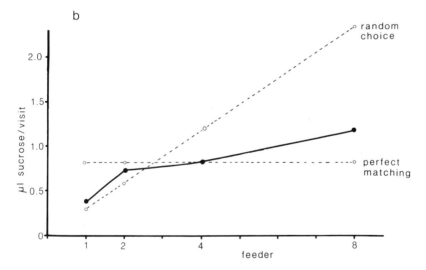

Figure 4.24. *a*. Choice matching under the conditions of a constant but different flow of sucrose solution in four feeders. The four feeders are numbered according to their relative flow rates (see text). The thick line indicates the choice matching for all choices. The dotted lines depict two extreme choice strategies discussed in the text, perfect matching and random choice. The latter is equivalent to a regular visit at any temporal pattern which leads to an average of equal choices at each feeder. The bars mark the choice behavior during return visits (RV) to the same feeder and alternate visits (AV) to one of the three remaining feeders. *b*. Average gain of reward (µl sucrose solution) per visit at the four feeders (see text). Dotted lines mark the two extreme strategies, perfect matching and random choice.

less frequently visited. Consequently, the reward gained per visit (Fig. 4.24b) is less for the low-reward feeders and higher for the highest-reward feeder than expected under perfect matching. From an informational point of view, the bee continuously collects information at the expense of imperfect matching. Perfect matching would provide the bee with a constant reward per visit but with no information about the differences among the four feeders. Only if the bee kept track of each choice over a long period of time would it be able to calculate the productivity of the feeder by dividing the total reward gain through the number of visits at each particular feeder. Obviously, the honey bee's memory content is insufficient for such a demanding job and so it adopts a different strategy.

Random choice, in contrast to perfect matching, would maximize information about the differences. However, in that case, the bee would either reduce its energy budget by visiting the four feeders at a frequency set by the flow rate of the high-reward feeders or it would leave excess amounts of reward in the high-reward feeders by visiting them at a frequency set by the low-reward feeders. In the latter case, it would risk other bees discovering the high-reward feeders. Given constraints on memory, the compromise between maximizing net energy gain and informational gain is manifested in suboptimal matching.

The experiment described in Figure 4.24 was repeated under several different conditions: vertically- vs.-horizontally-arranged feeders, shorter (30 cm) and longer (1.5 m) distances between neighboring feeders, equal or strongly different color signals around the tube entrance, other reward ratios (1:2:4:8), and different sucrose concentrations. Results appear to be independent of these parameters (Greggers and Menzel, in press). Interestingly, feeder No. 1 with the lowest reward was always chosen a bit more frequently than feeder No. 2. Within any given experiment, this difference was never significant but was consistent over all test conditions. This result suggests that honey bees are programmed to invest a certain amount of energy and time for probing, regardless of whether or not the low-reward conditions in less frequently visited feeders have changed. Information collection is thus as important as optimization of energy gain.

An analysis of behavioral sequences on a real-time scale gives us some hints about mechanisms of choice performance and the action of different forms of memory involved in the decision process. The bars in Figure 4.24 indicate that the total number of all return flights to the just-visited feeder (RV) match the reward distribution better than the total number of all AV flights (see also Figs. 4.21 and 4.26b). Since, on average, feeder No. 8 provides more reward per visit than feeders No. 1 and No. 2 (see Fig. 4.24a), the probability of an RV should depend on the amount of reward acquired during the last visit. Indeed, the probability of an RV increases with the amount of the last reward.

Correspondingly, the probability of AV is inversely related to the amount of reward acquired during the last visit. The choice during AVs depends strongly on the feeder at which the bee begins its AV flight (Fig. 4.25). Very good matching is found for AVs following a visit at feeders No. 1 and No. 2, while very poor matching after visits at flowers 4 and 8. This means that honey bees starting from low-reward feeders recruit a memory about reward distributions which contains more reliable information about the long-term experience in the patch than that of bees starting from a high-reward feeder. This "reference memory" is overridden shortly after

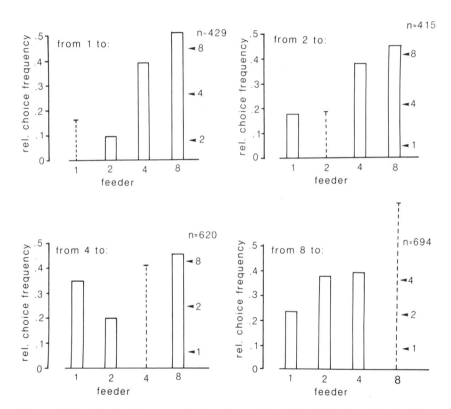

Figure 4.25. Choice frequency during alternating visits (AV), from feeder No. 1 (left upper graph), from feeder No. 2 (right upper), from feeder No. 4 (left lower), from feeder No. 8 (right lower). The dotted line gives the relative choice frequency of the corresponding RV flights. The arrowheads on the right side mark the relative choice proportions for perfect matching. n gives the number of choices of eight bees.

a high reward, irrespective of where the bee receives it. The sensitizing effect of a strong reward lasts in turn for only a short period of time (see Fig. 4.21). After that time (specifically, between 5 and 10 seconds after takeoff from the last-visited feeder, depending on average reward rate in the entire patch and thus on flight speed; see Greggers and Menzel, in press), choice is controlled primarily by the reference memory.

How is the information collected during a visit used to update memory? Theories on classical and instrumental conditioning favor the conclusion that the difference between the expected and the actually experienced US strength is the most important factor in learning (Rescorla, 1967, 1988; Rescorla and Wagner, 1972). Since flight time between visits correlates both with the average amount of reward in the whole patch and the amount of reward during the last visit, we can use flight time as an indicator of the expected next US and the flight time after that reward as a measure of the deviation between expected and experienced US. Furthermore, licking time depends primarily on the expected reward because the minute amount of sucrose solution is imbibed in less than 5 seconds of the average licking time. Table 4.1 gives the results for two categories of US strength: low reward (<0.4 μl sucrose solution), high reward (>0.4 μl). It is obvious that the US experienced at the "last" flower determines an expectancy for the US at the "actual" feeder, because a high "actual" reward induces a longer licking time after a high "last" reward than after a low "last" reward. The flight time after the "actual" reward for RV flights is short for the transition from low to high reward, long for the transition from high to low reward, and not different for successive rewards of the same amount (low-low, high-high). AV flights do not depend on those transitions (not shown).

It is obvious from these results that the dynamics of memory processes and the limited capacity of the bees' short-term memory have to be taken into account in explaining its foraging behavior. Bees, like all foraging animals, are not omniscient at any stage of their foraging cycle and are programmed to keep track of changes in food availability. The honey bee's foraging behavior reflects a compromise between food collection and information collection. Bees are hardly playing with a two-armed bandit for which they first would have to discover the probabilities of success in order to apply optimal rules (Houston et al., 1982; Krebs et al., 1978). Rather honey bees, like other animals, continuously collect and retrieve information in the process of making choices. The temporal dynamics of foraging, the structure of their memories, and the limited capacity of certain memory stages appear to be the framework in which proximate mechanisms account for ultimate goals.

The dynamic model developed on the basis of these results formulates the following two rules: (1) the informational capacity for updating the

Table 4.1. Results for two categories of US strength

Lick time at the actual feeder: (low < 0.4 μl, high > 0.4 μl):

		Reward "actual" feeder	
		Low	High
Reward "last" feeder	Low	7.1 sec ± 0.8 n = 693	14.9 sec ± 0.9 n = 1076
	High	7.0 sec ± 0.6 n = 721	18.5 sec ± 0.8 n = 769

Flight time of RV-flights after visiting the actual feeder:
(low < 0.4 μl, high > 0.4 μl):

		Reward "actual" feeder	
		Low	High
Reward "last" feeder	Low	6.0 sec ± 0.7 n = 231	4.3 sec ± 0.6 n = 253
	High	6.4 sec ± 0.8 n = 264	5.0 sec ± 0.7 n = 186

memory is proportional to the difference between the expected and the experienced amount of reward (i.e., a "difference rule") and (2) retrieval of the expected amount of reward activates either a short-lasting working memory or a long-term reference memory. Retrieval from working memory overrides retrieval from reference memory and dominates over short time periods immediately after a reward. Retrieval from working memory depends strongly on the amount of reward. Reference memory does not appear limited in time or capacity and stores the sum of all informational components resulting from the difference rule.

These rules predict certain peculiarities which were actually confirmed by the results. Figure 4.26 gives an example. One implication of the two rules is that the choice of feeders depends on the time interval between the last reward and the next choice. At very short intervals, the most choices are allocated to the high-reward feeders. At longer intervals, the highest-reward feeder, No. 8, loses its attractiveness, whereas feeders No. 1 and No. 4 become more attractive. This pattern is an outcome of the first rule,

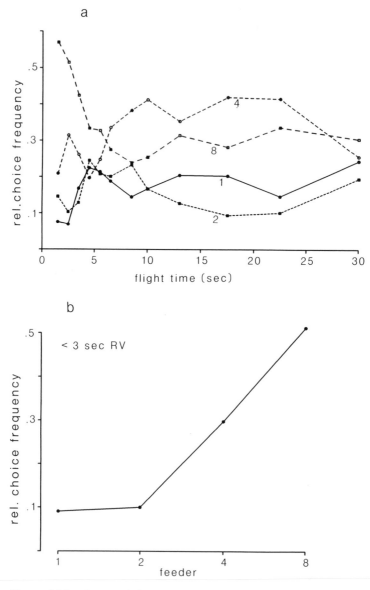

Figure 4.26. Temporal dynamics of the choice behavior for four feeders with the relative rates of sucrose flow 1, 2, 4, and 8 (see Figure 4.25 and 4.26). Part *a* gives the time dependence of the choice for intervals of 2–30 seconds between two successive visits for both RV and AV flights together. The numbers 1, 2, 4, and 8

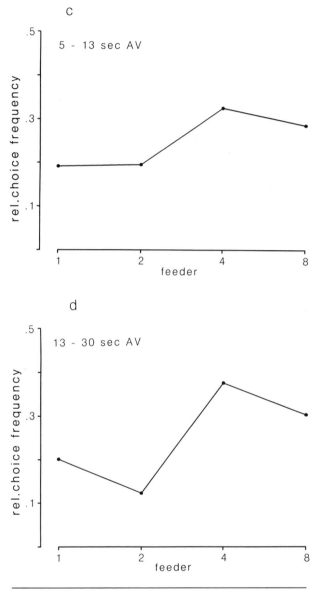

Figure 4.26. (Continued) beneath the curves indicate the choice for the respective feeder. Parts *b–d* plot the same results for three time horizons (*b*, < 3 seconds; *c*, 5–13 seconds; *d*, 17–30 seconds) and for the four different feeders (Nos. 1, 2, 4, 8 at the abscissa). Note the reduction in the choice of 8 and 2 and the enhancement of the choice of 1 and 4 at longer time intervals.

which leads to a relatively stronger long-term memory for feeders No. 1 and No. 4 and a relatively weaker memory for Nos. 2 and 8 (see Greggers and Menzel, in press).

Conclusion

Memory formation and retrieval is as highly dynamic and multiphasic a process in the honey bee as it is in animals with large brains. To the extent that memory dynamics reflect cellular and network properties underlying the different forms of plasticity in the nervous system, they are indicative of the existence of a series of information-storage mechanisms. Memory dynamics are also adapted to the particular needs of the animal in nature. Our results favor the conclusion that, for a honey bee working in a floral patch, the tight match between the expected sequence of food encounters and the programmed transitions between memory phases simplifies the decision-making task in a continuously fluctuating and highly unpredictable world. The honey bee's memory records neither the number nor the sequence of rewarded and unrewarded landings. It calculates neither an average of any sort nor a probability of positive or negative encounters. The decision rules applied by the bee are obviously different from that of a player at a two-armed bandit machine, yet optimization criteria are met both in the short term and the long term. The honey bee meets these criteria through the application of certain rules of thumb which relate the amount of reward, the time to next encounter, and the precision with which the most recent or the more remote memory is activated. These memory-retrieval mechanisms appear to produce expectations which may differ considerably among the different memories and permit the honey bee to adapt quickly to changes in environmental conditions. The signal for the next step of memory formation is the deviation between the expected and the experienced US or reward. The dynamics of memory processes protect the animal from being caught in a suboptimal patch of food distribution. At the same time, the limited capacity and duration of short-term memory prevent honey bees from accumulating information about the environment indefinitely. Lack of "knowledge," however, does not lead the bee to a probabilistic relationship between its actions and their consequences. The reason for this lies in the very nature of memory dynamics. If an individual honey bee experiences a higher reward than expected by retrieval from the long-term memory, short-term memory is triggered and keeps the individual within the patch, but only during the active status of the short-term memory which now carries an updated, higher expectation of reward. If a positive encounter is added within a short period of time, the long-term memory is not changed and the animal regulates its choice behavior according to the previous status of the long-term memory. If an even higher reward is experienced during the lifetime of the short-term memory, then the consolidation process updates long-term memory and maintains high expectations in short-term memory as well. It is obvious that such a system will work only if the temporal dynamics and the transfer properties between forms of memory match very well conditions of flower-foraging in nature. Since these can change quite drastically over the course of the year, one might expect additional long-term adaptations which account for the changing circumstances of the honey bee colony.

References

Bitterman, M.E. 1988. Vertebrate-invertebrate comparisons. NATO ASI Ser.-Intell. Evol. Biol. **17**:251–275.

Carew, T.J., Abrams, T.W., Hawkins, R.D. and Kandel, E.R. 1984. The use of simple invertebrate systems to explore psychological issues related to associative learning. In D.L. Alkon and J. Farley (eds.), Primary Neural Substrates of Learning and Behavioral Change. Cambridge University Press, New York, pp. 169–183.

Colwill, R.M., and Rescorla, R.A., 1988. The role of response-reinforcer associations increases throughout extended instrumental training. Anim. Learn. Behav. **16**:105–111.

Ebbinghaus, M. 1885. Über das Gedächtnis. K. Buehler, Leipzig.

Erber, J. 1972. The time-dependent storing of optical information in the honeybee. In R. Wehner (ed.), Information Processing in the Visual Systems of Arthropods. Springer, Berlin-Heidelberg-New York, pp. 309–314.

Erber, J. 1975a. The dynamics of learning in the honeybee (*Apis mellifera carnica*). I. The time dependence of the choice reaction. J. Comp. Physiol. **99**:231–242.

Erber, J. 1975b. The dynamics of learning in the honeybee (*Apis mellifica carnica*). II. Principles of information processing. J. Comp. Physiol. **99**:243–255.

Erber, J., Masuhr, T. and Menzel, R., 1980. Localization of short-term memory in the brain of the bee, *Apis mellifera*. Physiol. Entomol. **5**:343–358.

Fabre 1879. Souvenirs entomologiques. Paris.

Forel, A. 1910. Das Sinnesleben der Insekten. Reinhardt, München.

Frisch, K. v. 1967. The Dance Language and Orientation of Bees. Harvard University Press, Cambridge.

Garcia, J., and Koelling, R.A., 1966. Relation of cue to consequence in avoidance learning. Psychol. Sci **4**:124–124.

Gormezano, I. 1984. The study of associative learning with CS-CR paradigms. In D.L. Alkon and J. Farley (eds.), Primary Neural Substrates of Learning and Behavioral Change. Cambridge University Press, New York, pp. 5–24.

Gould, J.L., and Marler, P. 1984. Ethology and the natural history of learning. In P. Marler and H.S. Terrace (eds.), The Biology of Learning. Springer-Verlag, Berlin-Heidelberg-New York, pp. 47–74.

Greggers, U. 1989. Optimizing strategies in choice behavior in the honey bee. In J. Erber, R. Menzel, H.J. Pflüger, and D. Todt (eds.), Neural Mechanisms of Behavior—Proceedings of the 2nd International Congress of Neuroethology. Georg Thieme Verlag, Stuttgart-New York, p. 219.

Greggers, U., and Menzel, R. Memory dynamics and the foraging strategy of honey bees. Behav. Ecol. Sociobiol, in press.

Griffin, D.R. 1984. Animal Thinking. Harvard University Press, Harvard.

Grossman, K.E. 1971. Belohnungsverzögerung beim Erlernen einer Farbe an einer künstlichen Futterstelle durch Honigbienen. Z. Tierpsychol. **29**:28–41.

Hawkins, R.D. and Kandel, E.R. 1984. Is there a cell-biological alphabet for simple forms of learning? Psychol. Rev. **91**:375–391.

Heinrich, B. 1979. Resource heterogeneity and patterns of movement in foraging bumble bees. Oecologia **40**:235–245.

Heinrich, B. 1983. Do bumble bees forage optimally, and does it matter? Am. Zool. **23**:273–281.

Heinrich, B. 1984. Learning in Invertebrates. In P. Marler and H.S. Terrace (eds), The Biology of Learning. Dahlem Konferenzen. Springer-Verlag, New York, pp. 135–147.

Houston, A., Kacelnik, A., and McNamara, J. 1982. Some learning rules for acquiring information. In J. McFarland (ed.), Functional Ontogeny. Plenum Press, London, pp. 140–191.

Hull, C.L. 1934. Learning II: The factor of the conditioned reflex. In C. Murchison (ed.), Handbook of General Experimental Psychology. Clark University Press, Worcester.

Hull, C.L. 1943. Principles of Behaviour. Appleton-Century-Crofts, New York.

Kamin, L.J. 1969. Selective association and conditioning. In W.K. Honig (ed.), Fundamental Issues in Associative Learning. Dalhousie University Press, Halifax.

Kiltermann, R. 1969. Lern- und Vergessensprozesse bei der Honigbiene—aufgezeigt anhand von Duftdressuren. Z. Vergl. Physiol **63**:310–334.

Krebs, J.R. Kacelnik, A., and Taylor, P. 1978. Test of optimal sampling by foraging great tits. Nature **275**:27–31.

Kuwabara, M. 1957. Bildung des bedingten Reflexes von Pavlovs Typus bei der Honigbiene, *Apis mellifera*. J. Fac. Sci. Hokkaido Univ. Ser. VI Zool **13**:458–464.

Lindauer, M. 1955. Schwarmbienen auf Wohnungssuche. Z. Vergl. Physiol. **37**:263–324.

Lindauer, M. 1959. Angeborene und erlernte Komponenten in der Sonnenorientierung der Bienen. Z. Vergl. Physiol. **42**:43–62.

MacArthur, R.H., and Pianka, E.R. 1966. On optimal use of patchy environment. Am. Nat. **100**:603–609.

Maynard Smith, J. 1978. Optimization theory in evolution. Annu. Rev. Ecol. Syst. **9**:31–56.

Meineke, H. 1978. Umlernen einer Honigbiene zwischen gelb- und blau-Belohnung im Dauerversuch. J. Insect Physiol. **24**:155–163.

Menzel, R. 1967. Untersuchungen zum Erlernen von Spektralfarben durch die Honigbiene (*Apis mellifica*). Z. Vergl. Physiol. **56**:22–62.

Menzel, R. 1968. Das Gedächtnis der Honigbiene für Spektralfarben. I Kurzzeitiges und langzeitiges Behalten, Z. Vergl. Physiol. **60**:82–102.

Menzel, R. 1969. Das Gedächtnis der Honigbiene für Spektralfarben. II. Umlernen und Mehrfachlernen. Z. Vergl. Physiol. **63**:290–309.

Menzel, R. 1979. Behavioral access to short-term memory in bees. Nature **281**:368–369.

Menzel, R. 1984. Short-term memory in bees. In D.L. Alkon and I. Farley (eds.), Primary Neural Substrates of Learning and Behavioral Change. Cambridge University Press, Cambridge, pp. 259–274.

Menzel, R. 1985. Learning in honey bees in an ecological and behavioral context. In B. Hölldobler and M Lindauer (eds.), Experimental Behavioral Ecology. Gustav Fischer Verlag, Stuttgart, pp. 55–74.

Menzel, R. 1987. Memory traces in honeybees. In R. Menzel and A. Mercer (eds.), Neurobiology and Behavior of Honeybees. Springer, Berlin, pp. 310–325.

Menzel, R. 1990. Learning, memory and "cognition" in honey bees. In R.P. Kesner and D.S. Olten (eds.), Neurobiology of Comparative Cognition. Erlbaum, Hillsdale, NJ, pp. 237–292.

Menzel, R. and Erber, J. 1972. The influence of the quantity of reward on the learning performance in honeybees. Behavior **41**:27–42.

Menzel, R., Erber, J. and Masuhr, T. 1974. Learning and memory in the honey bee. In L. Barton-Browne (ed.), Experimental Analysis of Insect Behavior. Springer, Berlin, pp. 195–217.

Menzel, R. and Greggers, U. 1992. Temporal dynamics and foraging behavior in honeybees. In J. Billen (ed.), *Biology and Evolution of Social Insects*. Leuven University Press, Leuven.

Menzel, R. and Sugawa, M. 1986. Time course of short-term memory depends on associative events. Naturwissenschaften **73**:564–565.

Müller, G.E., and Pilzecker, A., 1900. Experimentelle Beiträge zur Lehre vom Gedächtnis. Z. Psychol. 1:1–288.

Pavlov, I.P. 1967. Lectures on conditioned reflexes. International Publishers, New York.

Pleasants, J.M. 1981. Bumblebee response to variation in nectar availability. Ecology 62:1648–1661.

Pyke, G.H. 1978a. Optimal foraging: movement patterns of bumble bees between inflorescences. Theor. Popul. Biol. 13:72–98.

Pyke, G.H. 1978b. Optimal foraging in bumblebees and coevolution with their plants. Oecologia 36:281–293.

Pyke, G.H. 1984. Optimal foraging theory: A critical review. Annu. Rev. Ecol. Syst. 15:523–575.

Rescorla, R.A. 1967. Pavlovian conditioning and its proper control procedures. Psychol Rev. 74:71–80.

Rescorla, R.A. 1988. Behavioral studies of pavlovian conditioning. Annu. Rev. Neurosci. 11:329–352.

Rescorla, R.A. and Holland, P.C. 1982. Behavioral studies of associative learning in animals. Annu. Rev. Psychol 33:265–308.

Rescorla, R.A. and Wagner, A.R. 1972. A theory of classical conditioning: Variations in the effectivness of reinforcement and non-reinforcement. In A.H. Black and W.F. Prokasy (eds.), Classical Conditioning. II: Current Research and Theory. Appleton-Century Crofts, New York, pp. 64–99.

Schmid-Hempel, P. 1984. The importance of handling time for the flight directionality in bees. Behav. Ecol. Sociobiol. 15:303–309.

Schmid-Hempel, P. 1985a. How do bees choose flight direction while foraging? Physiol. Entomol. 10:439–442.

Schmid-Hempel, P. and Schmid-Hempel, R. 1985b. Nectar-collecting bees use distant-sensitive movement rules. Anim. Behav. 34(2):605–607.

Schreurs, B.G. 1989. Classical conditioning of model systems: A behavioral review. Psychobiology 17:145–155.

Seeley, T.D. 1985. Honeybee ecology. A study of adaptation in social life. Princeton University Press, Princeton.

Smith, B., and Menzel, R. 1989. An analysis of variability in the feeding motor program of the honey bee: The role of learning in releasing a modal action pattern. Ethology 82:68–81.

Squire, L., and Cohen, N. 1982. Remote memory, retrograde amnesia and the neuropsychology of memory. In L.S. Cermak (ed.), Human Memory and Amnesia. Erlbaum, Hillsdale, NJ, pp. 275–303.

Vareschi, E. 1971. Duftunterscheidung bei der Honigbiene-Einzelzell-Ableitungen und Verhaltensreaktionen. Z. Vergl Physiol. 75:143–173.

Waddington, K.D. 1985. Cost-intake information used in foraging. J. Insect Physiol: **31**:891–889.

Waddington, K.D. Allen, T. and Heinrich, B. 1981. Floral preferences of bumble bees (*Bombus edwardsii*) in relation to intermittent versus continuous rewards. Anim. Behav. **29**:779–784.

Waddington, K.D., and Holden, L.R. 1979. Optimal foraging: On flower selection by bees. Am. Nat. **114**:179–196.

Weiskrantz, L. 1970. A long-term view of short-term memory in psychology. In G. Horn and R.A. Hinde (eds.), Short-Term Changes in Neural Activity and Behavior. Cambridge Univ. Press, London, pp. 63–74.

5

Merging Mechanism and Adaptation: An Ethological Approach to Learning and Generalization

Brian H. Smith

Introduction

Learning allows animals to infer correlations among stimuli in their environment in order to predict the future occurrence of resources or threats. Stimuli to which animals do not normally respond might be correlated over time with biologically important stimuli such as those related to food availability, mates, or predators. Once the correlation between stimuli is learned, detection of previously "neutral" stimuli might permit the animal to prepare for the biologically important event in optimal ways (Hollis, 1984, 1990). The robustness of such correlations can, however, change considerably both within and between generations. The ability to track this rapidly changing correlation with little or no concomitant genetic change is a way of coping with situations that change on a time scale shorter than that needed for genetic change. Models such as those proposed by Stephens (this volume) show how the time scale on which the environmental correlations change can affect the evolution of learning.

These ideas offer a potential explanation of why it is adaptive to be able to modify behavior based on experience. However, historically there have been considerable differences in the way different researchers have approached learning (Terrace, 1984). Ethologists focused on types of stimuli, termed sign stimuli, that innately release a behavioral action pattern (Tinbergen, 1951). Other types of stimuli might release a behavior because of a learned association with a biologically relevant sign stimulus. Animals might be predisposed to learn certain cues (e.g., during song learning in birds), or learning might be restricted to one life stage (e.g., imprinting). Psychologists eschewed working with these kinds of species-specific behaviors to focus on delineating general mechanisms or rules that would apply to any learning situation. The result was termed "general process theory," which held that learning behavior should not be subject to species-specific constraints (Terrace, 1984). Learning was viewed as a process with which an animal deals with any kind of uncertainty, regardless of whether that uncertainty was generated in a natural or unnatural setting.

The acceptance of the views summarized in the first paragraph contrasts with the widely different methodologies used to study learning. There still remains to be done a considerable amount of empirical and theoretical work to determine whether these methodologies can be synthesized (Shettleworth, 1984). Attempts at such a synthesis are becoming more and more common (Beecher, 1988; Bolles, 1988; Staddon, 1983), so a vast literature has now developed showing how ethological considerations can affect the interpretation of learning in defined psychological paradigms. A common finding in these later studies is that animals selectively associate some cues with items such as food, but other cues do not appear to be learned as well (Gould and Marler, 1984; Revusky, 1984). The converse of this issue, whether knowledge of mechanism can affect adaptive interpretation, is only beginning to be explored in detail (Hollis, 1984, 1990). The thesis of this chapter is that an understanding of both kinds of problems, that is, learning as adaptive phenomenon and learning as mechanism, are necessary to develop a more satisfying, predictive understanding of learning. Indeed, truly ethological analyses must emphasize, among other things, mechanism *and* adaptation (Barlow, 1989). To begin, learning mechanisms as defined in psychological studies will be briefly reviewed. One additional process will be specifically addressed—that of generalization from a learned to a novel stimulus. Finally, two examples at the end will stress the need to synthesize the study of adaptive behavioral modification with that of learning mechanism.

Learning Mechanisms

Psychological studies of animal learning and memory have provided a rich methodology for studying mechanism. There are potentially several different forms of learning, which have different consequences for modification of behavior. In essence, these mechanisms describe different logical rules for extraction of information from the environment. Historically there have been many different mechanisms proposed to explain behavioral modification, but animal-learning mechanisms have recently been subsumed into three basic categories (Macintosh, 1983; Rescorla, 1988): nonassociative, associative, and operant (instrumental) conditioning.

Nonassociative learning is tantamount to learning that a stimulus exists (Rescorla, 1988). The experimental procedure involves repeatedly exposing a subject to a stimulus without explicitly manipulating any other aspect of the conditioning situation. One type of response to such a conditioning procedure is *habituation*, which would lead to a decrease in a subject's initial response to that stimulus over repeated exposures. Thompson and Spencer (1966) have listed nine behavioral criteria that must be demonstrated in order to show habituation in an experimental paradigm. In contrast to habituation, the response to that stimulus and to other stimuli may be enhanced for some time after the initial exposure especially if the stimulus is a salient one such as food. This pattern in response is termed *sensitization*. Nonassociative mechanisms, which produce decrements or increments in a response to a stimulus, may interact with one another. For example, one criterion for proving habituation is to show *dishabituation*, in which an animal's response recovers through exposure to a sensitizing stimulus. Until recently, dishabituation was thought to result from sensitization superimposed upon habituation. However, recent studies with the gastropod mollusc *Aplysia* have shown that dishabituation arises at a different developmental stage than sensitization and thus may be a third nonassociative process (Rankin and Carew, 1988).

Associative (or Pavlovian) learning involves establishing a correlation between two or more stimuli. One of the stimuli, the unconditioned stimulus (US), elicits a powerful appetitive or aversive reaction in a properly motivated subject, as would a sign stimulus. A stimulus that elicits little or no prior response until association with the US is termed the conditioned stimulus (CS). (Cases where a prior response to the CS exists and is simply enhanced or otherwise altered by association with the US are termed α-conditioning; see Menzel et al., this volume, for a more extensive discussion of this phenomenon.) When the CS precedes the US, the procedure is called "forward pairing." When the US precedes the CS, it is termed "backward pairing." In both cases a correlation, established by temporal contiguity, exists between the two stimuli. If subjects learn that correlation, the expectation is that a response to the CS would be modified in a predictable manner. To assess this change in behavior, it is crucial to compare the behavior of subjects exposed to forward or backward pairing with that of a group of subjects that receive an explicitly unpaired exposure to the CS and US (Rescorla, 1988). In the latter treatment group, no correlation can be established and no change due to associative processes is expected in the subject's behavior. In addition to pairing relationships, factors such as the elapsed time between presentation of CS and US (interstimulus interval—ISI), and the time period that separates CS-US pairing trials (intertrial interval—ITI), affect the degree to which behavior changes. The ISI associated with optimal learning performance may depend on the ITI, a relation termed the "duty cycle" (Rescorla, 1988; Gallistel, 1990).

Operant (or instrumental) learning involves a subject learning the consequences of its own actions. To test whether learned behavior is due to an operant mechanism, the experimenter establishes a contingency between delivery of a reinforcing stimulus and the subject's behavior. Note that operant conditioning procedures differ from Pavlovian procedures in that no explicit CS is introduced by the experimenter. Thus when a subject performs a response in the course of its actions, it receives a US, which may be either appetitive or aversive. The behavioral measure of conditioning is an increase or decrease in the performance of the behavior with which the US is contiguous.

This discussion of conditioning mechanisms serves as an introduction to issues that relate to the examples outlined below. It is by no means a complete account of the wealth of procedures developed by psychologists to study learning (Rescorla and Holland, 1982; Rescorla, 1980, 1988). Three additional issues in the psychological literature are worth noting. First, the mechanisms outlined above are not necessarily independent of one another. Thus mechanisms of sensitization might be involved in conditioning, and any conditioning procedure is likely to involve an operant contingency between the subject's behavior and the US. These interrelationships can and must be addressed in a comprehensive study of learning. Second, a lack of a change in behavior as a result of a conditioning procedure can lead to problems in interpretation if proper experimental procedures are not followed (Rescorla, 1980; Terrace, 1984). For example, a variety of studies have shown that even though a subject does not demonstrate conditioning in the behavior that is being measured, it may still learn relations among stimuli. Several procedures have been developed to uncover unexpressed learning by using different response measures and different conditioning procedures (Rescorla and Holland, 1982; Rescorla, 1980, 1988). Thus a distinction must be carefully made between learning *ability* and the rules animals use to translate knowledge into *performance*. Third, temporal contiguity is not always sufficient to produce conditioning. In cases where a CS provides redundant information, a subject may not show conditioning even though an appropriate ISI and ITI have been used. Procedures have been developed to separate effects of *contingency* in the relationship between CS and US, that is, how reliably the CS predicts the US, from effects due to temporal *contiguity* (Rescorla, 1988).

Generalization as Error Reduction

Before proceeding to the examples, it will be necessary to briefly introduce one additional concept. A process that has received considerable attention in the psychological literature (Kalish, 1969), and which relates to a broader consideration of learning in an adaptive context, needs to be considered—stimulus generalization. Generalization refers to a subject's tendency to respond to stimuli that were not experienced during conditioning trials, but which vary from a learned CS along a defined perceptual dimension, such as shape or light wavelength. Superficially, generalization looks like a mistake because a subject responds to a stimulus that has never been associated with a reward. However, I contend that defining learning in statistical terms suggests a way in which generalization can be adaptive.

Having learned the pattern of sensory stimulation (the CS) that reliably predicts an event, a subject is faced with the task of identifying future occurrences of similar patterns and separating those patterns from ones that do not predict the event. Under natural conditions, however, it is unlikely that a complex sensory pattern from a CS will be exactly reproduced in the future (Shepard, 1987). So animals must have mechanisms for generalizing learned information. For example, floral odors that foraging honey bees use to identify resources such as nectar and pollen are complex mixtures of many individual odorants. Even among flowers that contain the resource, the exact composition of the odor signal may vary due to such factors as age, physiological condition, and genetic constitution. Under these conditions, a bee that constrains itself to search for the same complex of stimuli that it experienced from a flower at which it received a reward might, in the extreme, only revisit the flower it just depleted of resources. The bee would pass over many flowers of the same species that contain a reward but which vary slightly from the pattern given off by the flower at which the bee found nectar and/or pollen. In the other extreme, a bee that generalizes too broadly might visit every flower regardless of its species identity or reward potential; that is, it might not easily recognize stimuli that predict that a flower contains no pollen or nectar. The handling time involved in visitation of unrewarding flowers would be wasted (Schmid-Hempel, 1984).

Learning must be viewed in a broad sense as a means by which animals classify items in their environment, rather than simply as a means for identifying a single type of item (Shepard, 1987). Any complete study of learning must contain a description of the rules by which subjects generalize among stimuli. For example, among a class of items that possess a reward or pose a threat, how does a subject estimate and track the variability among items within that class? Generalization among stimuli enables a bee to sample a variety of items and learn which stimuli predict rewards and which do not. Animals learn to respond to stimuli that predict a reward, and, through mechanisms reviewed above, they learn to suppress responses to stimuli that do not contain a reward, or even produce aversive stimuli. Generalization mechanisms therefore provide animals with information that allows them to choose among a variety of stimuli that might be associated with reward. It might never be advantageous for animals to completely eliminate responding to stimuli that are unrewarding, because the reward value might change over time. But generalization and discrimination allow animals to track this change and to modulate the frequencies at which they respond to stimuli (see Mackintosh, 1983), for a review of discrimination conditioning and its relationship to generalization).

As with learning per se, several mechanisms can potentially account for generalization. Psychophysical models of response generalization predict how responses to novel stimuli will be graded according to their perceptual similarity to the sensory patterns produced by the CS (Kalish, 1969). Novel stimuli that are more similar to the CS will elicit stronger responses than novel stimuli that are less similar along a defined perceptual dimension (e.g., light wavelength or intensity, sound frequency, etc.). An animal might be able to track individual components of complex stimuli such as color, shape, and odor, and thus learn only those stimuli that most reliably predict a resource. Alternatively, worker honey bees, for example, process compounds of odors and colored visual cues as stimuli that are unique from either component (Couvillon and Bitterman, 1982). In that case, components cannot be tracked completely independently of one another.

In addition to perceptual mechanims that set the limits to novelty detection, the tendency to generalize learned information can be affected by other factors. Generalization may in some way reflect the value of the resource in terms of the risk of incurring such error costs as wasted handling time. Evolutionary pressures might act on a species' perceptual systems to change generalization and hence discrimination responses. For example, does one species, which occupies a different niche than other species, respond differently to error costs? A recent model of foraging in bumble bees has explained discrimination of reward frequencies based on risk-taking (Real, 1991). Physiological state might also alter an animal's perception of error costs. For example, might a honey bee on the brink of starvation for whom a small reward might be very important for ensuring survival over a short period generalize more broadly than a bee that is satiated?

The types of errors that a subject makes in generalization are roughly analogous to type I and II errors in statistical analyses (Sokal and Rohlf, 1981). A type I error occurs when a null hypothesis is rejected when in fact it should be accepted. This error might be analogous to a foraging bee accepting a flower as belonging to a rewarding class when in fact the flower possesses no reward. A type II error indicates acceptance of a null hypothesis when in fact it is false, which would be analogous to rejecting a flower when it possesses a reward. For statistics, setting a probability value of 5% usually minimizes the probability of making either type of error. However, there are situations in which animals might be inclined to minimize one type of error regardless of how the second type is affected (see Sokal and Rohlf, 1981, pp. 157–169). For example, a foraging honey bee might be much more likely to set an acceptance or rejection threshold based on its physiological state (or that of its colony), which would be analogous to adjusting a probability value to increase or decrease the risk of making one or the other type of error. A bee on the brink of starvation might set this threshold differently than a satiated bee, depending on the relative costs of making one or the other type of error.

We thus can make an argument that generalization is adaptive. What is needed now is a way of formalizing research in generalization.

Merging Mechanism and Adaptation

A crucial issue in the biology of learning is how, or even whether, species-specific learning in a natural setting can be accounted for by the traditional learning mechanisms. As already mentioned, working out the mechanisms that give rise to learned behavior requires a high degree of control over an array of experimental variables and procedures. This level of control is frequently impossible under field or more natural conditions, which would seem to limit the applicability of laboratory-based paradigms to studies of natural learning. However, behavioral ecologists often bring animals into controllable laboratory settings to study a variety of ecological problems, so it should not be inconceivable that controlled situations could be developed for studying ecologically based learning problems. Moreover, discounting psychological mechanisms as not being applicable to a natural learning situation is not a satisfying proposition because, as Terrace (1984) has pointed out, these mechanisms and models based on them (e.g., Rescorla and Wagner, 1972; Wagner, 1980), have accounted for a rich diversity of learning behavior. Obviously, these mechanisms are not an exhaustive account of all of the learning mechanisms that exist. For example, it is not clear whether natural-learning paradigms, such as bird song learning (Marler, 1991), can be accounted for by the mechanisms outlined above. Nevertheless, the two examples discussed below will illustrate how using those mechanisms to generate hypotheses about learning in a natural setting, and using ethological information to generate hypotheses about adaptive mechanism, can help us understand the structure and function of learning.

Recent developments have made some progress toward a synthesis of methodologies for studying learning in laboratory and natural settings. Most efforts are directed toward an integration of learning into the context of an organism whose behaviors are adapted to specific environments (Bateson, 1984; Bolles, 1988). In total, the developments are far too numerous and detailed to adequately review here, and readers should refer to recent reviews of current issues in animal learning (Terrace, 1984; Spear et al., 1990). However, one issue of that debate relates directly to the two examples outlined below—establishment of a predictable relationship between learning mechanism and adaptive role (Hollis, 1984, 1990). An understanding of learning mechanism should help in establishing and testing adaptive hypotheses in the same sense that a knowledge of natural history can help in interpretation of the expression of mechanism.

Some recent theoretical developments justify this type of research philosophy. Gallistel (1990) considers learning as a correspondence in the animal's central nervous system between stimulus correlations and a (innate) representation of the expected environment. For example, as we will see below, certain odors (pheromones) in bees carry innate meanings in specific behavioral contexts. Honey bees, moreover, have an innate neural representation of the meanings of those odors in particular that can influence how they respond to the odor in a learning paradigm (see Menzel et al., this volume). The concept of species-specific preparedness to learn certain associations between cues and motor routines is embodied in the behavior-systems approach to learning (Timberlake and Lucas, 1989; Timberlake, 1990). This approach is predicated on the notion that any animal is in possession of a set of innate behavioral motor routines, each of which relates to an evolved adaptive function such as mating or feeding. These behavioral routines are a sequence of activities that comprise a modal action pattern (MAP; Barlow, 1977) that is released by specific sign stimuli.

A sequence of defined behavioral acts is considered to be a MAP when the transition probability between those acts is high, that is, when performance of the actions is temporally correlated. For example, a sex pheromone may initially release a searching MAP in a male insect which then searches for visual stimuli that indicate the presence of a female. Once those visual stimuli are found, a different MAP, or a different part of the same MAP, that relates to contact and courtship may be set into motion. However, transition probabilities within a given MAP are almost never perfect (i.e., 1.0), and the transition among different MAPs is considerably lower. These low transition probabilities create the potential for learned modification in MAP expression. For example, the way in which several MAPs are linked in their sequential expression may depend on learning. The way that this control is expressed can in turn depend on the nature of the CS (e.g., visual vs. olfactory) and on the point in the expression of the MAP at which the CS is introduced. Some CSs might be better at controlling locomotory aspects of a MAP, such as those involved in search, whereas other cues might be more easily related to the endpoint of the MAP, such as mouthpart movements related to feeding (Timberlake, 1990).

We should therefore be able to use knowledge of natural history to interpret questions of mechanism and learning-performance differences. Furthermore, given that the MAP concept is so closely tied to adaptation, a knowledge of learning mechanism should aid the formulation of hypotheses regarding adaptation. Insects in particular have several advantages for studies that merge investigation of mechanism and adaptive value.

First, in order to combine a study of mechanism with a more natural, ethological study of learning, it is necessary to rear animals in controlled laboratory settings as well as to study behavior in a natural setting. Many insect species can be observed and manipulated in both situations and thus make ideal subjects for this kind of research. Second, cases in which analogous learning abilities have arisen in phylogenetically distinct groups may indicate a functional adaptation of the ability. Many insect groups have a high species and niche diversity that might place different requirements on learning ability and/or performance. For example, Shettleworth (1984) has proposed a research methodology for relating specialized vs. generalized learning abilities to the breadth of foraging niches that species occupy. Studying learning in species that differ in diet breadth may reveal a relationship between learning ability and niche. This type of methodology could be applied to insects, many groups of which show differences in niche breadth (Bernays, this volume).

Third, in emphasizing mechanism and adaptation I do not wish to understate the role of phylogeny. The ability to perform analyses on phylogenetically defined species is essential to sorting out the roles of phylogeny and adaptation. Working within an evolving lineage can shed much light on the origin and history of learning. Rosenheim (this volume) reviews methodological approaches to the comparative study of learning. It should be added that even in phylogenetic studies a consideration of mechanism is necessary. Behavioral outcomes of different learning mechanisms might be superficially similar, thus potentially leading a phylogeneticist astray. For example, appetitive Pavlovian conditioning and sensitization both increase responses to a stimulus. If proper control experiments are not performed, then two species, each expressing one of those mechanisms of learning, might be lumped together with regard to learning ability.

Learning in Social Bees

Studies of the mechanisms of animal learning can certainly benefit from an ethological perspective (see Menzel et al.'s and Gould's contributions to this volume). The thesis of the rest of this chapter is that ethological interpretations of behavioral modification, and any modeling efforts in that regard, can benefit equally from a consideration of mechanism as defined in psychological studies. That is, field studies of behavior under more natural conditions combined with controlled laboratory studies can complement each other. Two examples summarized below reflect this approach. The first example begins with field and laboratory studies of learning during mate choice and concludes by indicating how more complete knowledge of the learning mechanism is required to make predictions regarding the adaptive value of behavioral modification. The second example begins with a controlled laboratory-learning paradigm and concludes by indicating how ethological studies have been (and still are) needed to interpret differences in learning performance.

Learning About Mates in a Primitively Social Bee

Lasioglossum (Dialictus) zephyrum is a primitively social bee species in the family Halictidae (see Michener, 1974, for a detailed description of the life history). Queens overwinter without workers and emerge in the early spring in temperate climates. The queens then burrow into nearly vertical earthen banks along the south-facing slopes of streams and rivers. Such nesting burrows typically occur clustered in aggregations of dozens to thousands of nests within a restricted area. Queens are mostly solitary (a few join together in nesting activities) for the first 6 weeks of springtime, when they perform all of the nest activities involved in egg laying, defense, and nest construction themselves. The first generation of brood that emerges at the end of this period comprises mostly workers, which take over tasks such as guarding and foraging. In contrast to highly social honey bees, these workers can become fully functional queens if the original queen dies, and there are no major morphological differences between the workers and the queen. Dominance of workers by the queen is primarily behaviorally based (Brothers and Michener, 1974; Smith, 1987). There may be several worker generations throughout the summer. As the end of the summer approaches, though, an increasing percentage of the brood is comprised of males; by the end of the summer, new queens, who will overwinter, are produced along with males. As new queens emerge in late summer, they mate. The males and workers die with the first few frosts, and the queens return to soil burrows where they will spend the coming winter.

Shortly after emerging as adults, males leave their natal nests and begin patrolling just above the nest entrances within a confined area of the aggregation (Michener and Smith, 1987). Patrolling males typically pounce on any small, dark object on the ground beneath them; such objects may indicate the location of a female (Barrows, 1976). Most of the females encountered are workers, who for the most part do not mate and resist the males' mating attempts; only a small percentage of these attempts result in a successful mating (Fig. 5.1). Therefore, the most frequent type of encounter between the sexes involves a female struggling to free herself from mating attempts of a male, during which time she may repeatedly attempt to bite the male with her mandibles.

In laboratory studies of mate recognition, Barrows (1975a) found that males would make contact with (pounce upon) small black dots after in-

Figure 5.1. Mating pair of *Lasioglossum zephyrum*. Male is at right. In the laboratory, where this mating was photographed, matings last ca. 1 minute.

troduction of a tethered female to the cage. Such contacts in the presence of cues emitted from a female indicate increased response to visual stimuli that approximate the size and shape of a female. He obtained the same result by replacing the use of a tethered female with pieces of moist filter paper that had been kept with a female for 24 hours. No increase in flight activity or contact with the black dots was observed in the presence of control paper that had been kept for the same period but without contact with a female. He therefore deduced that a sex pheromone given off by the female was the cause of the increase in sexual activity (i.e., release of a MAP) on the part of the males.

After 1–2 minutes, responses to the pheromone-impregnated filter paper decreased significantly. Removal of the filter paper followed by immediate replacement with a different piece of filter paper impregnated with odors from the same female did not restore the males' arousal to the original level in spite of the disturbance of removal and replacement of the paper (Barrows, 1975b; Barrows et al., 1975). Thus the males were not marking the papers with a repellant odor of their own. It was also not the case that the decrease in response to the first paper was due to odor on it becoming less concentrated during exposure to the air. In either case, the second paper should have released levels of male activity equal to the first.

Experiments were then performed in which groups of males were given three successive exposures to female odors (Barrows, 1975b). The first two 1-minute exposures were, as above, to the odor of the same female, which led (as expected) to a significant decrease in contacts with the black dots between the first and second exposures. On the third presentation the odor of a second female was introduced, and the rate of contacts increased to a level that was not distinguishable from the first exposure. Therefore, males were not becoming fatigued by repeatedly flying at higher speeds during exposure to a sex pheromone or by repeated contacts with the black dots. Instead they appeared to be learning, through one or more of the mechanisms reviewed above, the odor of a given female. In other words, exposure to a female's odors modifies the probability that the MAP for mating is released by the those odors, and odors differ among individual females.

The adaptive value of mate learning in halictine bees was explained as follows. Learning odors of specific individuals could be adaptive if, by doing so, males avoid those females with which they have unsuccessfully attempted to mate. Such a mechanism would allow males to allocate time more efficiently to mating attempts if they patrol in small areas within nesting aggregations where they are likely to repeatedly contact the same females. Furthermore, if mating exposes males to risks, such as predation while they are stationary or injury during a struggle with an unreceptive female, then learning might help to minimize that risk. Learning would also enable males to focus on searching for novel females about which they have less information regarding receptiveness.

About the time that this hypothesis regarding adaptive value was first proposed, it was demonstrated that females in social colonies discriminate between conspecific female nestmates and non-nestmate conspecific females on the basis of individual odor. By breeding several different lines of *L (D.) zephyrum* in laboratory colonies, Greenberg (1979) was able to show that these odors have a heritable component. In other words, females who are genealogically related have more similar odors than females who are less related. Females who were unknown to the nestmates within a colony but were genealogically related to them were much more likely to be allowed entry into the colony than females unrelated to the colony nestmates.

Males also use the information on genealogical relationship contained in the odor signal (Smith, 1983). When males were exposed to an unreceptive female for 10 minutes, the number of copulatory attempts (pounces) with that female decreased dramatically at first and then more slowly as the exposure continued (Fig. 5.2, left). When the female was removed at the end of that period and immediately reintroduced, the rate of contacts with that female did not change. If a female who was genealogically un-

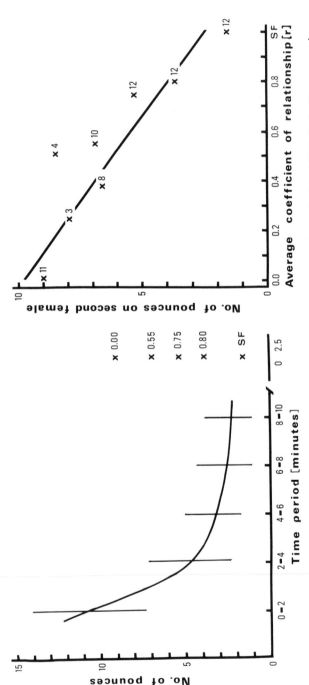

Figure 5.2. Learning and generalization of female odors by males in *L. zephyrum* (from Smith, 1983). Left: Decrement in copulatory attempts (pounces) by male bees on a female in successive 2-minute intervals over 10 minutes. Vertical lines are standard errors; the point at which the curve crosses the error bars indicates the mean attempts from three males in the cage with the female. The curve was fitted by eye. The break in the abscissa indicates removal of the female and reintroduction of either the same female (SF) or a different female of a known genealogical relationship (0.00–0.80) to the first. The *x*'s indicate the mean copulatory attempts by the same males on the reintroduced or new female. Right: Graph showing number of copulatory attempts on the reintroduced or new female as a function of genealogical relationship between the two females. Genealogical relationship for full, outbred sisters is 0.75 due to the haplodiploid mode of sex determination. Values higher than that indicate inbred sister or the same female, whereas values lower indicate degrees of cousins or females whose ancestors were collected from different nesting aggregations (i.e., *r* = 0.00). The straight line reflects a linear regression fitted to those points (see Smith, 1983), for statistical analyses of these results.)

related to the first was introduced instead, then the rate of copulatory attempts increased to a level that was not different from the first 2 minutes of exposure to the first female. These results confirmed Barrows' (1975a,b) studies that the decline in response was not due to disturbance or fatigue. However, reintroduction of females that were genealogically related to the first yielded results that were surprising.

Either inbred or outbred sisters of the original female elicited slightly more copulatory attempts than when the original female was simply reintroduced (Fig. 5.2, right). Cousins elicited even more copulatory attempts. In general, the number of copulatory attempts increased with decreasing genealogical relationship between the first and second females introduced to the males. Thus the odors used by males to recognize females have a heritable component, and the males use that information in making mating decisions. To use the terminology introduced in the first part of this chapter, males *generalize* the learned information about the odors of a female to her close relatives.

In light of the generalization results, Smith (1983) used the term "mistake" to point out an apparent problem with the original explanation proposed for this learning (outlined above). That explanation was based on males avoiding excessive time expenditures by learning a female's individual identification and her receptivity. The problem was that while odors are heritable, the association information that a given female will not mate does not necessarily apply more to her close relatives than to unrelated females. Thus a male, by rejecting a close relative of a female he has experienced, may make a mistake (called a type II error above) in that the second, related female might be receptive. Possibly, the generalization is a constraint on the male's perceptual system. However, the results indicated that alternative hypotheses need to be explored. A modified version of the original hypothesis which could account for the generalization was that males indeed avoid unreceptive females with whom they have had experience, partly to promote outbreeding (Bateson, 1983). Males learn the females from their small, and possibly inbred, aggregation and are preferentially, though not exclusively, attracted to a novel immigrant female who possesses a different genotype on average than females in the male's native locale.

In order to test this modified version of the hypothesis, the same sorts of testing procedures were run in field populations of a different species of halictine bee, *L. (Evylaeus) malachurum* (Smith and Ayasse, 1987). This species, which is found throughout temperate and Mediterranean regions of Europe, is similar to *L. (D.) zephyrum* in most aspects of its biology (Michener, 1974), except that nesting aggregations occur in horizontal clay soils in areas with little vegetation. On the day prior to testing, females were collected from two aggregations (termed HAG and WALD) that were separated by ca. 10 km. They were then frozen in the laboratory and held overnight until the following morning. The test consisted of pinning a female in one of two defined areas of an aggregation and counting the number of contacts by males each minute over 3 minutes. (Marked males tended to stay within a limited area of the larger aggregation.) These counts were then adjusted for variable levels of male flight activity. Each initial 3-minute test with a female was followed by an identical retest in a different situation. Some of the females were retested either in the same or in a different area within the aggregation. In other cases, the retest was performed in the same location with a different female collected either from the same nest, from a different nest, or from a different aggregation. Females collected from the same nest are on average more closely related than females collected from different nests (Crozier et al., 1987). Because of low male flight activity in the WALD population, these tests were performed only in HAG with both WALD and HAG females.

Male behavior in the field studies paralleled behaviors first described in the laboratory. In the first minute of testing, pinned females elicited high levels of copulatory activity (Fig. 5.3). This activity decreased over the next 2 minutes. When the same female was removed and then retested in the same location 1 minute later, significantly less copulatory activity was observed than in the first test (Fig. 5.3, top left). However, when the same female was retested in a different location within the same aggregation, where different males are patrolling, the copulatory activity recorded was significantly higher than when she was retested in the same location. Therefore, males in the field appeared to be learning the identity of the female. A retest in the same location, but with a nestmate (relative) of the first female, showed decreased attractiveness of the second female relative to the first (Fig. 5.3, top right). That is, the learning generalized to a relative. When a female collected from a different nest (not a close relative) within the same aggregation replaced the first female in the retest, the rate of copulatory attempts with the second female was not different from that with the first female (Fig. 5.3, bottom right). Therefore, males were not becoming fatigued. Wcislo (1987) has shown similar effects in field studies of mating in *L. (D.) zephyrum*.

If geographic separation led to some genetic differentiation between aggregations, then females from one aggregation should on average elicit more copulatory activity than females collected from the males' home aggregation. That is exactly what was observed. WALD females elicited significantly higher levels of activity across all 3 minutes than did females collected in the males' home aggregation, HAG (Fig. 5.3, bottom left).

In the mating studies reviewed above, focusing on stimulus generalization first helped to modify hypotheses about the adaptive basis for the behavior. The mechanism enabled formulation of further testable hypotheses about

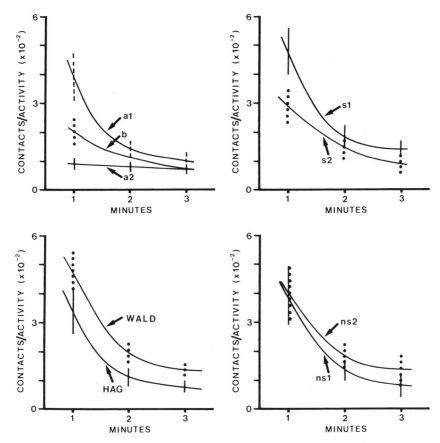

Figure 5.3. Field studies of learning and generalization of female odors by male *L. (Evylaeus) malachurum.* Curves show decrement in copulatory activity over three successive 1-minute intervals. Activity is measured by number of contacts males flying in the test area of the aggregation made with a female. Because differing numbers of males were flying on different days, or at different times of day, this activity was adjusted by dividing flight activity into the number of contacts (see Smith and Ayasse, 1987). Top left: A female (a1) was pinned in the test area for 3 minutes before removal. She was replaced into the same area (a2) or into a test area where different marked males were patrolling (b). Top right: A female (s1) was tested and then replaced by a female that had been collected from the same colony (s2), which presumably was a close relative (Crozier et al., 1987). Bottom right: A female (ns1) was tested and then replaced with a female from a different colony (ns2), which was presumably less related than a female from the same colony. Bottom left: Mean response on first tests by males from the HAG aggregation to females from HAG and to females from WALD, an aggregation some distance away (see text).

the behavior, which can now be interpreted as an outbreeding mechanism. Males generalize to close relatives of females they have experienced, the result of which is a heightened probability of detecting and mating with a novel female genotype. However, by generalizing too much, a male may risk mating with a female of another species that is found in the same locale but which produces a different pheromone mixture. Barrows (1975b) found that attractiveness decreased across species. Therefore, some optimum level of odor novelty may be preferred by the males, as has been shown for Japanese quail (Bateson, 1983).

Until now no mention has been made about the learning mechanism that gives rise to the curves in Figures 5.2 and 5.3. In studies that exposed males to live females, those females were always unreceptive. The learning curves (Figs. 5.2 and 5.3) always decrease through time in a nonlinear fashion—initially the decrease is rapid but is followed by a more gradual decrease after a few minutes. This type of learning curve would be expected if the males were habituating to female odors. However, the same type of curve might reflect associative conditioning assuming that an unreceptive female constitutes an aversive stimulus. This latter hypothesis would be reasonable given that the female's defensive reactions might injure the male. Therefore, although the studies reviewed above lead to the reasonable conclusion that learning takes place, they do not unambiguously identify the actual learning mechanism.

Future experiments that delineate the mechanism or mixture of mechanisms (e.g., nonassociative, associative, operant) could have an important impact on the adaptive interpretation of mate learning. The outbreeding hypothesis implicitly assumes that the primary mechanism is habituation. That is, exposure to females in the males' own locale should be sufficient to produce the learning. The quality of the experience, e.g., receptive or nonreceptive females, should not dramatically affect the habituation if this hypothesis is correct. However, males might associate qualities of the female (e.g., receptivity) with her odor, or a receptive female might sensitize a male's responses toward her and thus overcome habituation. If a female allows mating (all studies to date have used unreceptive or dead workers), does the male prefer odors of her close relatives? The outbreeding hypothesis would not easily account for these kinds of observations. Therefore, if future experiments document such an experienced-based preference, or at least a lack of response decrement after successful mating, then the outbreeding hypothesis would be brought into question. In any case, thinking about the exact learning mechanisms involved in mate choice will now guide future experimentation, while modeling studies (such as that reported in Roitberg et al., this volume) should help us to understand the adaptive consequences of different mechanisms.

We have only begun to understand male learning in halictine bees. Further work under a variety of conditions is now necessary to understand more fully the mechanism behind it as well as its adaptive role. Recent work has characterized in part the female odors to which males are reacting. (Smith et al., 1985, Smith and Wenzel, 1988) permitting use of synthetic conditioning stimuli. Several critieria can be used to evaluate whether the mechanism is due to habituation (Thompson and Spencer, 1966) or whether associative mechanisms might be involved. Furthermore, there are a variety of species of halictine bees that possess divergent life histories (Michener, 1974). Some bee species do not nest in aggregations, but instead disperse nests across large areas. Males of those species must search in areas where females can be most easily found (i.e., on flowers), though not in the numbers that commonly occur in aggregations of nests. A comparative study of the expression of mate learning in these different situations may lead to greater insight into the adaptive role of mate learning in general.

Learning About Odors During Feeding in Honey Bees

In order to establish general adaptive principles for learning, it will also be necessary to apply studies of learning and generalization to a variety of species in different contexts. Honey bees should be an ideal species in which to perform more detailed studies of the mechanisms of olfactory generalization. Foragers rapidly learn to respond to floral odors (von Frisch, 1967), and the learning processes can be easily studied in controlled situations (see Menzel et al., this volume).

Worker honey bees use several different kinds of odors (see literature cited in Winston, 1987). Many of these odors are pheromones, that is, odors produced and released by the honey bees themselves in order to communicate a variety of messages. In specific contexts, pheromones release MAPs. For example, alarm pheromones produced in the honey bee's sting potentiate worker bees' responses to a variety of visual, olfactory, and tactile stimuli that signal a threat to the colony. The honey bee's Nasonov pheromone is a lemony-smelling secretion released to attract other bees to an unscented foraging site, to the colony entrance in times of confusion, or to maintain the cohesiveness of a swarm of thousands of bees.

Many other kinds of odors do not release strong innate responses, because the correlation of those odors with any resource important to bees changes rapidly within a forager's lifetime. For example, foraging workers .must locate and learn to respond to floral stimuli, including odors, correlated with the presence of nectar and/or pollen. These resources can become depleted over very short time spans, which are much shorter than a worker bee's foraging lifetime (10–30 days). Depletion can be due to a variety of factors, which include short blooming cycles of some flower species and also competition for those resources with other colonies and species. In order to cope with this variability, honey bees associate a wide array of odors with floral rewards (von Frisch, 1967). Bees learn these associations rapidly and switch species depending on the availability and quality of other species (Seeley, 1985). The way in which foragers respond to floral odors must reflect adapations of the olfactory information-processing system (Koltermann, 1973).

Honey bees make ideal subjects for studies of learning mechanism under controlled and natural situations. A variety of conditioning procedures have been used, which range from use of freely flying subjects or walking subjects (von Frisch, 1967; Menzel, 1990) to use of subjects restrained in small harnesses where they can freely move antennae and mouthparts (Frings, 1944; Menzel, 1990; Menzel et al., 1974; Menzel and Bitterman, 1983; Fig. 5.4). The former procedure approximates natural settings, whereas the latter utilizes less-natural settings but has the advantage of better con-

Figure 5.4. Restrained worker honey bee whose antennae have just been touched by the sucrose-water droplet at the right. The subject has extended its proboscis in response to the sucrose (US) stimulation. Wires entering the bee's head from the rear are used for EMG recordings. The pointed structure at the tip of the proboscis is the glossa, extensions and retractions of which produce rhythmic bursts of spikes (licking movements) in the EMG. (Small labial palps stick out on each side of the glossa, but are not part of it).

trol of conditioning stimuli for elucidation of learning mechanisms (see Menzel et al., this volume and 1990, for an extensive review of proboscis extension [PE] conditioning of restrained bees and its use in studies of memory consolidation).

The PE procedure involves stimulation of olfactory and taste receptors of a subject's antennae with an odorant (CS) and sucrose (US). The latter is usually dissolved in water. Stimulation of the sucrose receptors on the antennae leads to extension of the proboscis (UR) in properly motivated subjects. Although there is usually a weak response to the CS in some subjects prior to any CS-US pairing, this response is always strengthened by forward pairing using specific ITIs (see Menzel et al.'s discussion of the α-response, this volume). There are many ways to quantify the extension response. One way is to compute the percentage of subjects in a population that show proboscis extension on a given trial. This measure allows for screening large numbers of subjects but does not accurately reflect the complexity of the feeding MAP.

In order to more accurately quantify the bee's feeding MAP, electromyogram recordings from the M-17 muscles in the bee's head can be used as the response measure (Snodgrass, 1956). Rhythmic contractions in these muscles during the PE response can be differentiated into three phases— extension, rhythmic glossal movements, and retraction (Rehder, 1987). Smith and Menzel (1989a) have used EMG recording to define independent parameters of the bee's feeding MAP and to test the ability of learned or novel stimuli to control expression of those parameters (Fig. 5.5). After subjects were conditioned to respond to an odor, they were tested under three different stimulus conditions to test the efficacy of different stimuli in releasing the MAP. These conditions were as follows: (1) Sucrose was applied to the antennae only, that is, no feeding occurred. (2) The odor to which subjects were conditioned was presented without the associated sucrose reward. (3) A novel odor (i.e., one that subjects did not experience during conditioning) was presented as in (2) to test for a generalization response. Only those subjects that responded with PE during test conditions were used for subsequent statistical analyses. Therefore, given that 100% of the subjects responded, no discrimination between test conditions would be possible using percentage data as the response measure. Differences in expression of the MAP among testing conditions must reflect differential ability of the different stimuli to control the MAP once it is released.

Only a subset of the response measures derived from EMG traces showed differences with conditioning experience, and in all cases the order of salience (high to low, respectively) for releasing and controlling MAP expression was sucrose (US), conditioning odor (CS), and novel odor. Parameters that described the length of the response (total spike count,

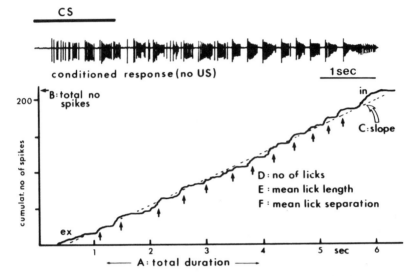

Figure 5.5. Top: EMG recording during and shortly after presentation of an olfactory conditioning stimulus. (The black bar indicates the time period during which the odor was presented). Long bursts of spikes at the beginning and ending of the recording correspond to proboscis extension and retraction, while rhythmic bursting in between corresponds to licking movements of the glossa (Rehder, 1987). Bottom: Graph of the cumulative number of spikes over the duration of recording, which shows different parameters that can be derived from the trace. These parameters are *duration* of the response; *total number* of spikes; the *slope* of the line through the cumulative spike curve, which indicates rate of responding; *number of licks* (bursts); *length* of bursts; *separation* of bursts; not shown is the *time* from beginning of the response to the first burst (from Smith and Menzel, 1989a).

duration of spiking activity, and number of "licks") all showed the highest values for tests with the sucrose US. Tests with the conditioning odorant showed an intermediate level, and tests with a novel odorant were lowest. Two other parameters that describe the "decisiveness" of the MAP (spike frequency or "speed" and number of interruptions) showed correlations with test condition. Tests with the US elicited the highest rate of spike activity, and this rate decreased from the CS to the novel odor. Tests with a novel odor were more likely to be interrupted one or more times than were tests with the US or the CS. Therefore, tests with the US and the CS elicited longer, faster, and more decisive responses than tests with a novel odor.

Other parameters did not vary. Neither the length of a spike burst nor the interburst interval, both of which are correlated to licking movements of the glossa, were different across testing conditions. The response latency, as measured from the start of the spike activity to detection of the first lick, was also not different across conditions.

Thus a CS can control some aspects of MAP expression but not others. This control is probably due to the buildup of an "associative strength" between neural representations for sucrose and for the CS (Wagner, 1980). Through proper temporal association, the CS can control feeding motor centers in the bee's CNS in a way that is similar to the innate control expressed by the sucrose US. Furthermore, a generalization response to novel odorants can be registered and is less than that to conditioning odorants. This response generalization arises in part from a gradient of perceptual similarity between the CS odor and other, novel odors that the subject had not experienced during conditioning. Further stimulus parameters leading to generalization still need to be established.

This latter problem is complicated because the perceptual dimensions along which odors vary are not clearly defined. For visual or acoustic stimuli, dimensions such as wavelength, size, movement, frequency, etc., can be clearly defined and experimentally manipulated. In contrast, odors vary according to carbon-chain length, presence of double bonds, functional groups, and many other parameters whose effect on sensory receptors cannot be described independent of a behavioral response. At best, these attributes of odors can be varied while others are held constant. Under those conditions, the bees' generalization tendencies can be quantified experimentally, and generalization gradients can be derived.

Accordingly, Smith and Menzel (1989b) defined olfactory similarity gradients according to parameters such as carbon-chain length, structure, and attached functional groups. Subjects were conditioned over several trials to associate an odor with a brief sucrose US. Those subjects that reached a given response criterion (at least three successive responses to the odor) were used in subsequent testing. A series of unrewarded tests were then performed with 21 novel odorants that varied in their structural features with respect to the conditioning odorant. (Several rewarded trials with the conditioning odorant were performed throughout the unrewarded series to prevent extinction of the response.) In addition, one unrewarded trial with the conditioning odor was used to assess the response to that odor during an unrewarded trial. Responses were quantified as the number of spikes in the EMG. All responses to novel odors that differed slightly in carbon-chain structure but contained a common functional group (aldehyde, ketone, alcohol, acetate, monoterpene alcohol, monoterpene aldehyde) were lumped because of a lack of difference in generalization responses to odors within those groups. The lack of difference in response to novel odors that differ only slightly in carbon-chain length indicates only that these odors are perceptually very similar. As a consequence, honey bees generalize strongly among them. This does not mean that they are indistinguishable to the bees.

After an equivalent amount of conditioning, there were significant differences in the abilities of odors that possess different oxygen moieties to release and control expression of a worker honey bee's feeding MAP (Fig. 5.6). Aldehydes, ketones, and alcohols all elicited relatively strong responses during extinction trials after conditioning with those odors. The

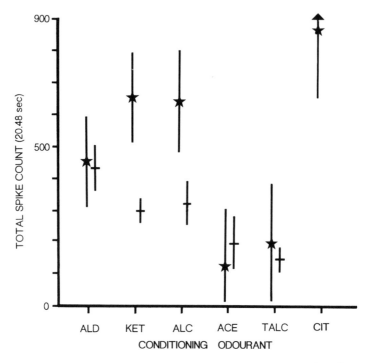

Figure 5.6. Responses of subjects to conditioning odors (stars) and to novel odors from the same chemical class as the conditioning odors (horizontal bars). Conditioning odors are listed as follows: ALD, aldehydes; KET, ketones; ALC, alcohols; ACE, acetates; TALC, monoterpene and sesquiterpene alcohols; CIT, the monoterpene aldehyde citral. ALD, KET, and ALC were all straight, saturated carbon chains of lengths five through eight or nine. KET and ALC contained the oxygen moiety on the second carbon in the chain. Three ACE compounds were used: isopropyl, isobutyl, and isopentyl. The TALC group was comprised of geraniol, nerol, and farnesol. All conditioning odors were pure compounds; means for each group of compounds are lumped response to each of the odors (from Smith and Menzel, 1989b).

monoterpene aldehyde citral was the most salient of the odors (Smith, 1991). In contrast, acetates and monoterpene alcohols were markedly less effective in regulating MAP expression. The acetates are components of the honey bee's alarm pheromone, which potentiates bees' responses to visual, olfactory, and tactile stimuli that indicate a threat to the colony. It might be the case that the neural pathways for alarm-pheromone processing release MAPs (e.g., stinging) that conflict with the expression of an appetitive MAP such as feeding; that is, learning performance is altered by arousal of conflicting motivational states. Alternatively, the sensory pathways that process alarm pheromone might not interact strongly with the neural pathways that produce the feeding MAP; in that case, storage of the learned information might not be as efficient as with other odors. Further experiments are necessary to determine whether alarm pheromones block learning altogether or whether they do not block learning, but instead block expression (Terrace, 1984; Rescorla, 1988; Spear et al., 1990). For example, changes in the concentration of the odor or in the motivational state of the subject might improve learning performance. Such experiments need to be performed. However, even if lower or higher quantities of the odor improved learning performance, it would not subvert the interpretation that specialized, adaptive stimulus-processing pathways affect the expression of learning mechanisms.

At first, the differences in responses to the terpene compounds did not seem to fit nicely into such a conclusion. Citral was strongly associated with rewards such as sucrose, whereas other terpenes were not. This appetitive predisposition to respond to citral also appeared in memory processing of this odor vs. processing after geraniol conditioning (Smith, 1991). The terpenoid compounds are all components of the bee's Nasonov pheromone, which elicits responding in a variety of natural situations (see Winston, 1987). In fact, geraniol, one of the alcohols used in the conditioning procedure, is the major component of the secretion (Pickett et al., 1980), yet it elicits significantly less responding during the conditioning procedure. However, closer scrutiny of the actual usage of the pheromone in natural situations leads to a potential explanation of differences in learning. Citral, although it is not the major component of the secretion, is the most attractive component in a variety of natural biological assays. Enzymes in the gland even convert the less-attractive alcohols into citral when the pheromone is released (Pickett et al., 1981). So citral may be the major component of the released pheromone. Therefore, the tendency for citral to elicit stronger responses in the conditioning assay (even though alcohols were tested at the same concentration) may simply reflect citral's significance in a natural context.

Figure 5.6 also shows generalization to novel odors that possess the same functional group (e.g., aldehyde, ketone, etc.). Among the aldehydes, generalization responses to novel aldehydes were as strong as were responses to the aldehyde to which bees had been conditioned. In contrast, generalization among ketones and alcohols of similar carbon chain lengths was different. In both of these chemical classes, novel odors (which differed only in carbon chain length) elicited significantly lower responses than the conditioning odors. These results are consistent with those obtained by Sass (1978), who measured responses of cockroach sensory receptors to short-chain alcohols of chain lengths comparable to those used by Smith and Menzel (1989a). He found that sensory receptors showed response differences across only a few carbon molecules that contribute to chain length; receptor types tended to have one "best" chain length, which elicited a peak response.

Generalization responses decreased across classes of compounds (Fig. 5.7). When subjects were conditioned to an aldehyde and then tested with a wide array of novel compounds, some of which were aldehydes of different carbon-chain length, generalization was strongest to the aldehydes relative to responses to other odors. The same pattern was observed for alcohols and ketones.

Thus generalization gradients can be described for odors in the honey bee. The most likely basis for gradients is perceptual similarity, in terms of coding processes at several levels in the olfactory system (Getz and Chapman, 1987), between certain odors and those experienced by the bees. Furthermore, honey bees show strong tendencies, as measured in learning performance to respond and generalize to certain odors (Smith and Menzel, 1989b; Smith, 1991). This response preference can be correlated with natural response tendencies toward pheromones that elicit strong appetitive response in a natural context. In short, honey bees show learning and generalization tendencies among odors that in some cases are biased by innate "meanings" of certain odors, much as occurs for different kinds of stimuli in many animals including humans (Shepard, 1987).

It is important to note that at least two other factors in addition to perceptual similarity affect the shape of these gradients. First, the generalization gradients reported in Figures 5.6 and 5.7 show mean levels of responding in populations of subjects under different testing conditions. Different subjects show different tendencies to generalize. No explanation for such individual differences exists. Subjects were collected for these studies as they departed from a colony. Bees depart colonies for a variety of reasons—foraging for pollen, nectar, or both, as well as removal of detritus, defense, and orientation flights. It may be that bees in different stages of development have different metabolic requirements. Thus a forager which uses its flight muscles on a regular basis may become depleted of energy reserves much faster than a bee that makes only occasional flights. A bee with a depleted energetic reserve might show different tendencies

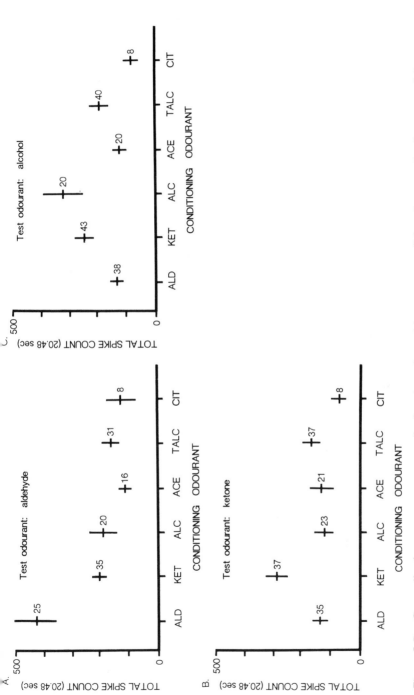

Figure 5.7. Generalization responses to a novel odor given that subjects were conditioned to an odor from one of the groups listed on the abscissa. Top left: Responses to a novel aldehyde from ALD. Top right: Responses to a novel alcohol from ALC. Bottom left: Responses to a novel ketone from KET. In all cases, the test odor was not one of the conditioning odors; thus when, for example, hexanal was the test odor, it was not one of the conditioning odors used in ALD on the abscissa. Subjects conditioned to an aldehyde from ALD were tested with the other aldehydes (from Smith and Menzel, 1989b).

to generalize than a bee that is closer to satiation. Alternatively, workers that are specialized for performing different tasks may preferentially learn different kinds of odors and their relationship to important environmental stimuli.

Second, aversive stimuli can influence the propensity to generalize and hence the shape of generalization gradients. If costs of generalization are increased, a subject may decrease its tendency to generalize, at least toward stimuli similar to ones that predict punishment. In natural foraging situations, flowers may possess no rewards. Furthermore, some nectars are bitter and elicit an aversive reaction from bees. In the extreme, some nectars contain sugars that are toxic to honey bees (Winston, 1987). Therefore, one would expect that honey bees have the capacity not only to learn to approach stimuli that predict floral rewards, but also to avoid stimuli that are associated with distasteful or toxic floral products (cf. Bernays, this volume).

Abramson (1986) used a conditioning paradigm in which freely flying foragers were trained to approach, land on, and feed from a target. While they were feeding, subjects received an aversive stimulus (shock or a burst of formic acid) associated with a vibratory CS. Foragers learned to avoid the stimulus by briefly flying off the target or by removal of their proboscides from contact with the sucrose solution. (The latter response broke contact between electrical leads that delivered the shock.) Smith et al. (1991) have recently extended this procedure to use of restrained subjects. In their procedure, subjects were differentially conditioned to two odors—one that predicted brief feeding on sucrose and another that predicted a 10-V AC shock if a PE response occurred upon sucrose stimulation of the antenna. Subjects could avoid shock if they withheld PE to the normally powerful releaser (the sucrose US) in the context of odor that predicts shock. A significant number of the subjects could indeed do that. But, as before, significant differences were observed among individuals in the extent to which they committed "errors."

These studies with honey bees show that an appeal to natural history can help to generate hypotheses regarding the expression of learning in defined conditioning paradigms. Variability in responses to odors can potentially be explained by the pheromonal nature of the odor and/or the motivational states of bees. In other experiments, awareness that naturally aversive stimuli occur (bitter and toxic nectars) led us to examine mechanisms of learned avoidance. However, most of the work, including studies of generalization, has focused more on elucidation of mechanism than on adaptive value. Although the experiments included pheromones among the conditioning odors in order to test for differences in processing from floral odors, the explanations regarding adaptive-processing mechanisms have been *ad hoc*; that is, they have been derived after the experiment was performed. It is now necessary to make predictions regarding adaptation and then choose the appropriate species in which to test them. For example, does pheromone processing in some cases interfere with storage and/or retrieval of learned information? What are the generalization responses of foragers when they and/or their colonies are stressed? In addition, there are phylogenetically unrelated species of bees that show flower preferences different from those of honey bees. The latter are flower generalists, which means that foragers visit a wide array of flowering plants. In contrast, a number of bee species specialize in collecting pollen from a narrow range of flowers. The life cycles of some bee species are tied to a single flowering plant. Is the expression of olfactory learning and generalization in these specialist species more restricted than that of honey bees? These sorts of studies should yield further insight into the relationship between learning mechanism, generalization, and adaptation.

Conclusions

Progress in the study of insect learning will take place on several fronts, as reflected in various chapters in this volume. It will require a variety of techniques including mathematical models as well as behavioral and physiological assays. Furthermore, researchers of insect behavior should seize the opportunity to apply these methods to a variety of species whose niches and phylogenetic relationships are now known. In this review, I have attempted to argue that a rich methodology for studying behavioral processes involved in learning needs to be exploited in this research. Neither of the projects outlined above is complete; each is now at a stage where more detailed questions need to be asked. However, increased use of controlled environments in general for studying insect learning should make these methods more effective. Coupled with studies of behavioral modification in natural contexts, well-controlled studies of mechanism can be powerful tools for understanding the adaptive significance of learning.

Acknowledgments

Throughout the studies reviewed above, the author has benefited from thoughtful interactions with a large number of colleagues. He wishes to thank in particular the following: C.I. Abramson, M. Ayasse, W.M. Getz, J.G. Hildebrand, R. Menzel, C.D. Michener, and J. Tengö. This work was supported in part by fellowships to the author from NSF/NATO and NIMH (NRSA award #5 F32 MH09783 to B.H.S.).

References

Abramson, C.I. 1986. Aversive conditioning in honey bees (*Apis mellifera*). J. Comp. Psychol. **100**:108–116.

Barlow, G.W. 1977. Modal action patterns. In T.A. Sebeok (ed.), How Animals Communicate. Indiana University Press, Bloomington, pp. 98–134.

Barlow, G.W. 1989. Has sociobiology killed ethology or revitalized it? In P.P.G. Bateson and P.H. Klopfer (eds.), Perspectives in Ethology: Whither Ethology?, Vol. 8. Plenum, New York, pp. 1–46.

Barrows, E.M. 1975a. Mating behavior in halictine bees (Hymenoptera: Halictidae): III. Copulatory behavior and olfactory communication. Insectes Soc. **22**:307–331.

Barrows, E.M. 1975b. Individually distinctive odors in an invertebrate. Behav. Biol. **15**:57–64.

Barrows, E.M. 1976. Mating behavior in halictine bees (Hymenoptera: Halictidae): I. Patrolling and age-specific behavior in males. J. Kansas Entomol. Soc. **49**:105–119.

Barrows, E.M., Bell, W.J., and Michener, C.D. 1975. Individual odor differences and their social function in insects. Proc. Natl. Acad. Sci. USA **72**:2824–2828.

Bateson, P.P.G. 1983. Optimal outbreeding. In P.P.G. Bateson (ed.), Mate Choice. Cambridge University Press, Cambridge, England, pp. 257–277.

Bateson, P.P.G. 1984. Genes, evolution, and learning. In P. Marler and H.S. Terrace (eds.), The Biology of Learning. Springer-Verlag, Berlin, pp. 75–88.

Beecher, M.D. 1988. Some comments on the adaptionist approach to learning. In, Evolution and Learning. R.C. Bolles and M.D. Beecher (eds.), Lawrence Erlbaum Associates, Hillsdale, NJ, pp. 239–248.

Bolles, R.C. 1988. Nativism, naturalism, and niches. In R.C. Bolles and M.D. Beecher (eds.), Evolution and Learning. Lawrence Erlbaum Associates, Hillsdale, NJ, pp. 1–15.

Brothers, D.J., and Michener, C.D. 1974. Interactions in colonies of primitively social bees, III. Ethometry and division of labor. J. Comp. Physiol. **90**:129–168.

Couvillon, P.A. and Bitterman, M.E. 1982. Compound conditioning in honeybees. J. Comp. Physiol. Psychol. **96**:192–199.

Crozier, R.H., Smith, B.H., and Crozier, Y.C. 1987. Relatedness and population structure of the primitively eusocial bee *Lasioglossum zephyrum* in Kansas. Evolution **41**:902–910.

Frings, H. 1944. The loci of olfactory end organs in the honey bee, *Apis mellifera* Linn. J. Exp. Zool. **97**:123–134.

Frisch, K. von 1967. The Dance Language and Orientation of Bees. Harvard University Press, Cambridge, MA.

Gallistel, C.R. 1990. The Organization of Learning. MIT Press, Cambridge, MA.

Getz, W.M., Chapman, R.F. 1987. An odor discrimination model with application to kin recognition in social insects. Int. S. Neurosci. **32**:963–967.

Gould, J.L., and Marler, P. 1984. Ethology and the natural history of learning. In P. Marler and H.S. Terrace (eds.), The Biology of Learning. Springer-Verlag, Berlin, pp. 47–74.

Greenberg, L. 1979. Genetic component of bee odor in kin recognition. Science **206**:1095–1097.

Hollis, K.L. 1984. The biological function of Pavlovian conditioning: The best defense is a good offense. J. Exp. Psychol: Anim. Behav. Proc. **10**:413–425.

Hollis, K.L. 1990. The role of Pavlovian conditioning in territorial aggression and reproduction. In D.A. Dewsbury (ed.), Contemporary Issues in Comparative Psychology. Sinauer Associates, Sunderland, MA, pp. 197–219.

Kalish, H.I. 1969. Stimulus generalization. In M.H. Marx (ed.), Learning Processes. MacMillan, New York, pp. 207–297.

Koltermann, R. 1973. Reasen-bzw. artspezifische Duftbewertung bei der Honigbiene und Ökologishe Adaptation. J. Comp. Physiol. **85**:327–360.

Mackintosh, N.J. 1983. Conditioning and Associative Learning. Clarenden Press, Oxford, England.

Marler, P. 1991. Song-learning behavior: The interface with neuroethology. TINS **14**:199–206.

Menzel, R. 1990. Learning, memory, and "cognition" in honey bees. In R.P. Kesner and D.S. Olton (eds.), Lawrence Erlbaum Associates, Hillsdale, NJ, pp. 237–292.

Menzel, R., and Bitterman, M.E. 1983. Learning by honey bees in an unnatural situation. In F. Huber and H. Markl (eds.), Neuroethology and Behavioral Phsyiology. Springer-Verlag, Berlin, pp. 206–215.

Menzel, R., Erber, J., and Mashur, T. 1974. Learning and memory in the honey bee. In L. Barton Browne (ed.), Experimental Analysis of Insect Behavior. Springer-Verlag, New York, pp. 195–217.

Michener, C.D. 1974. The Social Behavior of the Bees. Harvard University Press, Cambridge, MA.

Michener, C.D., and Smith, B.H. 1987. Kin recognition in primitively eusocial insects. In D.J.C. Fletcher and C.D. Michener (eds.), Kin Recognition in Animals. John Wiley, New York, pp. 209–242.

Pickett, J.A., Williams, I.H., Smith, M.C., and Martin, A.P. 1980. Nasonov pheromone of the honey bee *Apis mellifera* L. (Hymenoptera: Apidae), III. Chemical composition. J. Chem. Ecol. **6**:425–434.

Pickett, J.A., Williams, I.H., Martin, A.P., and Smith, M.C. 1981. Nasonov pheromone of the honey bee *Apis mellifera* L. (Hymenoptera: Apidae), III. Regulation of pheromone composition and production. J. Chem. Ecol. **7**:543–554.

Rankin, C.H., and Carew, T.J. 1988. Dishabituation and sensitization emerge as separate processes during development in *Aplysia*. J. Neurosci. **8**:197–211.

Real, L.A. 1991. Animal choice behavior and the evolution of cognitive architecture. Science **253**:980–986.

Rehder, V. 1987. Quantification of the honey bee's proboscis reflex by electromyogram recordings. J. Insect Physiol. **33**:501–507.

Rescorla, R.A. 1980. Pavlovian Second-Order Conditioning: Studies in Associative Learning. Lawrence Erlbaum Associates, Hillsdale, NJ.

Rescorla, R.A. 1988. Behavioral studies of Pavlovian conditioning. Annu. Rev. Neurosci. **11**:329–352.

Rescorla, R.A., and Holland, P.C. 1982. Behavioral studies of associative learning in animals. Annu. Rev. Psychol. **33**:256–308.

Rescorla, R.A., and Wagner, A.R. 1972. A theory of Pavlovian conditioning: Variations in the effectiveness of reinforcement and nonreinforcement. In A.H. Black and W.F. Prokasy (eds.), Classical Conditioning, Vol. 2. Appleton-Century-Crofts, New York.

Revusky, S. 1984. Associative predispositions. In P. Marler and H.S. Terrace (eds.), The Biology of Learning. Springer-Verlag, Berlin, pp. 447–460.

Sass, H. 1978. Olfactory receptors on the antenna of *Periplaneta*: Response constellations that encode food odors. J. Comp. Physiol. **128**:227–233.

Schmid-Hempel, P. 1984. The importance of handling time for the flight directionality in bees. Behav. Ecol. Sociobiol. **15**:303–309.

Seeley, T.D. 1985. Honeybee Ecology. Princeton University Press, Princeton, NJ.

Shepard, R.N. 1987. Toward a universal law of generalization for psychological science. Science **237**:1317–1323.

Shettleworth, S.J. 1984. Natural history and evolution of learning in nonhuman mammals. In P. Marler and H.S. Terrace (eds.), The Biology of Learning. Springer-Verlag, Berlin, pp. 419–434.

Smith, B.H. 1983. Recognition of female kin by male bees through olfactory signals. Proc. Natl. Acad. Sci. USA **80**:4551–4553.

Smith, B.H. 1987. Effects of genealogical relationship and colony age on the dominance hierarchy of the primitively eusocial bee *Lasioglossum zephyrum* (Hymenoptera: Halictidae). Anim. Behav. **35**:25–216.

Smith, B.H. 1991. The olfactory memory of the honey bee, *Apis mellifera*: I. Odorant-modulation of short- and intermediate-term retrieval after single trial conditioning. J. Exp. Biol. **161**:367–382.

Smith, B.H., Carlson, R.G., and Frazier, J. (1985). Identification and bioassay of the macrocyclic lactone sex pheromones of the halictic bee *Lasioglossum zephyrum* (Hymenoptera: Halictidae). J. Chem. Ecol. **11**:1447–1456.

Smith, B.H., and Ayasse, M. 1987. Kin-based male mating preferences in two species of halictine bees (Hymenoptera: Halictidae). Behav. Ecol. Sociobiol. **20**:313–318.

Smith, B.H., and Wenzel, J.W. 1988. Pheromonal correlation and kinship in the social bee *Lasioglossum zephyrum* (Hymenoptera: Halictidae). J. Chem. Ecol. **14**:87–94.

Smith, B.H., and Menzel, R. 1989a. The use of electromyogram recordings to quantify odorant discrimination in the honey bee, *Apis mellifera*. J. Insect Physiol. **35**:369–375.

Smith, B. H., and Menzel, R. 1989b. An analysis of variability in the feeding motor program of the honey bee: The role of learning in releasing a modal action pattern. Ethology **82**:68–81.

Smith, B.H., Abramson, C.I., and Tobin, T.R. 1991. Conditional withholding of proboscis extension in honey bees (*Apis mellifera*) during discriminative punishment. J. Comp. Psych. **105**:345–356.

Snodgrass, R.E. 1956. The Anatomy of the Honey Bee. Comstock, Ithaca, NY.

Sokal, R.R., and Rohlf, F.J. 1981. Biometry. Freeman, NY.

Spear, N.E., Miller, J.S., and Jagielo, J.A. 1990. Animal memory and learning. Annu. Rev. Psychol. **41**:169–211.

Staddon, J.E.R. 1983. Adaptive Behavior and Learning. Cambridge University Press, Cambridge, England.

Terrace, H.S. 1984. Animal learning, ethology, and biological constraints. In P. Marler and H.S. Terrace (eds.), The Biology of Learning. Springer-Verlag, Berlin, pp. 15–45.

Thompson, R.F., and Spencer, W.A. 1966. Habituation: A model phenomenon for the study of neuronal substrates of behavior. Psychol. Rev. **73**:16–43.

Timberlake, W., and Lucas, G.A. 1989. Behavior systems and learning: From misbehavior to general laws. In S.B. Klein and R.R. Mowrer (eds.), Contemporary Learning Theories: Instrumental Conditioning Theory and the Impact of Biological Constraints on Learning. Lawrence-Erlbaum Associates, Hillsdale, NJ, pp. 237–275.

Timberlake, W. 1990. Natural learning in laboratory paradigms. In D.A. Dewsbury (ed.), Contemporary Issues in Comparative Psychology. Sinauer Associates, Sunderland, MA, pp. 31–54.

Tinbergen, N. 1951. The Study of Instinct. Oxford University Clarenden Press, Oxford.

Wagner, A.R. 1980. SOP: A model of automatic memory processing in animal behavior. In N.E. Spear and R.R Miller (eds.), Information Processing In Animals: Memory Mechanism. Lawrence-Erlbaum Associates, Hillsdale, NJ, pp. 5–44.

Wcislo, W.T. 1987. The role of learning in the mating biology of the sweat bee *Lasioglossum zephyrum* (Hymenoptera: Halictidae) Behav. Ecol. Sociobiol. **20**:179–185.

Winston, M.L. 1987. The Biology of the Honey Bee. Harvard University Press, Cambridge, MA.

6

Motivation, Learning, and Motivated Learning

Marc Mangel

Introduction

As many of the papers in this volume illustrate, the behavior of an organism can be influenced by its own physiological state, by the state of the environment, and by the information that the organism has about the state of the environment. In this chapter, I develop a functional (i.e., evolutionary) approach that can be used to both separate and integrate physiology and environmental information, since each is connected with changes of behavior as a result of experience. The approach is based on dynamic, state-variable modeling (Mangel and Clark, 1988) which explicitly couples physiology and ecology within the framework of a Darwinian measure of fitness and thus responds to Kamil's (1983) call to integrate the "optimization approach" to behavior with other methods of ethology and psychology. Functional interpretations of learning require an assessment of the fitness, measured in terms of expected reproduction, of suites of behaviors. The technique used to determine fitness is called stochastic dynamic programming. Ward (1987) gives a simple example of stochastic dynamic programming for habitat acceptance; this example is in fact a special case of the methods developed by Mangel and Clark (1986).

Definitions

I modify Dudai's (1989, p. 6) definition: *Learning* is "an experience-dependent generation of enduring internal representations *of the external environment*, and/or experience-dependent lasting modification in such representations." In the language of neural networks (Edelman, 1987; Putters and Vonk, 1990) an "enduring internal representation" is a description of the external world based on connections between different neuronal groups and rules for modifying those connections. Learning represents changes in the pattern of connections or the rules for modifying those connections (e.g., Mangel, 1990a; Putters and Vonk, 1990). External cues and experience lead to modification of the pattern of connections and the rules for modifying those connections. With this definition, an organism cannot learn about its own internal state and learning consists of gathering information about the external environment and using that information to change the description of the environment. For example, if we want to study how oviposition behavior depends upon the rate of encounters with hosts, egg complement and age should be held constant as encounter rates are varied. By holding egg complement constant, we are able to understand how different encounter rates lead to different behaviors at a constant physiological state.

Motivation is a measure of physiological state directly related to the behavior of interest. Changes in motivation, via experience, can lead to changes in behavior, but this is not learning. For example, egg complement will increase over time if hosts are not encountered, and this can lead to a change in behavior (usually acceptance of an inferior host for oviposition). The objective in the host encounter study would be to separate changes in behavior due to increased egg complement (motivation) from changes in behavior due to changed descriptions of the world (learning) as encounter rates with hosts are varied.

The definition of learning that I adopt is narrower than "changes of behavior with experience." Alex Kacelnik (personal communication) has suggested the following analogy (modified for Central California): If I drive a car equipped with automatic transmission from Davis (elevation 19 m) to Lake Tahoe (elevation approximately 2,000 m), the car will change gears as the mountains are traversed. Although these changes of gear are determined by the "experience" of the automobile, they do not represent learning: gear changes are engineered responses to the state of the transmission.

The next two sections contain examples of learning and motivation separately, within the context of a functional determination of the value of information. In the third section, I show how the two can be combined.

Learning: Parasitoids and Patches of *Drosophila*

In this section, I model learning by a drosophilid parasitoid which is time, rather than egg, limited (e.g., van Alphen and Visser, 1990; Janssen, 1989) (see Table 6.1 for an explanation of parameters and their interpretations.) The assumption that the parasitoid is time limited means that the physiological variables (egg complement, nutritional status) can be ignored. In addition, I assume that patches of hosts consist of discrete clumps of rotting fruit which contain larvae of hosts and that the patches are hard to

Table 6.1. Parameters and their interpretations: parasitoids and patches of
Drosophila

Parameter	Interpretation
λ	Encounter rate of a parasitoid, once it is in a patch of hosts
α, ν	Parameters that describe the probability density of values of λ; in particular, the mean value of λ is $\dfrac{\nu}{\alpha}$ and the coefficient of variation is $\dfrac{1}{\sqrt{\nu}}$
α_0, ν_0	Initial values of the parameters, before a patch is visited; these correspond to "evolutionary information" concerning the distribution of possible values of λ
$f_0(\lambda)$	Prior probability density of λ, before a sampling is done
$\Gamma(\nu)$	Gamma function (for integers $\Gamma(\nu) = (\nu - 1)!$)
$f_p(\lambda|K, S)$	Posterior probability density of λ, given that K hosts were encountered in search time S
$F(\nu, \alpha, t)$	Expected (averaged over random encounters with hosts) accumulated ovipositions between t and T, given that the current values of the parameters describing the probability density of λ are ν and α
ρ	Probability that the parasitoid encounters a patch of hosts in a single period of search
μ	Probability that the parasitoid is killed during a single period
V_{leave}	Fitness value of leaving the current patch
N	Random number of hosts encountered in a single period of search, given that the parasitoid is in the patch
V_{stay}	Fitness value of staying in the current patch
m	Memory parameter used to weight past information

find. Since the parasitoid is not egg limited, when such patches are found there are fewer hosts available for oviposition than eggs. Patches of hosts, however, will vary in quality (number of hosts per unit volume, ratio of unparasitized to previously parasitized hosts) both over space and time (i.e., within the context of an individual's life) and over years (i.e., within the context of evolutionary time).

Here I adapt a model of learning by fishermen (Mangel and Clark, 1983; Mangel, 1990b) to describe learning by such parasitoids. For simplicity, I assume that patches are large enough such that depletion (see Mangel and Clark, 1983) and superparasitism (see Mangel, 1989, 1990b) can be ignored; these can be included in more complex models. In this case, the quality of a patch is determined solely by the encounter rate of hosts within that patch. The objective of the model is to provide a description for learning by the parasitoid as it encounters hosts.

Consider a parasitoid that has already found a patch of hosts. If hosts are randomly distributed in the patch, then we may assume that encounters with hosts in the patch follow a Poisson distribution (random encounters).

Pr {parasitoid encounters k hosts in time t given that the encounter rate is λ}

$$= \frac{e^{-\lambda t}(\lambda t)^k}{k!} \tag{1}$$

The encounter rate λ is not known to the parasitoid—it must be learned from experience in the particular patch. There is, however, a priori a probability distribution associated with different values of λ.

This *prior density* of possible values of λ represents an "internal representation" of the world in that the encounter rate in a particular patch is assumed to be randomly drawn from the probability density of λ. Experience (search and encounters with hosts) leads to modifications of this prior density and thus a change of the internal representation. The prior density provides a template for learning; the mechanism of learning still needs to be described.

A commonly used (e.g., DeGroot, 1970; Mangel, 1985) prior density is the gamma density

$$f_0(\lambda) = \frac{e^{-\alpha\lambda}\lambda^{\nu-1}\alpha^\nu}{\Gamma(\nu)} \tag{2}$$

That is, $f_0(\lambda)\Delta\lambda$ is the probability that the actual encounter rate is between λ and $\lambda + \Delta\lambda$. Here $\Gamma(\nu)$ is the gamma function. For integer values, $\Gamma(\nu) = (\nu - 1)!$; otherwise it can be viewed simply as part of the constant that ensures that the integral of $f_0(\lambda)$ over $0 \leq \lambda \leq \infty$ is equal to 1. The gamma density has two parameters, α and ν, that can be interpreted as follows. When λ has the density given by (2), its mean and coefficient of variation (standard deviation divided by the mean) are

$$E\{\lambda\} = \frac{\nu}{\alpha}$$

and

$$CV\{\lambda\} = \frac{1}{\sqrt{\nu}}$$

This form is convenient, because we can specify a mean encounter rate and then adjust the variability of this encounter rate by changing ν.

Combining (1) and (2) shows that

Pr {parasitoid encounters k hosts in time t}

$$= \int_0^\infty \frac{e^{-\lambda t}(\lambda t)^k}{k!} f_0(\lambda) \, d\lambda$$

$$= \frac{\Gamma(k + v)}{\Gamma(v)} \left(\frac{t}{\alpha + t}\right)^k \left(\frac{\alpha}{\alpha + t}\right)^v \tag{3}$$

This is a negative binomial distribution (Mangel, 1985) and can be put into the form more commonly used by ecologists (Southwood, 1966) in which the mean and overdispersion parameter are specified. One finds that the mean is $m = (v/\alpha)t$ and the overdispersion parameter is v. The mean number of encounters is m and the variance of the number of encounters is $m + (1/v)m^2$. Thus, when v is small, the variance in the encounters will greatly exceed the mean. Parasitoids will experience, once in patches, clumped encounters with hosts: in some patches many encounters will occur and in other patches very few encounters will occur.

Learning is the process of changing the description of the probability associated with different values of λ. We employ the methods of Bayesian updating (DeGroot, 1970). That is, suppose that the parasitoid has been in the patch for S units of time and has encountered K hosts. Learning modifies the prior density by the use of this information and produces a *posterior density of λ*

$f_p(\lambda \mid K, S)\Delta\lambda =$ Pr {encounter rate is between λ and $\lambda + \Delta\lambda$, given that K hosts were encountered in S units of time}

Applying Bayes's theorem shows that $f_p(\lambda|K, S)$ is again a gamma density with *updated parameters* $v + K$ and $\alpha + S$ (DeGroot, 1970). The Bayesian analysis provides an "updating rule" for the parameters:

$$v \rightarrow v + K$$

$$\alpha \rightarrow \alpha + S$$

Given the information concerning encounters, the posterior mean and coefficient of variation of λ are $E_p\{\lambda\} = (v + K)/(\alpha + S)$ and $CV_p\{\lambda\} = 1/\sqrt{v + K}$. These updated parameters represent a change in informational state (estimate of encounter rate distribution) caused by experience (actual encounters with hosts). The prior and posterior densities are "internal representations" which can be modified by experience.

We can compute the selective advantage of learning by relating learning to expected lifetime reproduction of the parasitoid. Assume that at emergence, the prior density of λ (the evolutionary template on which learning occurs) is given by (2) with parameters v_0 and α_0 and that the maximum reproductive life span of the parasitoid is T. As the parasitoid encounters patches and hosts within patches, the probability distribution of the encounter rate is described by the prior density or the current posterior density. We seek the behaviors that maximize expected lifetime reproduction. In this case, the behavior is particularly simple: the parasitoid can remain in the current patch or leave it and search for another patch. At any time between emergence and T, let

$$F(v, \alpha, t) = \text{maximum } E \text{ \{accumulated reproduction from}$$
$$\text{ovipositions between } t \text{ and } T \mid \text{current values of}$$
$$\text{parameters are } v \text{ and } \alpha\} \tag{4}$$

The "maximum" in (4) corresponds to a maximum over behavioral decisions (to remain in the current patch or leave) and the "E" denotes expectation over the random distribution of encounters. We can derive an equation for $F(v, \alpha, t)$ by considering the consequences of the two behavioral options.

First consider the value of leaving the current patch. If patches are randomly distributed and ρ is the probability that the parasitoid encounters a patch in a single period of search, then the probability that it takes s periods of search to find the next patch is $(1 - \rho)^{s-1}\rho$. If μ is the probability of death in a single period, then the probability that the parasitoid survives these s periods is $(1 - \mu)^s$. If encounter rates in patches are independent of each other, then the expected fitness upon encountering a patch after s units of search will be $F(v_0, \alpha_0, t + s)$. That is, since there is no information about the newly encountered patch, we assume that the probability distribution of λ is (2), with the initial parameters v_0 and α_0. The fitness value of leaving is thus

$$V_{\text{leave}} = \sum_{s=1}^{\infty} (1 - \rho)^{s-1}\rho(1 - \mu)^s F(v_0, \alpha_0, t + s) \tag{5}$$

Since T is the maximum time available for oviposition, if $t + s > T$ in (5), we replace $t + s$ by T.

If the parasitoid stays in the current patch, it may encounter any number of hosts in the next period. This number is a random variable N, with distribution given by (3). For simplicity, assume that superparasitisms are rare. This would occur, for example, if the parasitoid population is low and parasitoids systematically walk along the host patch. In this case, each encounter with a host increments lifetime fitness by an amount f. The value of staying is composed of two terms. The first is the expected fitness from hosts encountered in period $t + 1$. The second is the expected fitness from hosts encountered after period $t + 1$, taking into account the new information (i.e., that N hosts were encountered in one period of search in the patch). Hence we obtain

$$V \text{ stay} = \sum_{N=0}^{\infty} \{Nf + (1 - \mu)F(v + N, \alpha + 1, t + 1)\}$$

$$\times \left[\frac{\Gamma(N + v)}{\Gamma(v)} \left(\frac{1}{\alpha + 1}\right)^N \left(\frac{\alpha}{\alpha + 1}\right)^v \right] \tag{6}$$

The maximum expected fitness is then determined by comparing the value of leaving the patch and the value of staying in the patch:

$$F(v, \alpha, t) = \max\{V_{\text{leave}}, V_{\text{stay}}\} \tag{7}$$

Eq. (7) is called an equation of "stochastic dynamic programming" (Mangel and Clark, 1988). As seen from the derivation, it is simply a method of bookkeeping, augmented by the assumption that the parasitoid behaves to maximize expected reproduction.[1]

The solution of (7) determines values of v and α, as a function of time, for which the parasitoid should stay in the current patch and for which it should leave the current patch. (When $t = T - 1$, the optimal decision is obviously to stay in the patch, regardless of parameter values. This provides a check on the numerical solution).

A number of features emerge from the solution of (7). First (Mangel and Clark, 1983), even in the simplest case of $T = 2$ (so that at most two patches can be encountered), the value of acquiring information and updating parameters as described above can be considerable. For example, when $\rho = 1$ (so a patch is found with certainty in each period), $\mu = 0$ (so that the parasitoid survives each period up to T with certainty), and $\alpha = 0.1$ and $v = 1$ (so that the mean encounter rate is ten hosts per period

---

[1]Computing (5)–(7) is not completely trivial because the sums may involve many terms. This "curse of dimensionality" in dynamic programming is alleviated as better and faster computers allow us to deal with such problems more easily. There are two main difficulties. First, in principle at least, the value of s in (5) and N in (6) may be very large. The way around this difficulty is to choose maximum values of s and N that correspond to most of the cumulative probability (e.g., 99.9%) and restrict s and N to be less-than-or-equal-to those values. The second difficulty is that v can also become very large. Again, a simple solution is to restrict v to be less-than-or-equal-to some maximum value $v_{\max}$ in the sense that $N + v$ is never allowed to go above $v_{\max}$.

and the coefficient of variation is 100%), the expected reproduction in the period following sampling is nearly 40% higher than in expected reproduction in the period of sampling. This *value of information* is considerable, even though the parasitoid can only stay in the current patch or visit at most one more patch. A similar result is obtained if depletion is taken into account (Mangel and Clark, 1983).

Second (Mangel, 1990b), in a changing environment, there generally is an advantage (also see Stephens, 1991) to both updating parameters and to forgetting past information. A simple method for forgetting past information is to weight data collected p periods previously by e^{-mp} where m is a weighting parameter. When $m = 0$, all information collected is remembered but when $m > 0$, past information is forgotten at a rate which increases as m increases. The parameters for the probability density of λ are now functions $v(S)$ and $\alpha(S)$ of time spent in the patch, with $v(0) = v_0$ and $\alpha(0) = \alpha_0$. The updating rule becomes (Mangel, 1990b)

$$v(S + 1) = e^{-m}v(S) + K + (1 - e^{-m})v_0$$

$$\alpha(S + 1) = e^{-m}\alpha(S) + 1 + (1 - e^{-m})\alpha_0$$

This is a "linear operator" (Kacelnik et al., 1987) of the form commonly used in psychological studies of learning.

In conclusion,

- There is a selective advantage, in terms of lifetime reproduction, in acquiring information about the world and using that information to shape behavior.

- The selective advantage of such learning is a broad function of the amount of effort put into sampling so that precise point optima are not expected.

- In an environment which is changing, there is a selective advantage to "forgetting" past information. The fitness value of forgetting is also a relatively flat function of the rate at which the past is forgotten. Thus, there will be selective pressures for both sampling the environment (learning) and forgetting past information.

Motivation: Behavioral Changes Induced by Egg Maturation

In the previous section, experience (encounters with hosts) led to changes in behavior, without any concomitant change in physiological state. It may be, however, that changes in behavior occur solely because of changes in physiological state such as egg complement (Singer 1982, 1983, 1986; Singer et al., 1990). In such a case, we wish to develop a model in which behavior changes with experience, but the description of the external environment does not change.

A synovigenic insect (one which matures eggs over time) encounters host type i with probability λ_i in a single period of time. Unlike the previous case, I assume that no updating of the values of the λ_i occurs during the search for oviposition sites. Assume that $X(t)$ is the egg complement at time t and that when a host is accepted for oviposition, the entire current egg complement is laid (see Tatar, 1991; unpublished data). Laying a clutch of size x on a host of type i increases the mother's lifetime reproduction by $f_i x$ where f_i characterizes the "quality" of the host, from the perspective of the larvae. For simplicity no larval density dependence is assumed; this can be modified (cf. Roitberg et al., 1990). Oviposition of x eggs requires handling time $\tau(x) = \tau_0 + \tau_1 x$, where τ_0 is a fixed time needed for oviposition (e.g., host recognition and handling) and τ_1 is the variable time required per egg. The probability of mortality during search is μ_s and during oviposition is μ_{op}.

Expected lifetime reproduction is defined by

$$F(x, t, T) = \text{maximum } E \text{ \{reproduction accumulated from} \\ \text{ovipositions between } t \text{ and } T \mid X(t) = x\} \tag{8}$$

For simplicity, assume that there are only two host types and that $f_1 > f_2$ so that larval performance is superior on host type 1. Also assume that the time period is chosen so that the insect can mature one egg during each period and that the maximum egg complement is x_{max}. Under the assumption that acceptance of a host leads to oviposition of the entire egg complement (cf. Mangel, 1987), the equation that $F(x, t, T)$ satisfies is

$$F(x, t, T) = (1 - \lambda_1 - \lambda_2)(1 - \mu_s)F(x', t + 1, T)$$

$$+ \sum_{i=1}^{2} \lambda_i \max\{f_i x + (1 - \mu_{op})^{\tau(x)}F(\tau(x), t + \tau(x), T);$$

$$(1 - \mu_s)F(x', t + 1, T)\} \tag{9}$$

The first term on the right-hand side of (9) corresponds to the event that no host is encountered during period t. Otherwise, a host of type 1 or type 2 is encountered. When a host is encountered, the insect's behavior involves acceptance (with increment in lifetime reproduction) or rejection of the host. The first term following the "max" corresponds to acceptance of the host. In this case, lifetime fitness is incremented by $f_i x$, the elapsed time for oviposition is $\tau(x)$, the entire egg complement is laid, and $\tau(x)$ new eggs are matured during the process. The second term following the max corresponds to rejection of the host. In this case, one period of time is used, so egg complement is increased by one egg, up to the maximum egg complement. Hence $x' = \min(x_{\max}, x + 1)$.

As (9) is solved for lifetime reproduction, decisions $d_i(x, t)$ are generated. These are to either accept or reject for oviposition a host of type i encountered during period t when egg complement at the start of the period is x. Depending upon the egg complement at the time of acceptance of a host, different clutches will be realized over time and host types. As the difference $T - t$ increases, so that "end of life" effects are less important, the behaviors become independent of time and depend only upon egg complement. From such behaviors, it is possible to predict the results if an insect is presented with a two-type choice experiment (Fig. 6.1).

In such an experiment, an insect is allowed to oviposit. After an interval following the oviposition, she is presented with two hosts and her willingness to oviposit on each host is determined, usually by observing stereotypical preovipositional behavior, but she is not allowed to oviposit (see Singer, 1982, 1983, 1986; Singer et al., unpublished data). This procedure

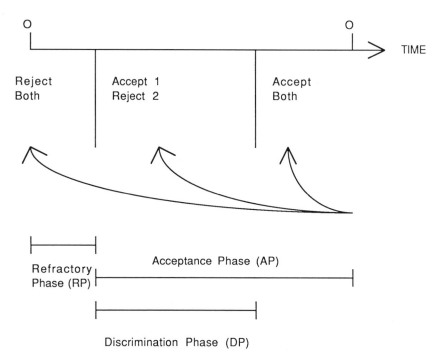

Figure 6.1. Results of the motivational model (9). O denotes an oviposition. See text for further details.

is repeated after another interval. The general prediction, which holds over a wide range of parameter values, for the time course of such an experiment is depicted in Figure 6.1. In the figure, O's denote ovipositions. After the first oviposition, the insect enters a "refractory phase" (RP) in which she rejects hosts, regardless of type. The fixed handling time τ_0 is the source of the refractory period. That is, when τ_0 is very small, we anticipate a small refractory period. Alternatively, when τ_0 is non-negligible, there is a "fixed survival cost" for any oviposition. This causes a delay in oviposition until egg load is such that a sizeable clutch can be laid. For example, laying a single egg requires time $\tau_0 + \tau_1$ and laying ten eggs requires time $\tau_0 + 10\tau_1$. If, for example, $\tau_0 = 10\tau_1$, then the relative risk in oviposition of ten eggs rather than one egg is about twice as great ($20\tau_1$ time units vs. $11\tau_1$ time units), but the relative fitness difference is tenfold if there is no larval density dependence.

As time continues, and egg load increases, there is a point at which the insect will accept type 1 but reject type 2. She is now "motivated" to oviposit and will continue to be so until the next oviposition. For a period defined as the "discrimination phase" (DP), she will be motivated but discriminate between host types 1 and 2 in that she will oviposit in host type 1 but not in host type 2. As time progresses (and no host type 1 is encountered), egg load continues to increase until the insect is both motivated and nonselective. During this "acceptance phase" (AP), the insect will oviposit on the next host presented. After that oviposition, depending upon values of parameters (Table 6.2), the insect may return to any of the three previous behavioral states (RP, DP, or AP). This insect, then exhibits a wide range of behaviors, and these change with experience, although no change of the description of the environment is involved. This is a case in which motivation changes as a result of experience and behavior changes as a result of motivation. However, learning, as defined above, does not occur.

An interesting, possibly counterintuitive, result emerges from this model. The superior host (type 1) will be accepted for oviposition over a wide range of egg complements. The inferior host (type 2) will be accepted for oviposition only for large egg complements, since it is better to oviposit on the inferior host than to simply waste eggs. Because of this, we predict that a range of clutches will be observed on host type 1, but only large clutches will be observed on host type 2. An investigator studying such an insect might reject the "optimality" model because the insect "puts only large clutches into the poor host and this is clearly not optimal." But this is completely consistent with the optimality model. Furthermore, if an investigator simply went to the field and measured clutch sizes as a function of plant quality, he ·or she could be misled concerning the relationship between preference and host quality or preference and performance. We

Table 6.2. Parameters and their interpretations: behavioral changes induced by egg maturation

Parameter	Interpretation
λ_i	Probability that the insect encounters host type i in a single period of search
f_i	Increment in lifetime reproduction from oviposition of one egg on host type i
$X(t)$	Egg complement at time t
x	Particular value of the egg complement
x_{max}	Maximum allowed value of the egg complement
$\tau(x)$	Handling time needed to lay a complement of x eggs; it is composed of a fixed time τ_0 and a variable time $\tau_1 x$
$F(x, t, T)$	Maximum expected accumulated reproduction from ovipositions between time t and T, given that the egg complement at time t is $X(t) = x$
μ_{op}	Probability of death during a period in which the insect is ovipositing
μ_s	Probability of death during a period in which the insect is searching

can only understand acceptance of poorer hosts in the context of life history, and not in the context of single host encounters.

We thus see that behavior changes with internal environment (egg complement), which itself changes according to the state of the external environment. This is, however, not learning in that parameters characterizing the external environment are not updated according to experience.

Combining Environmental Information and Physiology: Motivated Learning

The methods of the two previous sections can be combined to deal with motivated learning, i.e., situations in which both an informational state variable, which characterizes the external environment, and a physiological state variable, which characterizes the internal state of the insect, determine behavior. In this case, experience (e.g., host deprivation) changes both the information state (e.g., estimates of encounter rates with hosts) and the physiological state (e.g., egg complement). Mangel (1989) and Mangel and Roitberg (1989) describe two examples in which physiological and informational state variables are combined.

Mangel and Roitberg (1989) considered the so-called superparasitism behavior of the apple maggot *Rhagoletis pomonella*. Female apple maggots held in field cages usually oviposited in unparasitized fruit and on occasion would oviposit in previously parasitized fruit (i.e., "superparasitize" fruit). Because each fruit in the field cage was individually tagged and individual flies can be observed, the data could be collected according to the encounter history of the fly (the fraction FRAC of previously parasitized hosts in the last five encounters) and the time since the last oviposition (TSLO). Using methods (Iwasa et al., 1984; Mangel, 1987) similar to the ones described in this paper, the plane "TSLO-FRAC" can be divided into two regions. In one region, the next previously parasitized host encountered should be accepted and in the complementary region it should be rejected. When this theory was compared with the empirical results, however, 50% of the observed ovipositions fell into the "wrong" portion of the plane: the flies superparasitized when the theory suggested that they should reject the host. Adding an informational state changes the theoretical predictions. In particular, the TSLO-FRAC is now divided into three regions: one in which the next previously parasitized fruit should be accepted, one in which it should be rejected, and one in which it may be accepted or rejected depending upon the encounter history (information). All but two of the observed acceptances of previously parasitized fruit fell into the "accept" or "maybe" regions (Mangel and Roitberg, 1989: Fig. 4).

Mangel (1989) developed a model for the parasitization of sycamore aphids by *Monoctonus pseudoplatani* Marsh. In this case, the physiological-state variable was egg complement and the informational-state variable was the probability that an encountered aphid would be unparasitized. The patterns of parasitism predicted by the theory compared favorably with the observed patterns of parasitism.

Neither Mangel (1989) nor Mangel and Roitberg (1989) used a Bayesian model of the type described in the current paper. The combination of models of Bayesian updating and physiological variables is an open and fruitful area of research.

Discussion: Can Learning and Motivation be Separated?

In a sense the theory is now ahead of the experimental work, since the task of separating informational and motivational determinants of behavior remains a challenge to empiricists. Progress is being made. For example, Tatar (1991) suggests that informational state (seasonal host quality) interacts with physiological state (egg load) to influence clutch size in oviposition by a butterfly. Rosenheim and Rosen (1991), in an elegant study of the behavior of a parasitoid, show how informational and physiological states may be separated and how the predictions of models such as the ones developed in this paper can be tested. The physiological state variable (egg complement) was controlled through parasitoid size and ambient temperature. Informational state was controlled by encounter rates, holding egg complement relatively constant.

Roitberg et al. (1992) use photoperiod [the "closeness" of t to T, as in Eq. (4) or (9)] and encounter rates to provide cues about the state of the external environment in situations in which the physiological variable, energy reserves (rather than egg complement), determines survival. Theories similar to the ones developed here are used to predict the time on a patch and number of superparasitisms by a drosophilid parasitoid. The theoretical results are supported by empirical observation.

On the theoretical side, we still need models that effectively describe the "internal representations" in terms of neuronal groups. For example, it is unlikely that animals perform Bayesian updating in the manner described above. On the other hand, it is likely that a neural network which effectively performs the equivalent of Bayesian updating could be constructed and such networks need to be developed. Again, progress is being made (Putters and Vonk, 1990), but there is still much work to be done. The most progress will be made by developing theory and experiments in tandem, so that we will have practicable theories which can provide understanding of experiments on learning and motivation.

Acknowledgments

I have benefited from conversations with Mike Singer and Marc Tatar, both of whom also read the manuscript. In addition, Dan Papaj and an anonymous reviewer provided many comments that improved the presentation. This work was partially supported by NSF grants BSR 86-1073 and by the University of California.

References

van Alphen, J.J.M., and Visser, M.E., 1990. Superparasitism as an adaptive strategy. Annu. Rev. Entomol. **35**:59–79.

DeGroot, M.H. 1970. Optimal Statistical Decisions. McGraw-Hill, New York.

Dudai, Y. 1989. The Neurobiology of Memory. Oxford University Press, Oxford, England.

Edelman, G. 1987. Neural Darwinism. Basic Books, New York.

Iwasa, Y., Suzuki, Y., and Matsuda, H. 1984. The theory of oviposition strategy of parasitoids. I. Effect of mortality and limited egg number. Theor. Popul. Biol. **14**:205–227.

Janssen, A. 1989. Optimal host selection by *Drosophila* parasitoids in the field. Funct. Ecol. **3**:469–479.

Kacelnik, A., Krebs, J.R., and Ens., B. 1987. Foraging in a changing environment: An experiment with starlings (*Sturnus vulgaris*). In M.L. Commons, A. Kacelnik, and S.J. Shettleworth (ed.), Quantitative Analyses of Behavior, Vol. VI. Foraging. Lawrence Erlbaum Associates, Hillsdale, NJ, pp. 63–87.

Kamil, A.C. 1983. Optimal foraging theory and the psychology of learning. Am. Zool. **23**:291–302.

Mangel, M. 1985. Search models in fisheries and agriculture. Lec. Notes Biomath. **61**:105–138.

Mangel, M. 1987. Oviposition site selection and clutch size in insects. J. Math. Biol. **25**:1–22.

Mangel, M. 1989. An evolutionary interpretation of the "motivation to oviposit." J. Evol. Biol. **2**:157–172.

Mangel, M. 1990a. Evolutionary and neural network models of behavior. J. Math. Biol. **28**:237–256.

Mangel, M. 1990b. Dynamic information in uncertain and changing worlds. J. Theor. Biol. **146**:317–332.

Mangel, M., and Clark, C.W. 1983. Uncertainty, search and information in fisheries. J. Int. Council Explor. Seas **41**:93–103.

Mangel, M. and Clark, C.W. 1986. Towards a unified foraging theory. Ecology **67**:1127–1138.

Mangel, M., and Clark, C.W. 1988. Dynamic Modeling in Behavioral Ecology. Princeton University Press, Princeton, NJ.

Mangel, M. and Roitberg, B.D. 1989. Dynamic information and host acceptance by a tephritid fruit fly. Ecol. Entomol. **14**:181–189.

Putters, F., and Vonk, M. 1990. The structure-oriented approach in ethology: Network models and sex-ratio adjustments in parasitic wasps. Behavior **114**:148–160.

Roitberg, B.D., Mangel, M., and Tourigny, G. 1990. Density dependence in fruit flies. Ecology **71**:1871–1885.

Roitberg, B.D., Mangel, M., Lalonde, R.G., Roitberg, C.A., van Alphen, J.J.M., and Vet, L. 1992. Dynamic shifts in patch exploitation by a parasitic wasp. Behav. Ecol. **3**:156–165.

Rosenheim, J.A., and Rosen, D. 1991. Foraging and oviposition decisions in the parasitoid *Aphytis lingnanesis*: distinguishing the influence of egg load and experience. Anim. Ecol. **60**:873–893.

Singer, M.C. 1982. Quantification of host preference by manipulation of oviposition behavior in the butterfly *Euphydryas editha*. Oecologia **52**:224–229.

Singer, M.C. 1983. Determinants of multiple host use by a phytophagous insect population. Evolution **37**:389–403.

Singer, M.C. 1986. The definition and measurement of oviposition preference in plant-feeding insects. In J. Miller and T.A. Miller (eds.), Insect-Plant Interactions. Springer-Verlag, New York, pp. 65–94.

Southwood, T.R.E. 1966. Ecological Methods. Chapman and Hall, London.

Tatar, M. 1991. Clutch size in the swallowtail butterfly, *Battus philenor*: comparisons of behavior within and among seasonal flights in California. Behav. Ecol. Sociobiol. **28**:337–391.

Ward, S.A. 1987. Optimal habitat selection in time-limited dispersers. Am. Nat. **129**:568–579.

7

Choosing Hosts and Mates:
The Value of Learning

Bernard D. Roitberg, Mary L. Reid, and Chao Li

Introduction

Recent discussions of animal learning emphasize the importance of considering an animal's ecology when interpreting its learning abilities (Johnston, 1982; Bolles and Beecher, 1988; Staddon and Ettinger, 1989). The implementation of this approach has so far been directed primarily at a few well-studied species. As a result of some success in revealing correspondences between what an animal can learn and its lifestyle, it has been recommended that a species' ecology be thoroughly understood before learning experiments are conducted (Kamil and Mauldin, 1988). However, the risk here is that our understanding of learning will be limited to case studies that cannot be easily extended to other species. A useful companion approach would be to find general conditions favoring learning that can then be used as a framework for studying learning in individual species. As an example, the value of learning about a resource depends upon its variability or patchiness; if the resource is constant, then a fixed or innate response is favored, while increasing variability favors assessment and learning (Green 1980; see Stephens, this volume). Ideas such as this, generated from the functionalist's perspective without concern for particular mechanisms, may generate important organizing principles in the study of learning in animals.

In this chapter, we investigate another general condition that we believe might influence the value of learning in animal decision-making. We ask whether the frequency of the decision and the (inversely related) fitness payoff of each decision affect the optimal amount of information about a resource that an animal uses. Our model systems, representing the extremes of the frequency/magnitude decision trade-off, are a female parasitoid with many small eggs to lay individually in hosts and a female insect that can only mate a few times in her lifetime. Each egg contributes a small amount to the parasitoid's lifetime reproductive success, while each mate chosen affects a large proportion of a female's fitness. In each case, the female faces a patchy distribution of hosts or mates of various quality and must decide at each encounter whether to accept that host or mate. Our questions are: how much information should a female collect about local availability of high-quality hosts (mates) in order to realize the greatest fitness returns from her decisions, and is this dependent upon the frequency and magnitude of these reproductive decisions?

To answer this question, we use dynamic programming (Mangel and Clark, 1988) to model the life of a female insect. We vary the number of reproductive decisions she can make in her lifetime (and consequently the contribution each makes to her lifetime fitness) to determine how this variable affects the value of learning. Our index of learning is the length of the memory window, i.e., the number of past encounters with hosts or mates she remembers when making her decision. To determine the generality of the outcome, we also vary the patchiness of the environment, since this has been shown to be an important variable affecting the value of learning (Green, 1980). We discuss the model in more detail in the next section.

One goal of this work is to provide a common conceptual framework within which to examine two different life-history parameters. Typically, examination of host selection and mate choice have been studied in isolation of one another, although the problems may not be that dissimilar. As a telling example, an identical choice tactic was modeled independently by Mitchell (1975) for oviposition in a bean weevil and by Brown (1981) for mate choice by sculpins, with very similar conclusions. Other models of choice tactics (for mates) have been developed by Janetos (1980) and Real (1990). Like them, we employ an optimal sequential (one-step) search tactic. Unlike their models, ours makes explicit the value of learning and memory in this behavior. Real (1990) concluded that recall was not valuable in mate selection, but he also assumed that mate-seekers knew the quality distribution of available mates. We model such prior knowledge as well, but we also consider patchiness in the distribution that may alter an animal's actual encounters.

A Model for Learning to Accept Hosts and Mates

In order to evaluate the advantage of learning to accept hosts and mates one must do the following: (1) define goals for the relevant acceptance behavior, (2) establish a common currency through which one can compare the two processes, and (3) calculate payoffs for learning under various ecological conditions.

The first step in this evaluation process is to assume that natural selection acts to increase mean fitness of a population and that there is a correlation between heritable variation in the behavior under consideration and fitness (Endler, 1986). Thus, the goal of an acceptance behavior is that behavior which maximizes fitness. Second, fitness is indexed by lifetime reproductive success and as such, both host and mate selection can be evaluated by their contribution to lifetime reproductive success. This provides a common currency for both processes (McNamara and Houston, 1986). Third, we calculate payoffs for learning though the use of dynamic-state-variable models. Our choice of model is based upon the fact that learning alters the information state of individuals; dynamic models explicitly consider such changes in state (e.g., Mangel and Roitberg, 1989).

The System

To calculate payoffs for learning we will consider performance over time where the units of time are discrete. The calculations require that we characterize both females (i.e., decision-makers) and their environment.

Consider an environment where hosts or mates are distributed in some manner. Further, consider that hosts (mates) are of three types: good, moderate, and poor. By this we mean that acceptance of a good host provides a greater increment to an individual's lifetime reproductive success (f) than does a moderate or poor one. Thus, $f_g > f_m > f_p$ always holds.

Females can be characterized by several features. First, we assume that females harbor some limited resource that can be spent during each individual's lifetime. In the case of the host-choice decision-maker, that limiting resource is eggs. Thus we assume that at any time, t, an individual can be characterized by its egg load $\varepsilon(t)$. Each time that individual accepts a host (i.e., lays an egg) egg load decreases by one (so long as individuals harbor at least one egg—condition [1]). For example, assume that all individuals begin life with 30 mature eggs. By contrast, assume that a mate-choice decision-maker harbors a much smaller resource (e.g., spermatheca volume) such that the maximum number of matings that an individual can achieve in its lifetime is three (i.e., one magnitude fewer fitness-related events than for egg layers). Thus, each mating decreases the mating equivalent of ε by one and condition (1) holds.

Second, an individual can be characterized by its information state. By this, we mean that information obtained during encounters with hosts or mates is stored in memory and used in acceptance-rejection decisions. We employ the approach of Mangel and Roitberg (1989) and define the memory state as a memory vector $m_x(t)$ where positions in the vector describe, in chronological order, host types recently encountered (see Mangel, 1990, for alternative approaches to modeling information acquisition and processing). Thus, m_1 defines the host (mate) most recently encountered (with which the female is still in contact) and m_{max} defines the type of host encountered in the most distant past that the individual can still remember, where max is the size of the memory vector. Thus an individual with max = 5 will always remember the identity of the five hosts (mates) most recently encountered. By considering information in this way it is possible to compare the effects of storing and processing different amounts of information on lifetime fitness. As individuals encounter new hosts (mates), the identities of those individuals are entered into the memory vector with the most distant memory being lost (i.e., a sliding memory window *sensu* Cowie and Krebs, 1979). For example, an individual that is able to remember the previous three encounters with hosts may harbor the following sequence of encounters:

$$m(t) = \{g, p, m\} \tag{1}$$

where g, p, m refers to good, poor, and moderate host types, respectively. Now, suppose that the next host encountered is of the "good" type. Thus, $m_3(t)$ is replaced by $m_2(t)$ and so on such that

$$m(t + 1) = \{g, g, p\} \tag{2}$$

In this manner, the memory vector will be constantly updated as females encounter new hosts (mates). Note, however, that if a host (mate) is not encountered during a unit of time the memory vector does not change.

We considered memory lengths ranging from zero to five past events remembered. For zero memory units, the animal's view of the world is not influenced by experience. Instead, such females have a fixed expectation that the world contains proportions of the three host (mate) types. In our example we assume that the world harbors equal numbers of each host type and that zero-memory-unit females expect that the world is so. Individuals with one memory unit will only consider the host that they are currently in contact with to evaluate future chances of locating hosts (males).

Building an Optimal-Decision Matrix

One can employ the relationships outlined above to determine how females will respond to encounters with different types of hosts. Remember, we assume that decision-making has evolved to maximize lifetime fitness. Lifetime fitness can be defined as (Mangel and Clark, 1988):

$F(\varepsilon, m, t, T) = $ maximum expected fitness from host exploitation (or mating) between t and T when egg (mating) and memory states at time t are ε and m, respectively

Thus, fitness depends upon responses to hosts (mates). To calculate fitness several more parameters/assumptions must be defined:

1. Encounter rates: densities of each type of host are equal but because, in some situations, hosts are clumped, actual encounter rates with different host types might not be equal. Thus, the apparent (i.e., from the female's perspective) probability of encounter with a particular host i, λ_i, is dependent upon the proportional representation of that host type in the memory vector. For example, $\lambda_g = 0.4$ when $m(t) = \{g, m, p, p, g\}$.

2. Females cannot return to hosts (mates) after they have rejected them (i.e., hosts [mates] move and so are unavailable at the same location over time—see Li et al., 1992, for a discussion of spatial search and resource distributions).

3. The probability ρ that a female survives from one period to the next is independent of ε and m states. Since ρ is positive, however, there is an implicit assumption that future decisions are discounted by a less than 100% chance of survival. A lack of discounting in Janetos's (1980) mate-choice model has been strongly criticized by Real (1990) as being unrealistic.

4. No fitness is accrued after time T.

When the above assumptions hold, several mutually exclusive series of events can occur with the following consequences:

1. The female does not encounter any hosts (mates), with probability λ_0. In this case the future fitness of that individual can be defined by ρ multiplied by future reproductive fitness where neither ε nor m changes and time advances one unit to become $t + 1$.

2. The female encounters a host (mate) of type g. That host is always accepted such that $\varepsilon(t)$ becomes $\varepsilon - 1(t + 1)$ where the minimum egg condition of 0 holds. Further, the memory vector $m(t)$ becomes $m'(t + 1)$ where $m_1(t + 1)$ becomes g, m_{max} is removed, and all other $m_x(t)$ become $m_{x+1}(t)$. Finally, t advances one unit to become $t + 1$. (A "good" host will always be accepted because the female will always achieve a maximum fitness increment through acceptance of this best possible host.)

3. The female encounters a host (mate) of either moderate or poor quality. If that host is accepted then $\varepsilon(t)$ becomes $\varepsilon - 1 (t + 1)$ where the minimum egg condition of 0 holds. Further, the memory vector $m(t)$ becomes $m'(t + 1)$ where $m_1(t + 1)$ becomes m or p, respectively, m_{max} is removed, and all other $m_x(t)$ become $m_{x+1}(t)$. Finally, t advances one unit to become t + 1. If the female chooses

to reject the host (mate), ε does not change whereas both the memory vector and time become the same value as when the host is accepted.

The three cases can be summarized in the following equation:

$$F(\varepsilon, m, t, T) = \lambda_0 \rho F(\varepsilon, m, t + 1, T)$$

$$+ \lambda_g[f_g + \rho F(\varepsilon - 1, m', t + 1, T)]$$

$$+ \sum_{i=1}^{2} \lambda_i \max[f_i + \rho F(\varepsilon - 1, m', t + 1, T);$$

$$\rho F(\varepsilon, m', t + 1, T)] \tag{3}$$

The equation is solved "backward in time" starting with $t = T - 1$ and going until $t = 1$. The maximization terms in the equations indicate that the decision (e.g., accept or reject) is chosen that gives the highest expected fitness (see Mangel and Clark, 1988, for more details). Thus, for each combination of egg (spermatheca) and memory state, an optimal response to encounters with a host of a given type can be solved for individuals at a particular time in their lives. Those decisions (i.e., state-dependent accept/reject) can be assembled into a decision matrix that defines an individual's optimal response throughout its life for different egg and memory states. An example of one such decision matrix is shown in Figure 7.1. Clearly, all responses can be sensitive to all state variables.

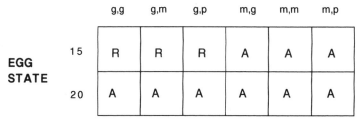

MEMORY STATE

		g,g	g,m	g,p	m,g	m,m	m,p
EGG STATE	15	R	R	R	A	A	A
	20	A	A	A	A	A	A

Figure 7.1. A decision matrix for an egg-laying insect that remembers its two most recent encounters with hosts. Rejection (R) and acceptance (A) decisions are shown for the insect at two different egg states and six different memory states. (Details are provided in text.) Other parameter values are $f_g = 1, f_m = 0.5, f_b = 0.25, \lambda_0 = 0.1, \lambda_g = 0.297, \lambda_m = 0.297, \lambda_b = 0.297, t = 135, T = 150, \rho = 0.999$.

An important feature of the model is that individuals estimate their future lifetime reproductive success from expected encounter rates with hosts. These expectations are based upon the composition of their memory vectors (i.e., learned information). It is this feature that allows us to evaluate the benefits of different degrees of learning. For example, if the sequence of the previous five encounters was g, b, g, g, g, an individual with a memory of only two units would base its decision on an expectation of encounter rates with good hosts to occur with 50% of the time (since $m = \{g, b\}$) while an individual with a memory of five units will expect to meet good hosts during 80% of all encounters. How important these decisions are depends upon the structure of events that an individual encounters in its lifetime.

Our model is very similar to that of Real's (1990) sequential mate-choice model in that a critical mate value (W_{crit}) is derived such that any potential passive mate that is encountered is accepted, if its value falls above W_{crit}. There is a fundamental difference between our two approaches, however, in that Real assumes that the searching females know the distribution of mate quality whereas the females that we model learn that distribution (except in the case where memory = 0). Furthermore, our goals are somewhat different in that Real's purpose was in developing a theory of mate choice whereas our principal interest lies in how different degrees of learning could affect mate (host) choice decisions. Finally, we do not attempt to find the best rule for acceptance as have Janetos (1980) and Real (1990) but rather try to evaluate use of varying amounts of information in mate- and host-choice decisions.

Evaluating Fitness Consequences of Learning

Once the decision matrix is completed we can evaluate the reproductive success of individuals using the decision matrix through Monte Carlo simulation. In so doing we "release" computer insects into environments that vary in host encounter rates and sequences. Stochasticity in encounters varies in two ways. First, individuals occasionally encounter patches of hosts (mates). We imagine that in some environments such patches would be very consistent (i.e., that most hosts would be similar to the first host encountered) while others would harbor a random mix of host types. Thus, there will be some correlation among host types encountered over time as follows:

$$\text{phost} = \exp(-\text{hslope} * t) \qquad (4)$$

phost is the probability that the next host encountered will be the same as the first host encountered in that patch; hslope is a parameter that determines the rate at which the correlation decays; and t is the number of time units that have elapsed from when the decision-maker first entered the patch. Clearly, if hslope is large then there will be little correlation between hosts encountered within a patch. (A more detailed discussion of this type of approach can be found in Roitberg et al., 1990, and Mangel and Roitberg, unpublished manuscript).

In addition to the consistency of events that occur within patches, the environment can also be characterized by the size or permanence of patches. In a similar manner to that described above, one can ask what the likelihood would be of an individual being able to continue searching in the same patch as time elapses. The following equation determines that likelihood:

$$ppatch = exp \, (-pslope * time)$$

Thus, when pslope is large, patches are either very small or transient. In biological terms, ppatch might describe the length of time over which a group of hosts (e.g., aphids) will continue to feed on a leaf before responding to the presence of a searching parasitoid or the clustering of potential mates according to male quality. [Phost and ppatch are analogous to Stephens's (this volume) within- and between-generation predictabilities.] We assume that females cannot recognize patch boundaries and simply wander through their environment encountering different patches by chance.

Effect of Memory Length

We varied the correlation among successively encountered hosts and the patchiness of the environment to determine the importance of environmental predictability to the types of decisions made. For each environment type, we calculated the lifetime fitness for females with memory windows ranging from zero to five. The six types, respectively, can be described as follows: (0) uses no information in estimation of its future fitness, (1) uses identity of current host (mate) to determine future encounter rates and future fitness, (2) uses identity of current and next most recent host (mate) to determine future encounter rates and future fitness, etc.

For each case, we "released" 500 females of each memory type into each environment and calculated the average fitness. To compare the fitness outcomes of different memory lengths, we standardized them by dividing by the maximum mean fitness in each environment. This procedure was conducted separately for females making many small decisions (e.g., seeking hosts) and females making few, major decisions (e.g., seeking mates). We can then compare the relative value of long memory for the two types of decision-makers.

Results

Some typical results obtained from our simulations are shown in Figure 2a–c wherein each section shows fitness as a function of memory length for a constant set of pslope and hslope values. Although exact shape and size of the curves varied among the different environments, two features were found throughout:

First, in almost all cases, a short memory (i.e., using only the identity of the current host or mate to determine future encounter rates and fitness) is the least favoured strategy (Fig. 7.2a–c). No learning at all (i.e., having fixed expectation) is better than a short memory, when the animal knows the global densities of the three host types, but a long memory (max = 5) is usually best of all. Individuals with short memories suffer the lowest

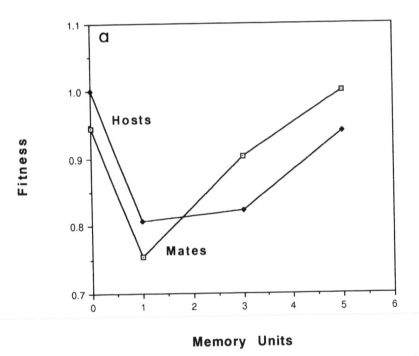

Memory Units

Figure 7.2. Mean lifetime fitness for 500 mate-choosing and 500 host-choosing computer insects that employ different numbers of memory units in their decision-making: (a) hslope = 0.1, pslope = 0.3; (b) hslope = 0.3, pslope = 0.1; (c) hslope = 0.3, pslope = 0.3. All other parameter values are the same as shown in Figure 7.1.

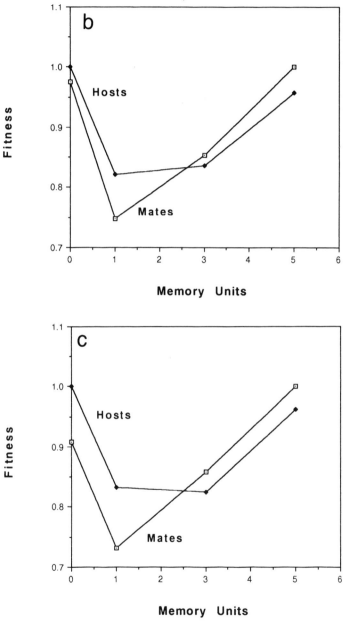

Figure 7.2. (*Continued*)

fitness because they accept all hosts (mates) regardless of quality. This is because such decision-makers perceive the availability of hosts in the future to be 100% for the host that they have just encountered (and entered into their limited memory bank) and 0% for the other two hosts. Under such conditions, it will always pay to accept the host that one has just encountered since it appears that all future hosts will be of the same type (i.e., there is no sense waiting for something better to come along). This insight does not arise in Real's (1990) paper because he assumed that females already know the distribution of hosts whereas our females sample to learn that distribution and then store it in their (sometimes very limited) memory.

Second, the exact manifestation of the pattern discussed in (1) varied greatly between females seeking hosts and those seeking mates, regardless of the nature of variability in the environment. Females choosing hosts (making many small decisions) experienced only a small decrease in fitness by having a short memory instead of no memory, while females choosing mates suffered a great reduction in fitness. In addition, the relationship among fitness, memory window, and the ppatch function will vary as hslope varies.

Discussion

Our model suggests that although a long memory provides a fitness advantage to both host- and mate-seeking females, the cost of having a short memory (i.e., using current encounters only to predict long-range encounter rates) is much greater relative to no memory for mate-seekers (those making few but major decisions). This conclusion is robust in that it is consistent across "worlds" differing in their changeability or patchiness. The significance of this difference to our real-life understanding of host-seekers and mate-seekers depends on several factors, some of which are related to the degree to which our model mimics reality. Three of our assumptions may be particularly influential. The first concerns our representation of memory as a sliding memory window (Cowie and Krebs, 1979). Whether this contains the essential elements of insect memory cannot be assessed with the current state of memory research, but Roitberg (personal observation) observed that host acceptance behavior of a tephritid fly (*Rhagoletis pomonella*) parasitizing fruits was best predicted by a sliding memory window of five units.

Our second assumption is that a female with no memory has an innate expectation of host quality that matches that which is found in the world (but not necessarily in any particular patch, or among those she actually encounters as might occur if, for example, host [mate] odor concentrations provided global information on host availability). This explains in part the relative success of a lack of memory relative to a short memory. However, the fact that a short memory in a mate-seeking female has such great fitness costs compared to zero memory suggests that even a somewhat incorrect innate expectation may still be better that having just a short memory.

Our third assumption is that the only cost of accepting an inferior host is that of lost opportunity should a high-quality host be encountered in the future. There may, however, be other costs to accepting any host or inferior hosts (e.g., Roitberg, 1989). For example, Li et al. (in preparation) observed female wasps, *Aphelinus asychis*, foraging for second-instar pea aphids, *Acyrthosiphon pisum* (one of their native hosts), on bean leafs. They found that the time taken to accept a host is much higher than that for rejection. Li et al.'s analysis of this apparently "cautious" acceptance suggests that several costs be considered, including (1) the high value of each egg, given the rather limited number of mature eggs available at any one time (20–30 maximum), (2) the limited time available for host search, (3) weak interspecific competitive abilities of *A. asychis* larvae, (4) energetic costs, and (5) host defensive behavior. These "other" costs may make testing of predictions from simple evolutionary models difficult at best because some or all of them can potentially drive evolutionary responses to host quality.

Finally, the analysis presented in this chapter is concerned solely with animals that forage in worlds where hosts (mates) are in relatively large supply. As hosts or mates become more rare one would expect foragers to be more willing to accept low-quality hosts (i.e., all memory types will begin to express similar behavior) regardless of their information state. Under such conditions, the value of information is low at best since it will have little opportunity to effect changes in behavior (Stephens 1989, this volume).

Assuming that our model captures the essence of reality, we can ask what our results predict about the amount of learning we should see in animals searching for hosts and those seeking mates. This depends on how the evolution of learning and memory occurs, which at this point is quite speculative at levels above the individual neuron (Quinn, 1984). One view of learning is that the brain is "prewired" to be sensitive to particular features of the environment that an animal should learn, enabling faster learning (Changeux et al., 1984). This model can explain aspects of honey bee learning, for example (Gould, 1984). If remembering more things requires a more complex assembly of "dedicated" neurons, then the evolution of a long memory would likely entail gradual genetic changes as complexity increases. It seems generally true in insects that the ability to learn one thing does not transfer to other tasks (Alloway 1978; but see Laverty, 1985, who does find some transfer), so there do not seem to be shortcuts to learning. In terms of the results of our model, this evolutionary route of a gradually incrementing memory would require passing through an adaptive valley (*sensu* Wright, 1965) in the region of one and two memory units (Figs. 7.2, 7.3) where fitness is reduced. Here the difference between making few, major decisions (choosing mates) and making many, smaller decisions (choosing hosts) is most pronounced; mate-seekers would pay a much higher fitness cost to evolve a short (at first) memory compared to host-seekers. So a novel prediction of our model is that animals choosing mates will in general be less likely to learn about their local environment than will host-seekers. On the other hand, the payoff curves for host and mate choice are qualitatively similar and thus give us confidence that one can evaluate searching for hosts and mates within a similar framework using the same currency.

Before evaluating this prediction with a review of the literature, we must mention some qualifications. First, we reiterate that we have been referring to host-seekers and mate-seekers as a convenient shorthand for situations where animals make many small fitness decisions or few large decisions, respectively. There are certainly cases in which ovipositing females lay all their eggs in one basket, so to speak, and our model predicts that these females should have an innate rather than learned preference for basket quality. Conversely, some animals will mate very frequently through their life, and here the cost of a short memory will be less severe. Consequently, in our review of learning about hosts and mates, we try to consider learning in a particular species in light of the frequency of ovipositing or mating. Second, because there are many types of learning, we must specify what is relevant to our model. We are concerned with whether an animal adjusts its preference for host or mate quality according to local availability, where local availability is determined through individual encounters. Changes in individual preferences due to factors unrelated to host or mate availability, (e.g., predation, Sih, 1988) are not within the scope of our chapter.

For the remainder of this chapter we will review the literature with regard to the theory that we developed earlier in this chapter. Due to the coverage given to host-related decisions in other sections of this volume (cf. Mangel, this volume), we will concentrate on aspects of mate choice.

Learning About Mates

The existence of mate choice is no longer in doubt (Bateson, 1983), but there is still much to know about why it exists and how the preferences develop (Bradbury and Andersson, 1987). Fitness advantages to mate choice may occur indirectly in theory, by becoming linked to traits that confer a mating advantage to males (Arnold, 1985), but direct benefits to mate choice are also increasingly being identified in nature (e.g., Watt et al., 1986; Phelan and Baker, 1986; Simmons, 1987; Dussourd et al., 1989;

McLain et al., 1990). Consequently we feel safe in our model's assumption that mate choice can significantly influence female lifetime reproductive success.

The nature of mating preferences, such as whether they are fixed or relative, has not been widely studied. Two studies have reported fixed threshold preferences in mate choice, as we predict from our model. Naive female cockroaches (*Nauphoeta cinerea*) prefer socially dominant males, and the strength of this preference is apparently the same whether the mate is alone or compared to a subordinate male (Moore and Moore, 1988; Moore and Breed, 1986). The conclusion that females use only fixed threshold criteria and do not compare males would have been strengthened if female preferences for solitary and paired males were measured in the same study, but clearly there is a strong innate preference for dominant males. Female cockroaches produce few broods at relatively long intervals (Breed, 1983), which suggests that mating decisions are few and of major consequence, the condition under which we predict innate preferences to occur.

Fixed threshold preferences have also been found in red jungle fowl (*Gallus gallus*, Zuk et al., 1990). In this study, individual females were given free access to pairs of males which could not interact with each other. In cases where both males exhibited poorly developed secondary sexual characteristics (combs), females chose randomly, while in trials in which one male had large combs (while differing from his partner a similar amount to other trials), females mated with this male. This indicates that females do not simply mate with the larger-combed male. When the same female encountered a pair of males both with large combs and another pair of males (at a different time) both having small combs, lack of choice occurred disproportionately with the pair of small-combed males. In this species, females copulate frequently, but typically with a small number of males and few dominant roosters (Zuk et al., 1990), which is congruent with our prediction that learning is not expected to occur when females make few mating decisions.

In milkweed beetles (*Tetraopes tetraophthalmus*), it is the male who exhibits mate choice, preferring larger females (Lawrence, 1986). In this species, males (and females) mate repeatedly and males provide no resources to females (Lawrence, 1986). So, each mating likely contributes a small proportion to a male's lifetime fitness, and under these circumstances, learning may more readily evolve according to our model. To test whether male preferences depended on local conditions, we used Lawrence's (1986) data on mating times to examine whether mating duration with a preferred female depended on whether there was another, less-preferred female also available. We found that males spent significantly more time mating with a medium-sized female when she was alone than when a small female was also present ($T = 3.385$, $N = 52$, $p < 0.01$), and males also spent more time mating with a single large female than with a large female in the presence of a mid-sized female ($T = 2.558$, $N = 52$, $p < 0.02$). Comparisons between single large females and large with small females and between large females with either a medium or small female were not significant. Because females do not compete with each other (Lawrence, 1986), it can be concluded that males adjust their mating behavior toward a given-sized female according to the presence of other females. Whether the effect is simply due to the presence of another female or the size of the other female cannot be distinguished with these data.

Females bark beetles, *Ips pini*, make very few mating decisions over their lifetimes (probably rarely more than three), and these decisions may have large fitness consequences because males provide help and protection to their offspring in addition to genes (Reid and Roitberg, 1992a). Reid and Roitberg (1992b) tested whether female *Ips pini* had innate or learned preferences, or both. Naive females were placed in enclosed arenas of either all large or all small males. Females tended to persist longer at the entrance holes to male nuptial chambers when the male inside the chamber was larger rather than smaller. Because the females were naive, this was considered an indication of their innate preferences. After experiencing either large or small males in this manner, each female was then immediately placed with a male of intermediate size. Females that had previously experienced only smaller males tended to persist longer at the midsized male than females which had encountered large males. Thus, female response to the same male depended on their experience with larger or smaller males. However, this design does not distinguish whether females gained information about the local availability of males through individual encounters or from the general pheromone blend in each arena. The latter possibility seems plausible, since pheromones are the means by which males attract females, at least from a distance (Byers, 1989), and pheromone quantity and quality may provide information about male quality (Gries et al., 1990). The dense local aggregations of males in nature would provide a blend of pheromones that probably provides females with a great deal of information about the availability of quality males at that site.

However, there is some evidence from field observations that female *I. pini* may reduce their acceptance thresholds as a result of experience (Reid, unpublished). Females accepted less preferable males when they had previously encountered similar males and when they had visited (unsuccessfully) more males. Also, some females accepted a male who was much less preferable than a previously rejected male, but if the difference in quality was very large, the better male had also been visited much earlier in her sequence of encounters. This observation is compatible with a sliding memory window, in which the existence of high-quality males is eventually forgotten. Although these results are from uncontrolled field observations, they suggest that mate preferences of female *I. pini* are affected by individual experience. This is at odds with our model's predictions. Perhaps the distribution of males is much more patchy than pheromone blends can indicate, though preliminary observations of male distributions do not suggest extreme patchiness. The cost of having a short memory is reduced in very patchy worlds, so learning could evolve more readily. This and other possible explanations are quite speculative without further information about *I. pini* and our model's assumptions, so we leave this case unresolved.

Learned female mate preferences have been implicated in the rare male effect. Here, the general observation is that when a male genotype or phenotype is rare, it obtains disproportionately more matings, and when it is in the majority, it gets fewer-than-expected matings (Knoppien, 1985). Most studies reporting the effect conclude female choice, not male competition, is responsible for this pattern. The nature of female preferences, particularly with regard to whether they are learned or not, is usually less clear. For example, Spiess (1987, 1989) interprets his observations of a rare male effect in *Drosophila* as the result of a female preference for male phenotypes different from the phenotype that first courted them. The strongest support for this idea is found in Spiess and Kruckeberg's (1980) study in which this pattern was found even when males of the two phenotypes were in equal proportions. This suggests it is the sequence of encounters that determines the preference for rare males. However, subsequent studies have not found such clean results, and significant rare male effects are found only when cases in which females mate after only being courted by one type of male are removed from analysis (Partridge, 1989).

An alternative explanation for these observations is that some females have fixed preferences for one type of male (O'Donald, 1980; Knoppien, 1985; Partridge, 1989). Genetic models involving fixed female preferences can also explain the rare male mating advantage in ladybird beetles (*Adalia bipunctatus*; Marjerus et al., 1982), and a parasitoid wasp (*Nasonia vitripennis*, O'Donald, 1989). Genetic preferences have been demonstrated in cockroaches (Moore, 1989), ladybird beetles (Marjerus et al., 1982), and guppies (*Poecilia reticulata*; Houde, 1988). However, the existence of genetic preferences does not exclude the possibility that local availability of males may influence preferences as well. Distinguishing fixed preference differences among females of the same phenotype may be difficult. If females do not demonstrate virtually unanimous preferences in an experimental situation, then individual females should be given the opportunity to express a preference in more than one distribution of male qualities to control for individual differences. Omitting "deviant" females from analysis (close to half of *Drosophila* mating trails can be omitted in some studies

(Spiess, 1987)) may obscure the processes by which frequency-dependent mating success in males is achieved (Partridge, 1989).

Some experiments have demonstrated that male pheromone blends, as discussed previously for *Ips pini*, can influence the expression of minority male advantage. The experimental design involves a mating chamber in which females have access to males of two phenotypes in unequal proportions, plus an adjacent but separated chamber containing more males of one of the two phenotypes. Odors from the adjacent males can be introduced into the mating chamber to change the proportion of odors from the two male phenotypes, but not the actual numbers of each male type that females can encounter. Using this technique, frequency-dependent mating could be induced in *Drosophila pseudoobscura* when males were in equal proportions and reduced when males differed in their frequencies in the mating chamber (Ehrman, 1972). A similar result was found for the parasitoid *Nasonia vitripennis*; the mating advantage of a rare male genotype could be eliminated when odors from males of that phenotype were added to the mating chamber to compensate for the majority male odor advantage there (White and Grant, 1977). For these species, then, female mating preferences do appear to depend on the local availability of males. However, learning clearly occurs by assessing pheromones as a single summarizing cue, and not by sequential encounters of males as our model considers.

To summarize this discussion of learning in mate choice, we suggest that lack of learning in cockroaches and jungle fowl, and learning in male milkweed beetles, are in accordance with our prediction that learning about the availability of mates is more likely to occur when mating decisions are many and each has minor fitness consequences. However, this prediction is based on learning through sequential encounters. It appears that in many insects, animals can assess the local distributions of male types by pheromone blends. This provides an accurate description of what is available without sampling, and presumably with no more neuronal circuitry than would be able to detect the pheromones of one individual. Because more accurate local knowledge provides higher fitness than innate knowledge (Figs. 7.2, 7.3), we would expect animals to use this information whenever the evolutionary route to that stage does not require intermediate stages of short memory. Insects choosing mates from aggregations seem particularly suitable for using pheromone blends, as indeed the ones described here appeared to do. Similar "summarizing" variables of male availability that influence female preferences include hunger in species where males provide nuptial gifts (Steele, 1986; Thornhill, 1984) and spermatheca fullness (Rutowski 1980). If differences in learning ability associated with the fitness consequences of the decisions are to be found, we should first look for species in which summarizing cues are not available. Sequences of encounters can then be varied experimentally, with controls for individual differences in preferences, to determine whether previous experience affects mate preferences.

Acknowledgments

This work was supported by an NSERC (Canada) grant to B.D.R., by a Science Council of B.C. grant to M.L.R. and by a Macmillan Family Fund Scholarship to C.L. We thank Marc Mangel and an anonymous reviewer for comments on an earlier version of this paper.

References

Alloway, T.M. 1978. Learning in insects except Apoidea. In W.C. Corning, J.A. Dyal, and A.O.D. Willows (eds.), Invertebrate Learning, Vol. 2. Plenum Press, New York, pp. 131–171.

Arnold, S.J. 1985. Quantitative genetic models of sexual selection. Experientia **41**:1296–1310.

Bateson, P. 1983. Mate Choice. Cambridge University Press, Cambridge, England.

Bolles, R.C., and Beecher, M.S. (eds.). 1988. Evolution and Learning. Lawrence Erlbaum Associates, Hillsdale, NJ, 263 pp.

Bradbury, J.W., and Andersson, M.B. (eds.). 1987. Sexual Selection: Testing the Alternatives. John Wiley, Chichester, England.

Breed, M.D. 1983. Cockroach mating systems. In D.T. Gwynne and G.K. Morris (eds.), Orthopteran Mating Systems. Westview Press Inc., Boulder, CO, pp. 268–284.

Brown, L. 1981. Patterns of female choice in mottled sculpins. Anim. Behav. **29**:375–382.

Byers, J.A. 1989. Chemical ecology in bark beetles. Experientia **45**:271–283.

Changeux, J.-P., Heidmann, T., and Patte, P. 1984. Learning by selection. In P. Marler and H.S. Terrace (eds.), The Biology of Learning. Springer-Verlag, Berlin, pp. 115–133.

Cowie, R., and Krebs, J.R. 1979. Optimal foraging in patchy environments. In R.M. Anderson, B.D. Turner, and L.R. Turner (eds.), Population Dynamics. Blackwell Press, London, pp. 183–205.

Dussourd, D.E., Harvis, C.A., Meinwald, J., and Eisner, T. 1989. Paternal allocation of sequestered plant pyrrolizidine alkaloid to eggs in the danaine butterfly, *Danaus gilippus*. Experientia **45**:896–898.

Ehrman, L. 1972. A factor influencing the rare male mating advantage in *Drosophila*. Behav. Genet. **2**:69–78.

Endler, J. 1986. Natural Selection in the Wild. Princeton University Press, Princeton, NJ.

Gould, J.L. 1984. Natural history of honey bee learning. In P. Marler and H.S. Terrace (eds.), The Biology of Learning. Springer-Verlag, Berlin, pp. 149–180.

Green, R.F. 1980. Bayesian birds: A simple example of Oaten's stochastic model of optimal foraging. Theor. Popul. Biol. **18**:244–256.

Gries, G., Bowers, W.W., Gries, R., Noble, M. and Borden, J.H. 1990. Pheromone production by the pine engraver *Ips pini* following flight and starvation. J. Insect. Physiol. **36**:819–824.

Houde, A.E. 1988. Genetic differences in female choice between two guppy populations. Anim. Behav. **36**:510–516.

Janetos, A.C. 1980. Strategies of female choice: A theoretical analysis. Behav. Ecol. Sociobiol. **7**:107–112.

Johnston, T.D. 1982. Selective costs and benefits in the evolution of learning. Adv. Study Behav. **12**:65–106.

Kamil, A.C., and Mauldin, J.E. 1988. A comparative-ecological approach to the study of learning. In R.C. Bolles and M.D. Beecher (eds.), Evolution and Learning. Lawrence Erlbaum Associates, Hillsdale, NJ, pp. 117–133.

Knoppien, P. 1985. Rare male mating advantage: A review. Biol. Rev. **60**:81–117.

Lawrence, W.S. 1986. Male choice and competition in *Tetraopes tetraophthalmus:* Effects of local sex ratio variation. Behav. Ecol. Sociobiol. **18**:289–296.

Li, C., Roitberg, B., and Mackauer, M. 1992. The search pattern of a parasitoid wasp, *Aphelinus asychis*, for its host. Oikos, in press.

Mangel, M. 1990. Dynamic information in uncertain and changing worlds. J. Theor. Biol. **146**:317–332.

Mangel, M., and Clark, C. 1988. Dynamic modeling in behavioral ecology. Princeton University Press, Princeton, NJ.

Mangel, M., and Roitberg, B.D. 1989. Dynamic information and oviposition decisions in a fruit fly. Ecol. Entomol. **14**:181–189.

Marjerus, M.E.N., O'Donald, P., and Weir, J. 1982. Female mating preference is genetic. Nature **300**:521–523.

McLain, D.K., Hancock, M.B., and Hope, M.A. 1990. Fitness effects of nonrandom mating in the ragwort seed bug, *Neacoryphus bicrucis* (Hemiptera Lygaeidae). Ethol. Ecol. Evol. **2**:253–262.

McNamara, J., and Houston, A.I. 1986. The common currency for behavioural decisions. Am. Nat. **127**:358–378.

Mitchell, R. 1975. The evolution of oviposition tactics in the bean weevil, *Callosobruchus maculatus* (F.) Ecology **56**:696–702.

Moore, A. 1989. Sexual selection in *Nauphoeta cinerea:* inherited mating preference? Behav. Genet. **19**:717–724.

Moore, A.J., and Breed, M.D. 1986. Mate assessment in a cockroach, *Nauphoeta cinerea*. Anim. Behav. **34**:1160–1165.

Moore, A., and Moore, P. 1988. Female strategy during mate choice: threshold assessment. Evolution **42**:387–391.

O'Donald, P. 1980. Genetic Models of Sexual Selection. Cambridge University Press, Cambridge, England.

Partridge, L. 1989. Frequency-dependent mating in female fruitflies? Behav. Genet. **19**:725–728.

Phelen, P.L., and Baker, T.C. 1986. Male-size-related courtship success and intersexual selection in the tobacco moth, *Ephestia elutella*. Experientia **42**:1291–1293.

Quinn, W.G. 1984. Work in invertebrates on the mechanisms underlying learning. In P. Marler and H.S. Terrace (eds.), The Biology of Learning. Springer-Verlag, Berlin, pp. 197–246.

Real, L. 1990. Search theory and mate choice. I. Models of single-sex discrimination. Am. Nat. **136**:376–404.

Reid, M.L., and Roitberg, B.D. 1992a. Benefits of prolonged male residence with mates and brood in pine engravers (*Coleoptera: Scolytidae*). Behav. Ecol. Sociobiol., submitted.

Reid, M.L., and Roitberg, B.D. 1992b. Innate and learned mate preferences in pine engravers. (*Coleoptera: Scolytidae*). Anim. Behav., submitted.

Roitberg, B.D. 1989. The cost of reproduction in roseship flies: Eggs are time. Evol. Ecol. **3**:183–188.

Roitberg, B.D., Mangel, M., and Tourigny, G. 1990. Density dependence in fruit flies. Ecology **71**:1871–1885.

Rutowski, R.L. 1980. Courtship solicitation by females of the checkered white butterfly, *Pieris protodice*. Behav. Ecol. Sociobiol. **7**:113–117.

Sih, A. 1988. The effect of predators on habitat use, activity and mating behavior of a semi-aquatic bug. Anim. Behav. **36**:1846–1848.

Simmons, L.W. 1987. Female choice contributes to offspring fitness in the field cricket, *Gryllus bimaculatus* (De Greer), Behav. Ecol. Sociobiol. **21**:313–321.

Spiess, E.B. 1987. Discrimination among prospective mates in *Drosophila*. In D.J.C. Fletcher and C.D. Michener (eds.), Kin Recognition in Animals. John Wiley, Chichester, England, pp. 75–119.

Spiess, E.B. 1989. Comments on some criticisms of minority mating advantage experiments in *Drosophila*. Behav. Genet. **19**:729–733.

Spiess, E.B., and Kruckeberg, J.F. 1980. Minority advantage of certain eye color mutants of *Drosophila melanogaster*. II. A behavioral basis. Am. Nat. **115**:307–327.

Staddon, J.E.R., and Ettinger, R.H. (eds.). 1989. Learning: An Introduction to the Principles of Adaptive Behavior. Harcourt Brace Jovanovich, Orlando, FL, 436 pp.

Steele, R.H. 1986. Courtship feeding in *Drosophila subobscura*. II. Courtship feeding by males influences female mate choice. Anim. Behav. **34**:1099–1108.

Stephens, D.W. 1989. Variance and the value of information. Am. Nat. **134**:128–140.

Thornhill, R. 1984. Alternative female choice tactics in the scorpionfly *Hylobittacus apicalis* (Mecoptera) and their implications. Am. Zoo. **24**:367–383.

Watt, W.B., Carter, P.A., and Donohue, K. 1986. Females' choice of "good genotypes" as mates is promoted by an insect mating system. Science **233**:1187–1190.

White, H.C., and Grant, B. 1977. Olfactory cues as a factor in frequency-dependent mate selection in *Mormoniella vitripennis*. Evolution **31**:829–835.

Wright, S. 1965. Factor interaction and linkage in evolution. Proc. R. Soc. Lond. B. **162**:80–104.

Zuk, M., Johnson, K., Thornhill, R., and Ligon, J.D. 1990. Mechanisms of female mate choice in red jungle fowl. Evolution **44**:477–485.

8

Learning and Behavioral Ecology: Incomplete Information and Environmental Predictability

David W. Stephens

The Incomplete Information Problem

When a foraging notonectid bug extracts the juices from a prey item, it will, at some point, stop extracting and begin to search for another prey item. Behavioral ecologists have found that predators, like my hypothetical notonectid, will extract a high proportion of the available resources from a given prey, taking a long time to do it, when prey are scarce, but the same predator will extract less and give up more quickly if prey are abundant.

, To the behavioral ecologist this pattern is evolutionary economics at work. The notonectid gives up because the process of resource extraction shows diminishing returns: the forager obtains fewer resources for the second unit of time it spends exploiting a resource than it obtained for the first, and so on for each successive unit of time spent exploiting a given resource. This pattern of diminishing returns leads immediately to the tendency for overall prey abundance to influence extraction time. When prey are abundant a forager should tolerate a smaller reduction in its rate of return before giving up, simply because it can do quite well elsewhere.

More generally, the behavioral ecologist supposes that the relative costs and benefits of alternative tactics are *somehow* evaluated. Behavioral ecologists are careful not to claim that their models imply any particular mechanism of evaluation. (It may be real evaluation by some cognitive process— learning, memory, problem-solving—or in some cases natural selection may have "evaluated" the alternatives so that the animal actually implements simple rules of thumb.) Typically models in behavioral ecology claim that the organism acts *as if* it knows the relevant environmental features. In the notonectid example, the modeler might suppose that the notonectid can act as if it knows the average time it will take to catch another prey item. This assumption is sometimes called the "complete information" assumption: the animal of interest can act as if it has complete information about the relevant features of its environment.

The reader will probably agree that the complete information assumption cannot be true in general, but consider its flaws when applied to our hypothetical foraging notonectid. If we adhere to the complete information assumption we effectively suppose that the alternatives of (1) continuing to extract for one more time unit or (2) initiating a search for a new item can be evaluated solely in terms of time and food value (usually energy in foraging models). The complete information assumption, therefore, prevents us from considering the possibility that continuing to extract for a few seconds might reveal that the present prey item is of exceptionally high quality. There must be many natural situations in which foraging animals obtain not only food but also information about the nature of their environment. Of course, this problem is not restricted to models of foraging behavior; it applies equally to models of mate choice or territoriality; however, it has been most widely studied in the context of foraging models.

Behavioral ecologists recognize these difficulties as the so-called "problem of incomplete information." There is a large theoretical literature (Estabrook and Jespersen, 1974; Bobisud and Potratz, 1976; Oaten, 1977; Arnold, 1978; Green, 1980, 1984; McNamara and Houston, 1980; Pulliam and Dunford, 1980; Iwasa et al., 1981; Orians, 1981; Pulliam, 1981; Houston et al., 1982; McNamara, 1982; Stephens, 1989) and a smaller empirical literature (Lima, 1983, 1985; Krebs et al., 1978; Shettleworth et al., 1988; Tamm, 1987) that discuss these difficulties.

A Typical Incomplete Information Problem

To understand the components of a typical incomplete information problem consider the theoretical work of Green (1980) and the complementary experimental study of Lima (1985). Green models a patch exploitation problem in which patches resemble egg cartons in that food can only be found in a limited number of discrete sites or holes. Patches are externally identical, but they vary in the proportion of holes that contain food. Green's forager is incompletely informed because it cannot know what proportion of a patch's holes contain food before it exploits a given patch. Hence exploiting one more hole may yield food, but it may also yield information about the quality of the present patch. Following earlier foraging models, Green solved for the patch-leaving behavior that maximized the rate of prey capture.

Like many authors who have studied incomplete information problems, Green was interested in the conditions under which violating the assumption of complete information makes the biggest difference. In answer to this question, he found that taking account of the information value of alternative tactics seems to be most important when the distribution of patch qualities is highly variable. Many students of incomplete information models have reported results similar to this "variance effect," in special cases (like Green's egg-carton patch model), but the generality of this result has remained unclear (Stephens and Krebs, 1986).

Lima (1985) studied a similar situation experimentally using downy woodpeckers (*Picoides pubescens*) feeding on short lengths of ash log. Lima drilled 24 holes in each log, and he placed pieces of sunflower seed in some holes and covered all holes, whether full or empty, with a piece of opaque masking tape. The woodpeckers easily learned to peck through the tape to obtain food. Half of Lima's logs contained no seeds, while in the remaining half a proportion p of the holes were filled: that is, there were logs with no food and logs with some food. The proportion p of holes containing food in the "some food" logs was Lima's independent variable. The prediction here, based on a simplified version of Green's model, is that the woodpeckers should stay longer when p is smaller, because when p is small, more experience is required to discover whether a patch is full or empty. The woodpeckers' observed behavior was consistent with this prediction. Lima's study was not designed to investigate the "variance effect" predicted by Green.

The literature of incomplete information problems has produced few generalizations; we are certainly still a long way from anything that might reasonably be called "a solution" to the problem of incomplete information (see Stephens and Krebs, 1986, for review). Despite this disappointment, the problem of incomplete information continues to spark excitement in behavioral ecology, primarily because it is seen as a way to ask evolutionary questions about animal learning (Shettleworth, 1984; Kacelnik and Krebs, 1985; Kamil, 1983, 1987). According to this view, learning is information gathering. An animal's experience reveals some previously unknown attribute of the environment and this presumably leads to a change in behavior. This paper presents two models that treat learning and the incomplete information problem from a behavioral ecological perspective, although neither model is a typical incomplete information model. The first model attacks the incomplete information problem directly, by asking the deceptively simple question: what makes information valuable? The second model uses population genetics to unravel the effects of environmental unpredictability and predictability on the evolution of learning.

The Value of Information

This section reviews a model, presented by Stephens (1989), that asks, "What makes information valuable?" This model differs from typical incomplete information models in behavioral ecology (as outlined above) because it is not intended to model a particular experimental situation. Instead it is an attempt to understand what makes information valuable in an abstract way, with the goal of guiding our thinking about the role of information in behavior ecology.

This analysis is based on a definition of the value of information proposed by the economist Gould (1974). The details of Gould's definition reveal a great deal about the basic structure of incomplete information problems. Suppose that some feature of the environment can be in any one of k states called $s_1, s_2, s_3, \ldots s_k$. The decision-maker knows the probabilities of the various states (it knows that state s_1 occurs with probability p_1, for example), but it does not know which of the k states is the true state of the environment. The decision-maker must adopt a course of action represented by an algebraic variable called t (for example, the decision-maker might have to decide how much time to devote to a given activity, and t might represent this time). Further suppose that there exists a function $H(t|s_i)$ that represents the payoff received if the decision-maker chooses to play t given that the true state of the environment is s_i. Biologically, an ideal $H(t|s_i)$ would be a function that represents the *fitness consequences* of adopting behavior t when the environment is in state s_i.

If the decision-maker knows that state s_i is true, then it can adopt the behavior (a value of t) that is most appropriate for the known state, and this would be the t value that yields the highest payoff in this state. In symbols, we can write

$$H(t_i^* \mid s_i) = \max_t H(t \mid s_i) \tag{1}$$

where I call t_i^* the *state-appropriate behavior* for state s_i, and it represents the best option available to the decision-maker in state s_i. However, if the decision-maker does not know which state applies, then it must pick one t value that represents a compromise between the k possible outcomes. The best behavior available to the uninformed decision-maker can be specified formally as

$$\sum_{i=1}^{k} p_i H(t'|s_i) = \max_t \sum_{i=1}^{k} p_i H(t \mid s_i) \tag{2}$$

In words, t' is the t value that is best "on average."

Now suppose that some omniscient profiteer offers to tell the decision-maker the true state of the environment for a price. How much should the

decision-maker pay? The answer is that the decision-maker should be willing to pay no more than

$$\underbrace{\sum_{i=1}^{k} p_i H(t_i^*|s_i)}_{A} - \underbrace{\sum_{i=1}^{k} p_i H(t'|s_i)}_{B} \tag{3}$$

the difference between (A), the mean value of being informed (and hence
being able to pick the behavior that is appropriate for a given state), and
(B), the mean value of acting without any further information. This expression (3) is Gould's "value of information."

Notice that the conditional payoff functions (the $H(t|s_i)$, functions that
specify the "economic" consequences of different behaviors in different
states), play a central role in this definition, and this makes this definition
different from some other measures of information (e.g., the Shannon
index) that place equal weight on any reduction in ambiguity. In Gould's
definition, information is valued in terms of how it can be put to use.

Unfortunately, this intuitively appealing property also makes it difficult
to extract further generalizations from Gould's definition, because the H
functions are unknown and, in general, they may be quite different from
one problem to the next. One fairly reasonable form for the H functions
is an inverted parabola, as shown in Figure 8.1. The equation of an inverted
parabola can be written as

$$H(t \mid s_i) = H(t_i^* \mid s_i) - c(t - t_i^*)^2 \tag{4}$$

curve that reaches its highest point $(H(t_i^*|s_i))$ at $t = t_i^*$, and where c
represents the curvature of the function; a high c value defines a strongly
peaked function, while a low c value defines a relatively flat function. This
shape can be justified (via Taylor's theorem) as an approximation to many
unimodal functions, but its most appealing feature is that it captures the
intuitively reasonable idea that payoffs decrease smoothly as the actual
decision deviates from the optimal value (t_i^*).

Moreover, substituting conditional payoff functions of this form (4) into
Gould's definition yields an especially simple result. When the payoff functions have this inverted parabola form, then the best thing an uninformed
decision-maker can do is to set t' equal to the average of the state-appropriate behaviors (the t_i^*'s); that is

$$t' = \overline{t^*} = \sum_{i=1}^{k} p_i t_i^* \tag{5}$$

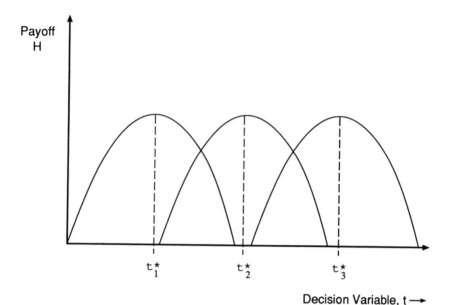

Figure 8.1. Three conditional payoff functions represented by inverted parabolae. Each parabola can be represented by t_i^*, the value of the decision variable t at which the peak payoff occurs, and a flatness parameter c that measures how strongly peaked the function is.

(Note: the calculations that lead to this result will be familiar to students of least-square estimators.) So the value of information in this case is

$$\sum_{i=1}^{k} p_i H(t_i^*|s_i) - \sum_{i=1}^{k} p_i [H(t_i^*|s_i) - c(\overline{t^*} - t_i^*)^2] \tag{6}$$

and simplifying this leads to

$$c \sum_{i=1}^{k} p_i (\overline{t^*} - t_i^*)^2 \tag{7}$$

In words, the value of information is the product of c, a term that measures the flatness of the payoff functions, and the variance of the state-appropriate behaviors. The central position of the variance term suggests that other things being equal; information is most valuable *when it has the greatest potential to change behavior*. A simple and general example of this effect is the case where the best behavior is the same regardless of which state is true: in any such case information has no value. Information that cannot change behavior is useless! When payoff functions have this simple inverted parabola form one can say even more: the more information can change behavior, the more useful it is.

Note that this effect is subtly different from the variance effect in traditional incomplete information models, because those models emphasize the variance in states (in Green's model it is the variance in the proportion of full holes per patch). For example, Green's variance effect is due to an orderly relationship between states and state-appropriate behaviors; an increase in the variance of states also increases the variance in state-appropriate behaviors. However, this need not be true. Changes in the distribution of states may or may not affect the distribution of state-appropriate behaviors.

The importance of the flatness parameter c also reflects the importance of behavior. When c is a small number the payoff functions are flat and, although there is still a "best" behavior for any given state, small departures from this optimum are relatively less costly when c is small. However, when c is large, the payoff functions are strongly peaked and a small departure from the state-appropriate behavior can have disastrous consequences. Moreover, it is the relative costs of departing from the optimum that affect the value of information; the absolute costs and benefits are not important. For example, the height of the payoff function's peaks can be shifted up and down arbitrarily *without* changing the value of information.

This model emphasizes behavioral change, but it also deemphasizes environmental states: the most significant thing about an environmental state is the behavior it implies. This is a little surprising, because many students initially think of the incomplete information problem as a problem in "statistical estimation." This model makes it clear, however, that animals faced with incomplete information have the problem of deciding how to behave and that determining (or estimating) the current state is only peripherally important.

These results provide a curious link between the view of learning as information acquisition that behavioral ecologists often adopt and the well-worn definition of learning as "the modification of behavior by experience," because both ideas focus our attention on behavioral *change*. When viewed in terms of this model the modification-of-behavior-by-experience definition seems especially apt, because this model suggests that the types of experience that are most useful to attend to are those that have the greatest potential to change behavior.

Learning and Predictability

In the previous section I have analyzed the economics of information acquisition in the most direct way that I can imagine. The arguments of the previous section bear on learning to the extent that one views learning as a process of information acquisition. In this section, I take a different direction.

What is the ultimate, evolutionary reason that animals learn? A quick survey of the literature will yield many answers of the form: learning has evolved because animals live in unpredictable environments (Thorpe, 1963; Mackintosh, 1983; Alcock, 1979; Plotkin and Odling-Smee, 1979; Johnston and Turvey, 1980; Gray, 1981; Johnston, 1982; Shettleworth, 1984). More-over, it is easy to see the importance of unpredictability by imagining an absolutely fixed (and hence absolutely predictable) environment. Surely one would not expect learning to be useful in such an environment; one would expect an absolutely fixed environment to be met with fixed behavior.

The reader may already see that this argument can be turned on its head. Consider a completely unpredictable environment, where today's state bears no predictable relationship to tomorrow's. Surely one would not expect learning here. Indeed, a more careful reading of the literature shows that some authors have emphasized the importance of predictability in the evolution of learning (Slobodkin and Rapoport, 1974; Staddon, 1983; Johnston and Turvey, 1980; Stephens, 1987). Of course, there is plenty of room between the extremes of absolute predictability and absolute unpredicta-bility, so one might argue that learning has evolved in response to inter-mediate levels of predictability (cf. Slobodkin and Rapoport, 1974; John-ston and Turvey, 1980).

This claim seems reasonable, but consider the here-and-now economics of learning as discussed in the previous section. Suppose that some ex-perience confirms that a particular environmental state is true, and you change your behavior appropriately. How will the benefits of this behav-ioral change differ if this environmental state remains the same for 1 min-ute, for 1 day, for 1 year? My guess is the longer the better; indeed, the formal model of Stephens (1987) suggests just that in a more elaborate case. It would seem that being able to behave appropriately for a long time is better than being able to behave appropriately for a short time. So where do these intermediate levels of predictability come in? The answer, I believe, is that thinking of learning as an adaptation to intermediate levels of predictability is somewhat misleading; it is a special case of a more explanatory model.

In this section I review an alternative model, originally presented by Stephens (1991), that considers the effects of two components of environ-mental predictability on the evolution of learning. To see my argument consider an animal, like a temperate-zone insect, with nonoverlapping generations, and further consider two successive generations, a parental and an offspring generation. Now suppose that we can distinguish between the processes that relate the environment's state in the parental generation to the environment's state in the offspring generation and those processes that relate the environment's state at earlier periods *within* the offspring generation to the later periods. That is, I distinguish between *between-generation* predictability and *within-generation* predictability. These two terms may be different for any number of reasons. For example, if offspring disperse a great distance from the parental habitat, then between-generation predictability may be lower than within-generation predictability.

Table 8.1 shows how these two components of overall predictability can be combined to provide an alternative to the intermediate predictability argument. In the first column, we would expect no learning based on the absolute unpredictability argument mentioned above; similarly, the "highest-highest" cell corresponds to the absolute fixity argument, and we would not expect learning here. It does, however, make sense to learn when the environment changes unpredictably between generations but is predictable within generations. One satisfying feature of this table is that it helps to reconcile the contradictory claims of some students of animal learning. For example, we have Johnston, 1982, p. 74) asserting that adaptation to environmental unpredictability is "the primary selective benefit" of learning, but Staddon (1983, p. 407) clearly stresses predictability when he writes that "the evolutionary function of . . . learning is to detect regularities." Table 8.1 makes it clear that both views are reasonable depending on one's assumptions. For example, a hypothetical advocate of unpredictability might assume that within-generation predictability is high and focus on the comparison between the two cells in the second column of Table 8.1; an advocate of predictability, by contrast, might assume that between-generation predictability is low and focus on the comparison between the two cells in the first row of Table 8.1.

Table 8.1 hints that separating environmental predictability into within- and between-generation components could help us understand how predictability affects the evolution of learning. However, Table 8.1 has the defect that it only applies at extreme values of the two predictability terms (lowest vs. highest). In an attempt to correct this difficulty I developed a

Table 8.1. Combining between- and within-generation predictability terms

		Within-generation predictability	
		Lowest	*Highest*
Between-generation predictability	*Lowest*	Ignore experience	Learn
	Highest	Ignore experience	Ignore experience

Reproduced from Stephens (1991).

simple model that allows both the within- and between-generation predictability terms to vary along a continuum. This model represents a first step in the analysis of these two components of environmental predictability in that it is the simplest model I can imagine that considers both predictability terms.

Assumptions of the Model

The model considers an animal with discrete generations. This animal's environment consists of two alternative resources. One resource is stable and the other varies between good and bad states. If an individual exploits the varying resource it gains one unit if the varying resource is in the good state, but it gains b if the varying resource is in the bad state. An individual exploiting the stable resource always obtains s. The model assumes that the stable resource is intermediate in quality between the good and bad states $(1 > s > b)$. (The assumption that the good state of the varying resources yields one unit can be made without losing generality, because any three values can be rescaled to make this true.)

Perhaps the most significant simplification in this model is the assumption that there are exactly two periods in each generation. The model treats a special type of environmental predictability, more precisely called persistence (Colwell, 1974). One parameter ω represents the within-generation persistence and is defined as the probability that the state in the second period of a given generation will be the same as the state in the first period. A second parameter β represents the between-generation persistence and is defined as the probability that the state in the beginning of a given offspring generation will be the same as the state at the end of the parental generation. Both persistence terms vary between one-half and one, with one-half corresponding to the least predictable case (the previous state provides no information) and one corresponding to the most predictable case. Figure 8.2 illustrates these assumptions.

The model is further simplified by supposing that the animal of interest is haploid (thereby avoiding the problems of dominance), and that three alternative alleles exist in the population. Individuals carrying allele V always exploit the varying resource regardless of their experience, while individuals carrying allele S always exploit the stable resource regardless of experience. Both alleles V and S represent non-learning alternatives. An individual carrying the third allele, L, exploits the varying resource in the first period of its life and bases its behavior in the second period on what it experiences in the first period: if it observes a good state in the first period, then it exploits the varying resource again in the second period, but if it observes a bad state in the first period it exploits the stable resource in the second period. To account for the possibility that the learning tactic

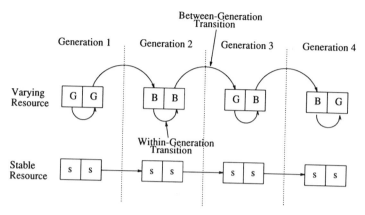

Figure 8.2. Diagram of the model's assumptions. An individual may exploit (1) a varying resource that varies between good (marked G) and bad (marked B) states or (2) a stable resource (marked S). The value of the stable resource is between the values of good and bad states of the varying resource. Two distinct Markov transition processes control the state of the varying resource. The first Markov process, governed by parameter ω, controls within-generation transitions, while the second Markov process, governed by parameter β, controls between-generation transitions.

may be more expensive to implement than the other two tactics (e.g., the higher costs of building the required cognitive equipment) a cost of learning term, c, is used. This cost term is measured in the same units as the two benefit terms s and b.

Putting these assumptions together, we see that four types of generations can occur as follows: GG, the varying resource may be in the good state in both the first and second periods; GB, the varying resource may be good in the first period, but bad in the second period; BB, the varying resource may be bad in both periods; and finally, BG, the varying resource may be bad in the first period and good in the second. Table 8.2 shows the fitness consequence for each genotype in each generation type. For a fixed generation type it is easy to specify how allele frequencies will change using the fitnesses in Table 8.2 and standard population genetics equations. The problem is that the generation type is not fixed, but changes from one generation to the next according to the within- and between-generation persistence parameters ω and β. Unfortunately, this stochastic change presents significant mathematical difficulties.

Table 8.2. The fitness consequences for the three genotypes in all possible
generation types

Allele: Tactic	Generation type			
	GG	GB	BB	BG
V: Fix on V	2	$1 + b$	$2b$	$1 + b$
S: Fix on S	$2s$	$2s$	$2s$	$2s$
L: Learn	$2 - c$	$1 + b - c$	$b + s - c$	$b + s - c$

Simulation Results

A simple way to overcome these mathematical difficulties is to perform
computer simulations of this evolving system. The difficulty with this numerical approach is that one must run many simulations at many different
combinations of the cost-and-benefit parameters (b, s, and c) to have any
confidence in the generality of one's results. I have performed a large
number of these simulations, and Figure 8.3 shows two typical results. In
rough outline, the simulation results agree with Table 8.1: learning does
not evolve when there is no within-generation persistence (the first column
of Table 8.1), nor does learning evolve when both persistence values equal
one. (I call this the absolute-fixity point and it corresponds with the
"highest-highest" cell of Table 8.1.) However, the details between these
two extremes are surprising. Specifically, excluding the absolute-fixity point
and its neighborhood, the pattern seems to depend only on the within-
generation persistence ω, while the between-generation term has almost
no effect.

Approximate Analytic Results

The domination of this problem by the within-generation term ω can
also be justified using an analytical argument. The stochastic transitions
that govern the states of the varying resource in a given generation (i.e.,
GG, GB, BB, or BG) can be formulated as a mathematical structure called
a Markov chain. The importance of this fact is that the long-term behavior
of Markov chains can be characterized easily. Under certain conditions,
Markov chains approach what is called a stationary distribution. In this
case, the stationary distribution allows us to say that in the long-run the
varying resource will be (1) good in both periods in proportion $\omega/2$ of all
generations, (2) good in the first period but bad in the second $(1 - \omega)/2$
of all generations, (3) bad in both periods in $\omega/2$ of all generations, and
(4) bad in the first but good in the second in $(1 - \omega)/2$ of all generations.
The surprising thing about this stationary distribution is that it does not
depend on the between-generation persistence, β. Although β can have a

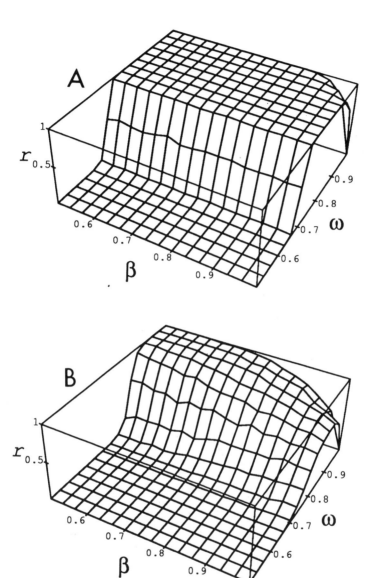

Figure 8.3. Typical simulation results. Both panels plot the
mean frequency of an allele representing the learning genotype r
as a function of the two terms that determine environmental
predictability: ω, the within-generation persistence term, and β
the between-generation persistence term. The r values are the
means of 500 simulations run either until fixation of any allele or
3,000 total generations. The two panels differ because different
cost-benefit parameters were assumed. In panel A, $s = 0.3$, $b =
0.15$, $c = 0$. In panel B, $s = 0.7$, $b = 0.35$, $c = 0$. Reproduced
from Stephens (1991).

strong effect on the sequence of generation types (e.g., whether GG generations tend to be followed by BB generations), it does not influence the relative frequency of generation types in the long run.

As mentioned above, the stationary distributions apply as long as certain conditions are satisfied. In this case, the required condition is that at least one of the two persistence values must be less than one. Put another way, the stationary distribution, with its independence from β, applies given that the environment is not absolutely fixed. In practice the stationary distribution is guaranteed to hold given infinite time, and the closer we are to the absolute-fixity point, the longer it will take to reach the stationary distribution. The results shown in Figure 8.3 illustrate this clearly. We see two qualitatively different regions. The first region is the neighborhood of absolute fixity. In this region, the stationary distribution does not apply and increasing environmental change promotes the evolution of learning. Hence, in the absolute-fixity region the old idea that increasing unpredictability promotes learning holds, but it holds in a curious and limited way: unpredictability promotes learning in an environment that changes very little. The second region is the rest of the parameter space, outside of the absolute-fixity neighborhood. In this region, increasing within-generation persistence (ω) promotes the evolution of learning and between-generation persistence has no effect. Put another way, given that there is some change, increasing within-generation persistence promotes learning.

Returning to the question of whether predictability or unpredictability is *the* driving force behind the evolution of learning, we see that this is a nonsense question. Both predictability and unpredictability need to be considered in any complete theory of the evolution of learning. Indeed, the importance of having both predictability and unpredictability in the model helps to explain the asymmetric effects of the within- and between-generation persistence terms. The within-generation term dominates this model's results for the following reason: the unpredictability required to promote learning can come either between or within generations, but the predictability required to promote learning must come within generations.

Finally, if for some reason the two persistence terms were always the same, then we could choose either term to represent overall environmental predictability. In this case, we can use Figure 8.3 to understand the effects of this overall persistence term by drawing an imaginary straight line from the point $(\beta,\omega) = (0,0)$ to the point $(\beta,\omega) = (1,1)$. If we trace out the frequency of the learning allele r as we move along this line from $(0,0)$ to $(1,1)$, then we see that r first increases and then decreases as we enter the absolute-fixity neighborhood. In short, if there was only one persistence term that measured both within- and between-generation persistence then we would expect learning to be most likely to evolve at an intermediate level of overall environmental persistence. Hence the argument that learn-

ing evolves in response to intermediate levels of environmental predictability (Slobodkin and Rapoport, 1974; Johnston and Turvey, 1980) can be
viewed as a special case of the present model that depends on the existence
of a special relationship between the two persistence terms.

General Conclusions

What's Learning

Before considering whether any general conclusions emerge from these
two models, I should clarify which aspect of learning I am addressing. I
focus on behavioral change caused by experience, and I am not concerned
with the *mechanisms* that an animal uses to change its behavior in light of
experience; experience can be used to change behavior via some complex
cognitive process or via some extremely simple stimulus-response scheme
(see Papaj and Prokopy, 1989; Shettleworth 1984; and Staddon, 1983, for
more on the limitations of the "modification of behavior by experience"
definition of learning). Strictly speaking, my interest is in behavioral phenotypic plasticity. I view learning as an instance of the more general phenomenon of phenotypic plasticity and, from a biological perspective, these
models can be viewed as part of the general literature of the origins of
phenotypic plasticity (Stearns, 1989; West-Eberhard, 1989; Levins, 1968;
Via and Lande, 1985). A reader who adopts a narrow and mechanistic
view of learning may prefer to substitute the phrase "behavioral phenotypic
plasticity" for the word "learning" in the following discussion.

Behavioral Ecology Meets Psychology

The models presented here, and behavioral ecology itself, offer an evolutionary perspective on animal learning. The importance of taking an
evolutionary approach to learning can be most clearly understood in terms
of the adamantly nonevolutionary perspective of "traditional" learning
theory. The difference between "the behavioral ecology of learning" and
"learning theory" is partly a distinction between evolutionary and mechanistic approaches, but this is a very incomplete characterization of the
difference. For example, the mechanistic approach of learning theorists is
not the study of the physiological and neurobiological basis of learning, as
one might suppose. Instead, the mechanisms that interest learning theorists
are those that can be deduced solely from observing behavior (to use a
crude analogy, learning theorists are interested in algorithms, not hardware). Moreover, learning theorists have studied an extremely narrow
range of species (e.g., rats and pigeons) in fairly limited situations (e.g.,
Skinner boxes and similar apparatus).

Two related ideas, the general process view and the principle of equipotentiality, are the subtext behind traditional studies of animal learning, and understanding these ideas will help the reader to see what behavioral ecology offers to the study of animal learning. According to the general process view, "all instances of associative learning involve the same basic underlying mechanism or process" (Roper 1983, p. 180). (Recall that the word "mechanism" does not mean a physiological mechanism). Although this statement may seem like a precise claim about nature, its interpretation varies widely. To some the general process view is the strong claim that there is one and only one learning mechanism for all species in all situations. To others it is nothing more than the mild claim that there exist *some* generalizations about learning. The so-called principle of equipotentiality is the much less ambiguous claim that "all pairs of events $E1$ and $E2$ can be associated with equal ease, in any species. . . . According to this view, when learning fails to occur . . . this is because of limitations on the species' sensory or motor capacity, rather than because of limitations on its ability to learn." (Roper, 1983, p. 184).

What makes these two ideas important is that they have been central to the experimental and theoretical paradigms adopted by learning theorists; for example, they have been used to justify studying such a narrow range of species. The focus has been on the search for features of learning that are common to all organisms and types of learning. In particular, there has been a strong bias against looking for learning specializations. In contrast, behavioral ecology and other evolutionary approaches to learning are well equipped to evaluate learning specializations.

Why is it important to consider learning specializations? Consider a less-abstract biological system like locomotion. There certainly are general features of locomotion: e.g., most forms of locomotion involve muscular contraction; all locomotion obeys Newton's laws of motion. However, if we restricted our attention to these aspects of locomotion we would have a superficial view of the problem. For example, we would overlook how the locomotion and locomotory apparatus of aquatic organisms differ from those of terrestrial organisms. Whether one finds the general or the special features of locomotion (or learning) most interesting is a matter of taste, but no one can doubt that both aspects must be considered in any attempt to achieve a complete understanding.

Although many psychologists still insist on the primacy of "general processes" in learning (Bitterman, 1975; Macphail, 1985), many others have been persuaded of the importance of looking for and understanding learning specializations (Kamil, 1987; Roper, 1983; Shettleworth, 1984). These psychologists have largely been persuaded of the need for an evolutionary approach by results from within animal psychology such as the Garcia effect. In this well-known challenge to the principle of equipotentiality,

Garcia and his colleagues (Garcia and Koelling, 1966; Garcia et al., 1966) showed that rats could readily learn an association between a taste and a gastric illness; but they learned an association between a "light-sound" stimulus and illness very poorly, or not at all. Similar results are now fairly common in the animal-learning literature (Roper, 1983; Seligman and Hager, 1972; Shettleworth, 1975, 1978, 1981). These results are encouraging to behavioral ecologists interested in learning, because the association between taste and gastric illness seems more "natural" than the association between sound and gastric illness. But, one would like to understand the diversity of learning abilities in terms that are more precise than "naturalness." The models presented here represent initial attempts to tell students of learning adaptations where to look, in terms of attributes that are less vague than naturalness. In the next two sections, I discuss the two models presented in this paper in terms of what they say about learning adaptations.

The Value-of-Information Model

The connection between this model and learning is not as explicit as learning's connection with the second model; indeed the reader may ask whether there is any connection. The answer is yes, but it requires some explanation. Recall that the model considers the value of attending to potential sources of information. The only way that an animal can obtain information is via its senses, or equivalently via *experience*. Ultimately, information must come from experience. It's not unreasonable, therefore, to think that this model is as much about the value of experience (see appendix of Stephens, 1989) as about the value of information.

When viewed in this way the main result of this model is that the types of experience that are most useful to attend to are those that have the greatest potential to change behavior. Of course, experience does not "change behavior" in itself; this expression is shorthand for the idea that there is an economically defined relationship between experiences and the "best behavior": e.g., given experience A the best behavior is to stay in the present patch for 10 seconds. This model leads us to expect that animals should attend to types of experience that make a big difference in behavioral outcomes and, although the behavioral outcomes ought to be quite different, the specific experiences that lead to these different outcomes may be similar.

Can Behavior Modify Experience?

What is curious about this model's relationship to learning is that the relationship between experience and behavior is backward. In learning, we naturally think of an experience causing a change in behavior, but in this model it is the *potential to change behavior* that determines the value of experience. When viewed in this way, the model suggests that the nature of an organism's behavioral abilities can affect the fitness value of its sensory abilities. Figure 8.4 illustrates this point in a hypothetical case. The two unimodal curves represent payoff curves that would apply in two possible states of the environment, say state 1 and state 2. The peaks of these curves show the "unconstrained" behavioral optima that the animal would choose if it "knew" which state was true. Suppose, however, that a behavioral constraint exists such that animals cannot adopt a t value greater than t_{max} (t might be a search speed, and t_{max} the fastest search speed that is physically possible). In the case illustrated in the figure, the best the animal can do given the behavioral constraint will be the same (set $t = t_{max}$) in both states; and this means that there is *no advantage* to the ability to distinguish between the two states. To restate the point, limitations on behavioral abilities can affect the fitness value of sensory abilities and it seems reasonable to hypothesize that over evolutionary time behavioral abilities have strongly influenced sensory abilities.

The idea that there may be an evolutionary link between behavioral and sensory abilities may help us to understand the failings of the principle of

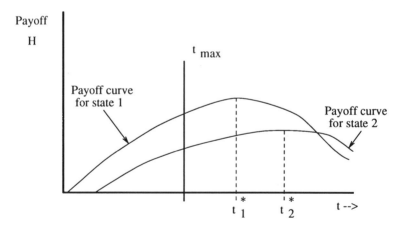

Figure 8.4. The effect of a behavioral limitation on the fitness value of the ability to discriminate two states. The two curves show hypothetical relationships between fitness grains (called "payoffs") and the value of some behavioral variable t. If there is a constraint that prevents the value of t from being larger than t_{max}, then the best behavior is the same (set $t = t_{max}$) regardless of which state is true, and according to the model discussed in the text this means that there is no benefit to discriminating between these two states. However, there would be some benefit to discriminating the states if t were unconstrained.

equipotentiality. This model does not preclude the possibility that some general process exists that can link arbitrarily chosen "experience-behavior" pairs. However, it does provide a hypothesis to guide further studies of the relationships between sensory and behavioral abilities.

The Learning-and-Predictability Model

This model suggests that we should be surprised to find learning in a case where states change randomly within a generation, or to find learning in a case of *absolute* environmental fixity. However, it would be less surprising to find learning in an absolute-fixity case for two reasons: (1) only a little change is required to maintain learning (for example, one change every 50 generations may be enough), so many putative cases of absolute fixity may actually have enough change to maintain learning; and (2) if the direct costs of learning are very low, then the forces selecting against learning in some kinds of fixed environments may be quite small (Stephens, 1991). The extreme sensitivity of the absolute-fixity case may help to explain the mixed results of those who have looked for the absence of learning abilities in relatively constant environments (Gray, 1981; Greenberg, 1985; Papaj, 1986).

Although learning may occur under many other conditions, there is never any case where learning is more likely than when the environment changes between generations but is constant within-generations. Hence, the discovery of a case where learning does not occur under these conditions, but does occur under some other regime of change, would strongly contradict this model.

Although it seems natural to think of these predictions in the context of between-species comparisons, I am not too hopeful about this enterprise. One difficulty is the well-known problem of establishing equivalent "experience" for different species (Kamil, 1987). Even setting this problem aside, consider how we might apply the hypotheses above to a between-species comparison. We would not expect the "general ability to learn" in a species that lives in a completely random environment, for example, but we might well expect the ability to learn in a species inhabiting an environment with high within-generation persistence. But what exactly is a *completely* random environment, and how does one measure *general* learning abilities? Environments have many attributes (each of which may or may not be learned "about"), and it is naive to suppose that one could meaningfully describe any natural habitat as completely random. Similarly, it is quite likely that a given species may be good at learning one sort of thing, and bad at another; and so to measure a generalized learning ability may be as pointless an exercise as trying to measure overall environmental randomness.

It seems more reasonable, in light of the comments above, to look for evidence bearing on my predictions using within-species comparisons. We might expect a given species to be able to learn about those attributes of its environment that have high within-generation predictability, but not about those attributes that vary randomly. An example of how such an endeavor might proceed exists in the so-called "biological constraints on learning" literature (Roper, 1983; Garcia and Koelling, 1966; Seligman and Hager, 1972), in which there are several demonstrations that the same species (indeed the same individual) learns more readily in one context than in another. For example, in a series of experiments Shettleworth (1975, 1978, 1981) found that hamsters would easily learn to dig to obtain food, but that attempting to reward scent-marking with food produced no evidence of conditioning. In view of the present model, one might argue that digging/food associations are relatively persistent in the hamster's natural habitat, while scent-marking/food associations, while they may occasionally occur in nature, are usually short-lived. Of course, this is speculation, the missing link in such studies being evidence bearing on the statistical persistence of particular states (or associations) in the animal's natural habitat.

Caveats and New Directions

While animal psychologists are in danger of missing significant aspects of learning by overemphasizing "general properties," behavioral ecologists may be in danger of under-emphasizing general properties. There are two reasons for this. First, it is relatively easy to build an evolutionary model of learning in a special case—such as my simple-predictability model— but it is more difficult to build an evolutionary model of general learning abilities. The second problem is that it is difficult to imagine how one might meaningfully test for a "general learning ability." However, these two points do not diminish the possibility that an animal might use a given learning mechanism in several ways. To take a crude example, an animal might use the same mechanisms in learning about its social environment that it uses to learn about food qualities. This is a general problem for the behavioral ecology approach to learning because behavioral ecologists are trained to focus on particular fitness consequences of behavioral traits.

I should emphasize that behavioral ecology offers several other points of view about the evolution of learning beyond those discussed here. I have only scratched the surface of the growing literature that attempts to integrate ideas from animal psychology with behavioral ecology (Kamil,

1987; Shettleworth, 1984; Fantino and Abarca, 1985; Logue, 1988). The papers by Kamil (1987) and Shettleworth (1984) are especially recommended. Another noteworthy theme is the use of numerical dynamic programming in the study of learning. Learning is fundamentally a dynamic problem because actions taken in the light of today's experience may influence how an organism should respond to tomorrow's experience. Until recently the mathematics of solving any reasonable dynamic learning problem were daunting; however, the techniques popularized by Mangel and Clark (1988) make it possible to gain insight into many realistic problems. See the chapters in this volume by Roitberg et al. and Mangel for examples of their techniques.

Summary

Behavioral ecologists have viewed learning as a problem in obtaining information about unknown environmental states, and they have built models that consider the effects of uncertainty on behavior. This paper reviews two models in this vein. The first model directly asks: what makes information valuable? The answer is that the most valuable types of information are those with the potential to produce large changes in behavior. The second model considers the effects of environmental predictability on the evolution of learning by dividing overall environmental predictability into within- and between-generation components. The model concludes that the environmental regularity necessary to promote the evolution of individual learning must come within generations, but that the environmental change required to promote the evolution of learning can come either within or between generations. Moreover, the model concludes that relatively little change is required for learning to evolve. Finally, some general implications of these two models are discussed.

Acknowledgments

I am grateful to Al Kamil for many helpful discussions about learning. The comments of Tony Joern, Bill Mitchell, and Dan Papaj greatly improved the manuscript. I am grateful to the National Science Foundation for their support via grant number BNS-8958228.

References

Alcock, J. 1979. Animal Behavior: An Evolutionary Approach. Sinauer Associates, Sunderland, MA.

Arnold, S.J. 1978. The evolution of a special class of modifiable behaviors in relation to environmental pattern. Am. Nat. **112**:415–427.

Bitterman, M.E. 1975. The comparative analysis of learning. Science **188**:699–709.

Bobisud, L.I., and Potratz, C.J. 1976. One-trial versus multi-trial learning for a predator encountering a model-mimic system. Am. Nat. **110**:121–128.

Colwell, R.K. 1974. Predictability, constancy and contingency of periodic phenomena. Ecology **55**:1148–1153.

Estabrook G.F., and Jespersen, D.C. 1974. The strategy for a predator encountering a model-mimic system. Am. Nat. **108**:443–457.

Fantino, E., and Abarca, N. 1985. Choice, optimal foraging, and the delay reduction hypothesis. Behav. Brain Sci. **8**:315–362.

Garcia, J., Ervin, F.R., and Koelling, R.A. 1966. Learning with prolonged delay of reinforcement. Psychonom. Sci. **5**:121–122.

Garcia, J., and Koelling, R.A. 1966. Relation of cue to consequence in avoidance learning. Psychonom. Sci. **4**:123–124.

Gould, J.P. 1974. Risk, stochastic preference, and the value of information. J. Econ. Theory **8**:64–84.

Gray, L. 1981. Genetic and experiential differences affect foraging behavior. In A.C. Kamil and T.D. Sargent (eds.), Foraging Behavior: Ecological, Ethological and Psychological Approaches. Garland STPM Press, New York, pp. 455–473.

Green, R.F. 1980. Bayesian birds: a simple example of Oaten's stochastic model of optimal foraging. Theor. Popul. Biol. **18**:244–256.

Green, R.F. 1984. Stopping rules for optimal foragers. Am. Nat. **123**:30–40.

Greenberg, R. 1985. A comparison of foliage discrimination learning in a specialist and a generalist species of migrant wood warbler (Aves: Parulidae). Can. J. Zool. **63**:773–776.

Houston, A.I., Kacelnik, A., and McNamara, J.M. 1982. Some learning rules for acquiring information. In D.J. McFarland (ed.), Functional Ontogeny. Pitman Books, London, pp. 140–191.

Iwasa, Y., Higashi, M., and Yamamura, N. 1981. Prey distribution as a factor determining the choice of optimal foraging strategy. Am. Nat. **117**:710–723.

Johnston, T.D. 1982. The selective costs and benefits of learning: An evolutionary analysis. Adv. Study Behav. **12**:65–106.

Johnston, T.D., and Turvey, M.T. 1980. An ecological metatheory for theories of learning. In G.H. Bower (ed.), The Psychology of Learning and Motivation: Advances in Research and Theory, Vol. 14, Academic Press, New York, pp. 147–205.

Kacelnik, A., and Krebs, J.R. 1985. Learning to exploit patchily distributed food. In R.M. Sibly and R.H. Smith (eds.), Behavioral Ecology, 25th Symposium of the British Ecological Society. Blackwell Scientific Publications, Oxford, England, pp. 189–205.

Kamil, A.C. 1983. Optimal foraging theory and the psychology of learning. Am. Zool. **23**:291–302.

Kamil, A.C. 1987. A synthetic approach to the study of animal intelligence. In D.W. Leger (ed.), Comparative Perspectives in Modern Psychology. University of Nebraska Press, Lincoln, NE, pp. 257–308.

Krebs, J.R., Kacelnik, A., and Taylor, P. 1978. Test of optimal sampling by foraging great tits. Nature **275**:27–31.

Levins, R. 1968. Evolution in Changing Environments. Princeton University Press, Princeton, NJ.

Lima, S.L. 1983. Downy woodpecker foraging behavior: Efficient sampling in simple stochastic environments. Ecology **65**:166–174.

Lima, S.L. 1985. Sampling behavior of starlings foraging in simple patch environments. Behav. Ecol. Sociobiol. **16**:135–142.

Logue, A.W. 1988. Research on self-control: an integrating framework. Behav. Brain Sci. **11**:665–709.

Mackintosh, N.J. 1983. General principles of learning. In T. Halliday and P.J.B. Slater (eds.), Animal Behaviour. Vol. 3: Genes, Development and Learning. W.H. Freeman, New York, pp. 149–177.

Macphail, E.M. 1985. Vertebrate intelligence: The null hypothesis. In L. Weiskrantz (ed.), Animal Intelligence. Clarendon Press, Oxford, England, pp. 37–50.

Mangel, M., and Clark, C.W. 1988. Dynamic Modeling in Behavior Ecology. Princeton University Press, Princeton, NJ.

McNamara, J.M. 1982. Optimal patch use in a stochastic environment. Theor. Popul. Biol. **21**:269–288.

McNamara, J.M., and Houston, A.I. 1980. The application of statistical decision theory to animal behaviour. J. Theor. Biol. **85**:673–690.

Oaten, A. 1977. Optimal foraging in patches: A case for stochasticity. Theor. Popul. Biol. **12**:263–285.

Orians, G.H. 1981. Foraging behavior and the evolution of discriminatory abilities. In A.C. Kamil and T.D. Sargent (eds.), Foraging Behavior: Ecological, Ethological and Psychological Approaches. Garland STPM Press, New York, pp. 389–405.

Papaj, D.R. 1986. Interpopulation differences in host preference and the evolution of learning in the butterfly, *Battus philenor*. Evolution **40**:518–530.

Papaj, D.R., and Prokopy, R.J. 1989. Ecological and evolutionary aspects of learning in phytophagous insects. Annu. Rev. Entomol. **34**:315–350.

Plotkin, H.C., and Odling-Smee, F.J. 1979. Learning, change, and evolution: An enquiry into the teleonomy of learning. Adv. Study Behav. **10**:1–41.

Pulliam, H.R. 1981. Learning to forage optimally. In A.C. Kamil and T.D. Sargent (eds.), Foraging Behavior: Ecological, Ethological and Psychological Approaches. Garland STPM Press, New York, pp. 379–332.

Pulliam, H.R, and Dunford, C. 1980. Programmed to learn: an essay on the evolution of culture. Columbia University Press, New York.

Roper, T.J. 1983. Learning as a biological phenomenon. In T.R. Halliday and P.J.B. Slater (eds.), Genes, Development and Learning. W.H. Freeman, New York, pp. 178–212.

Seligman, M.E.P., and Hager, J.L. 1972. Biological Boundaries of Learning. Appleton-Century-Crofts, New York.

Shettleworth, S.J. 1975. Reinforcement and the organization of behavior in golden hamsters: hunger, environment, and food reinforcement. J. Exp. Psych. Anim. Behav. Proc. **1**:56–87.

Shettleworth, S.J. 1978. Reinforcement and the organization of behavior in golden hamsters: sunflower seed and nest paper reinforcers. Anim. Learn. Behav. **6**:352–362.

Shettleworth, S.J. 1981. Reinforcement and the organization of behavior in golden hamsters: differential overshadowing of a CS by different responses. Q. J. Exp. Pysch. **33B**:241–256.

Shettleworth, S.J. 1984. Learning and behavioural ecology. In J.R. Krebs and N.B. Davies, (eds.), Behavioural Ecology: An Evolutionary Approach, 2nd ed. Blackwell Scientific Publications, Oxford, England, pp. 170–194.

Shettleworth, S.J., Krebs, J.R., Stephens, D.W, and Gibbon, J. 1988. Tracking a fluctuating environment: A study of sampling. Anim. Behav. **36**:87–105.

Slobodkin, L.B., and Rapoport, A. 1974. An optimal strategy of evolution. Q. Rev. Biol. **49**:181–200.

Staddon, J.E.R. 1983. Adaptive Behavior and Learning. Cambridge University Press, New York.

Stearns, S.C. 1989. The evolutionary significance of phenotypic plasticity. Bioscience **39**:436–444.

Stephens, D.W. 1987. On economically tracking a variable environment. Theor. Popul. Biol. **32**:15–25.

Stephens, D.W. 1989. Variance and the value of information. Am. Nat. **134**:128–140.

Stephens, D.W. 1991. Change, regularity and value in the evolution of animal learning. Behav. Ecol. **2**:77–89.

Stephens, D.W., and Krebs, J.R. 1986. Foraging Theory. Princeton University Press, Princeton, NJ.

Tamm, S. 1987. Tracking varying environments: sampling by hummingbirds. Anim. Behav. **35**:1725–1734.

Thorpe, W.H. 1963. Learning and Instinct in Animals. Methuen, London.

Via, S., and Lande R. 1985. Genotype-environment interaction and the evolution of phenotypic plasticity. Evolution **39**:505–522.

West-Eberhard, M.J. 1989. Phenotypic plasticity and the origins of diversity. Annu. Rev. Ecol. Syst. **20**:249–278.

9

Learning and the Evolution of Resources: Pollinators and Flower Morphology

Alcinda C. Lewis

Learning can enable an animal to find and use its resources more efficiently (Papaj and Prokopy, 1989), but what are the consequences of learning for the evolution of the resource organism? If the resource loses fitness, selection should favor protective or escape mechanisms. If the resource benefits from use, selection should favor traits that make it easier to find and use. An examination of a classic mutualism—the flowering plants and their pollinators—reveals that these simple scenarios do not always hold: flowers advertise their presence, as one would expect, but they hide their rewards. Why?

In this chapter I present an argument following Laverty (1985; see also Laverty and Plowright, 1988) that hidden rewards may be explained in part as a response to limitations on the ability of pollinators to learn to handle flowers. In outline, the argument is this. (1) Flowering plants benefit when pollinators exhibit visit consistency (or flower constancy), that is, when pollinators tend to restrict their visits to a single flower species. (2) While floral structures that require unique pollinator morphology would promote visit consistency, these are not common, perhaps because of the disadvantage of restricting the number of potential pollinating species. (3) Requiring pollinators to learn how to extract rewards would also encourage visit consistency, as Darwin (1895, p. 419) suggested, particularly if pollinators have limited ability to learn how to handle more than one species of flower. (4) Some pollinators who show visit consistency learn how to handle flowers and have limitations on the ability to learn to handle more than one species. (5) More complex flowers are usually harder to learn. Therefore (6), flower morphology can be seen as shaped in part by the advantages conferred when pollinators must make a substantial investment in learning to handle a flower. In summary, I argue that plants should evolve to attract generalist pollinators which then become facultative specialists because of influences by the flower on learning. The chapter concludes with consideration of some open questions about pollinator learning and its relationship to flower morphology.

In focusing on learning to handle flowers and its possible implications for flower morphology, I will not be considering other ways in which learning figures in pollinator-flower interactions, such as learning of flower cues. Also, pollinators may learn about the magnitudes of rewards offered by different flower species, and perhaps even by different kinds of individuals within species (Cresswell and Galen, 1991). This kind of learning no doubt influences the selective pressures on flowers, but is less likely to affect the details of floral structure.

Flowering Plants Benefit When Pollinators Tend to Restrict Their Visits to a Single Flower Species

A flower visitor that confines its visits to plants of one species is usually described as being "flower constant." But as Waser (1986) points out, this term has acquired multiple meanings, reflecting the fact that the behavior can have several causes. Constancy may be fixed or labile, genetically determined or learned. To avoid confusion, I use the term visit consistency (Lewis, 1989). An insect is consistent in its visits if it is more likely to visit a flower of the same species as the one just visited than one of a different species, whatever the cause of this behavior may be. This broad definition allows for pollinators who visit more than one flower species, or for those whose visit likelihoods for various species change over time.

The value of visit consistency to the plant as a way of ensuring correct pollen transfer has long been recognized (reviewed in Baker, 1983). Outcrossed progeny are often superior to selfed progeny when fitness is measured in various ways (e.g., Schemske, 1983; Shoen, 1983). Consistent insects avoid the pollen wastage of indiscriminate visitors or wind pollination. They may also lessen receipt of competing heterospecific pollen (Waser and Price, 1983). Campbell (1985) showed that two plant species that flower simultaneously and are visited indiscriminately by pollinators suffer reductions in the amount of pollen received, in pollen dispersal distance, and in outcrossing.

While Floral Structures That Require Unique Pollinator Morphology Would Promote Visit Consistency, These Are Not Common, Perhaps Because of the Disadvantage of Restricting the Number of Potential Pollinating Species

If visit consistency is advantageous to the plant, selection should favor adaptations that increase its likelihood. Some plant traits promote visit consistency by making rewards available only to a narrow group of pollinators whose adaptations make them relatively less able to extract or exploit rewards from other species. These plant traits include morphology as well as nectar constituents geared to physiological needs of particular taxa, flowering phenology, and signaling devices (color, pattern, and scent) (reviewed by Waser, 1983; Baker and Baker, 1983; Roubik and Buchmann, 1984; Armbruster, 1990).

Morphology can promote visit consistency in various ways. The structure of a flower may physically restrict access to insects with the appropriate morphology. A well-known example is the long corolla tube of certain lilies, allowing only hawkmoth species with equally long tongues to reach the reward (Nilsson, 1988). Such cases now appear to be the exception rather than the rule. More common are cases where there are general limits to use. For instance, many bees seem to partition flower species roughly by tongue length (Harder, 1985; Johnson, 1986). But insects can bypass these restrictions by adopting behaviors suited to particular flower sizes and shapes so that specialized flower morphology does not necessarily imply pollination by specialist vectors (Schemske, 1983).

Despite the compelling notion that plants have unique pollinators, there is increasing recognition that "general use is the general case" (Howe, 1984). Muller (1883), Macior (1974), and others looked for specific relationships between plant and insect and instead found generalist pollinators. Feinsinger (1983) and Howe (1984) also document numerous such examples, including orchid species lacking the specialized pollinators long assumed. Recent studies confirm the existence of generalized pollination systems in several additional taxa (Herrera, 1988; Roubik, 1989).

If visit consistency is valuable to plants, why is it not achieved by specialized pollinators? Several hypotheses have been put forward. Feinsinger (1983) maintains that the risks of specialization are too great for both plant and pollinator: should populations of either partner become too small, individuals of the partner species are less likely to reproduce. Competition among plants for pollinators might increase with specialization if pollinators are rare (Ratchke, 1988). Inbreeding depression might result from the short distance flown between flowers by specialist bees, leading Herrera (1987) to speculate that selection would favor mechanisms to attract additional pollinators with longer flight distances. Furthermore, there might be limits to selection for specialization even if it were advantageous. Horvitz and Schemske (1990) argue that variation in populations of potential pollinators over time and space would constrain selection for specialization. Herrera (1987) points out that the most common pollinators among the 34 on lavender are themselves morphologically diverse, also constraining evolution of morphological specialization by the plant (see also Galen and Blau, 1988).

Requiring Pollinators to Learn How to Extract Rewards Would Also Encourage Visit Consistency, as Darwin Suggested, Particularly if Pollinators Have Limited Ability to Learn to Handle More Than One Species of Flower

An often-quoted but little-tested hypothesis of Darwin (1895, p. 419) suggests another way in which visit consistency might arise:

> That insects should visit the flowers of the same species for as long as they can is of great significance to the plant, as it favors cross fertilization of distinct individuals of the same species; but no one will suppose that insects act in this manner for the good of the plant. The cause probably lies in insects being thus enabled to work quicker; they have just learned how to stand in the best position on the flower, and how far and in what direction to insert their proboscides.

In this argument, visit consistency is advantageous to the insect because learning to extract the reward from a flower of a new species is costly. Part of this cost is energetic: Heinrich (1984) calculated a significant energetic cost in time to master flower handling by bumble bees. An additional cost might be increased susceptibility to predation while immobile on a flower. Of course, these costs of learning to handle a new flower should be balanced against the costs of extended search needed to find flowers of an already-learned species.

There is an additional cost not discussed by Darwin but one with important consequences for visit consistency: the cost of switching between species is greater if learning to handle a new species interferes with the insect's ability to handle an already-learned species (Waser, 1983). Clearly an insect whose memory could hold only the amount of information needed to handle one flower species would pay a high price for switching among species. The insect would not only have to bear the cost of learning a new species but also the cost of relearning species previously learned.

Darwin's learning-centered mechanism for visit consistency can be seen as addressing the difficulties associated with specialist pollinators, noted in the previous section. In effect, learning requirements turn generalist pollinators into specialists, providing the benefits to the plant associated with specialists without limiting the pool of pollinator species, and without requiring pollinator species to rely exclusively on any one host.

To test Darwin's hypothesis, it is necessary to show that insects are consistent in their visits. Then it must be shown that insects learn the location and extraction of nectar within the flower. Additionally, it should be determined whether insects suffer a cost in relearning when switching between species.

Some Pollinators That Show Visit Consistency Learn How to Handle Flowers and Have Limitations on Their Ability to Learn to Handle More Than One Species of Flower

I have examined the behaviors linked by Darwin's hypothesis (i.e., visit consistency, learning to handle flowers, and relearning when switching among flower species) in the cabbage butterfly, *Pieris rapae*. In this section, I summarize findings for *P. rapae* and for other species for which data are available.

Visit Consistency

Measuring visit consistency is fraught with problems, and some approaches have led to questionable conclusions (see Waddington, 1983a). For example, a common method of determining consistency, identifying flower species from pollen loads, gives no information about the sequence of visits.

I attempted to avoid the common pitfalls by using the following procedure. I chose an insect known to visit many flower species for nectar, *P. rapae*. I followed males and nonovipositing females in morning nectaring flights in a species-rich habitat. I recorded the flowers visited and the flowers skipped. I compared the likelihood that an encountered flower would be visited when the last flower visited was the same or a different species (Fig. 9.1). The difference in likelihood I termed a history effect. The probability of visiting any individual flower the insect encounters is 0.09 but if the flower is the same species as last visited the probability increases to 0.74. Subsequent analyses confirmed that the history effect was not due to differences in abundances among flower species, flower distribution patterns, or pooling of observations across individuals or time periods, all of which are potential sources of misleading results.

These field results were supplemented by data from laboratory studies permitting more complete control. I put naive butterflies in a large cage with equal numbers of two flowers: vetch, *Vicia cracca*, and trefoil, *Lotus corniculatus*, in the first experiment and trefoil and bellflower, *Campanula rotundifolia*, in the second experiment. I recorded the species chosen by the butterflies for their first and second feed (Table 9.1). Butterflies apparently prefer vetch to trefoil and trefoil to bellflower, but in both cases they continue to feed from the flower from which they first fed. Thus laboratory and field observations agree in showing visit consistency in *P. rapae*.

Possible Differences in Visit Consistency Across Taxa

Since there has been no generally accepted measure of visit consistency, it is not possible to rank taxa quantitatively in terms of consistency. In

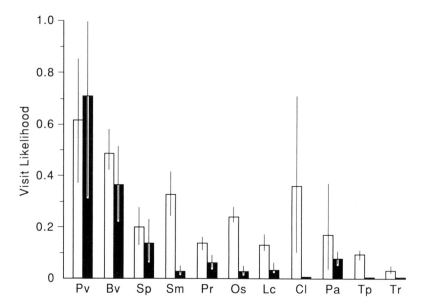

Figure 9.1. Visit likelihoods (visits/encounters) with 95% confidence limits when the encountered species was the same as the last species visited (open bar) and different from the last visited (closed bar). Species and number of encountered inflorescences for both bars: Pv, *Prunella vulgaris* (23, 7), Bv, *Barbarea vulgaris* (220, 59), Sp, *Specularia perfoliata* (136, 66), Sm, *Stellaria media* (123, 473), Pr, *Potentilla recta* (488, 174), Os, *Oxalis stricta* (691, 663), Lc, *Lotus corniculatus* (728, 812), Cl, *Chrysanthemum leucanthemum* (11, 529), Pa, *Potentilla argentata* (18, 263), Tp, *Trifolium pratense* (992, 3,330), Tr *Trifolium repens* (312, 4,600) (from Lewis, A.C. 1989. *J. Anim. Ecol.* 58:1–13.)

general, honey bees appear to be more consistent than butterflies or bumble bees (reviewed in Lewis and Lipani, 1990). This may be an indirect consequence of sociality. Foraging honey bees need not engage in other behaviors such as mate search or egg laying which may interfere with memory of flower cues and handling (Stanton, 1984). In addition, honey bees may have greater learning ability than the other insects. Dukas (1987) found that honey bees, but not solitary bees, could remember overnight a learned distinction between rewarding male flowers and unrewarding female flowers of the same species.

Roubik (1989) speculates that strict constancy would be rare in tropical bees, as compared with bees of the temperate zone, because of the large number of nectarless flowers. He also notes that bees in the tropics often have mixed pollen loads, although this does not preclude a substantial degree of visit consistency. In fact, increased competition for resources

Table 9.1. Test of constancy. Frequencies of initial and second feedings of butterflies given binary choices. Tests 1 and 2, X^2 with Yates's correction $p < 0.001$[a]

First feeding	Second feeding		
	V. cracca	*L. corniculatus*	*C. rotundifolia*
		Test 1	
V. cracca	67	5	
L. corniculatus	7	21	
		Test 2	
C. rotundifolia		8	24
L. corniculatus		68	0

[a]Reprinted with permission from Lewis. 1986. *Science* 232:863–865.

and hyperdispersion of many plant species in the tropics might increase rather than decrease the probability of visit consistency (Lewis and Lipani, 1990). For example, the neotropical *Heiconius* butterflies appear to trapline pollen sources and to vigorously defend individual plants from competing butterflies during droughts. These butterflies are long-lived, large-brained insects that may well be capable of learning the location and appearance of adult resources (Murawski and Gilbert, 1986; reviewed in Lewis and Lipani, 1990). Measuring the degree of visit consistency could lead to a testable prediction of learning ability of this insect, as well as shedding light on Roubik's assertion.

Learning Flower Handling

Darwin hypothesized that insects learn "how to stand in the best position on the flower and how far and in what direction to insert their proboscides." Such learning should be widespread but has been studied in only a few species, mainly the generalist honey bee and bumble bees (Heinrich, 1976, 1979; Laverty, 1980; Laverty and Plowright, 1988; see also Kevan and Lane, 1985). Both Heinrich (1979) and Laverty (1980) found that the time required by inexperienced bumble bees of various species to extract nectar or pollen decreased with experience on several flower species. Inexperienced bees landed on incorrect areas of the flower, and once landed, probed incorrect areas. In contrast is the narrow specialist *Bombus consobrinus*, which requires much less time for nectar extraction on monkshood, *Aconitum*, then generalist bumble bees (Laverty and Plowright, 1988). The specialist has an unlearned behavior, probing initially near the nectar, absent in the generalists that more often probed in the area of the stamens and sepal margins where nectar is more commonly found in other flower species. Nevertheless, the specialist also shows some improvement with experience, indicating that learning occurs. The reason for this is unknown but it may reflect slight variation in nectar location among flowers, discussed below. Buzz pollination, in which the anthers are vibrated, may also be learned despite specialized morphological adaptations for the behavior in some bumble bees (Buchmann, 1983).

I tested for learning of flower handling in *P. rapae*. Since the butterfly shows visit consistency, by Darwin's hypothesis it should also learn flower handling. I placed naive butterflies in cages with flowers of bellflower or trefoil and observed their behavior. When butterflies land on a flower, they walk around the flower, searching the sepals and corolla with their proboscides, eventually finding the nectar and beginning to drink. I term the time from landing on a flower to finding nectar in any flower *discovery time*. Note that discovery times may therefore include investigation of more than one flower. Discovery times for bellflower and trefoil are given in Figure 9.2 for the first eight attempts (median time between attempts, 1 second; first and third quartiles, 1 and 10 second). The discovery times show an improvement over attempts, one criterion of learning (Staddon, 1983). The curves are well fit by a power law, the function which best fits typical learning curves (Newell and Rosenblum, 1981; median and first and third quartiles of the percentage variance accounted for by the model: bellflower 82% [90–71%], trefoil 84% [93–63%]). The rates of learning, measured as the model estimate of the exponent, did not differ significantly between species. The initial times among butterflies for both groups are variable, as with bees, but the variance decreased with attempts.

Two butterflies on bellflower gave up before contacting nectar. Two individuals on trefoil had more erratic performance: both approached the flower from the back of the corolla and contacted nectar from this unusual position but were then apparently unable to find an approach that consistently led to nectar. These failures suggest that the learning task posed by some flowers can be too difficult for some insects, a point to which I return later.

Interference From Later Learning

As discussed above, Darwin's hypothesis gains added force if flower-visiting insects cannot remember the handling of more than one species simultaneously (Waser, 1983, 1986), or if there is a cost in doing so. Several authors have suggested that memory limitations for either flower appearance or handling may be important in the behavior of pollinators but there have been few tests of the hypothesis (Nilsson et al., 1987; Dukas, 1987; Laverty and Plowright, 1988; Real et al., 1990).

Laverty (1985) measured handling times for bumble bees foraging in a mixed patch of flowers. Individuals that switched to one flower species from another took more handling time, and made more handling errors,

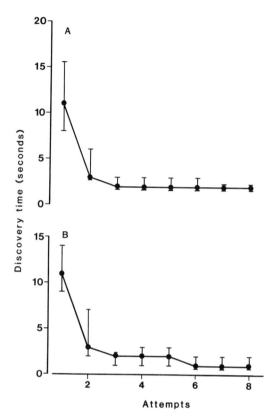

Figure 9.2. Discovery times (medians with first and third quartiles) for butterflies given either (A) *Campanula rotundifolia* (n = 24) or (B) *Lotus corniculatus* (n = 18). See text for details (from Lewis, A.C. 1986. *Science* 232:863–865).

than individuals that remained constant to that species, when the species differed in morphology.

Waser (1986) predicted that if visit consistency was due to memory limitations, it should be more pronounced when flowers in a mixture differed in appearance. He found that constancy by bumble bees increased as differences among test flowers increased. However, it is not entirely clear whether differences in appearance should increase or decrease memory demands, as I consider below.

Honey bees may have very persistent memory of flower cues after as few as three visits although the memory is susceptible to interference immediately after learning and 3 minutes later (Menzel et al., 1974, this volume; Menzel, 1985). If a different rewarding flower is offered at the same time on a subsequent day, the memory of the first one is erased (Bogdany, 1978). Gould (this volume) reports evidence indicating that bees can remember two different sets of cues for feeders at two different times of day. Memory for handling artificial flowers is also linked to time of day in honey bees (Gould, 1987). If this memory is similar to ones for flower cues, learning a second flower at the same time of day that a first was learned may interfere with the ability to recall the first.

I tested the ability of *P. rapae* to remember the handling of more than one flower species by giving two groups of butterflies bellflower to learn until a minimum of five and a maximum of ten successive discovery times did not exceed three seconds each. The experimental group was then given trefoil to learn, while the control group was held in the cage without flowers for 20 minutes, the maximum time for experimental butterflies to reach criterion on trefoil. Both groups were then offered bellflower again. The initial discovery times on the second offering of bellflower were compared with the final times on the first (Fig. 9.3). Butterflies given a second species to learn had to relearn the first while control butterflies did not: learning a second species interfered with the ability to remember the first. Discovery times of the control group may have been shortened by hunger, but this seems unlikely. Those butterflies in the experimental group that had the option of feeding on trefoil but did not, possibly because they were not hungry, had final and test discovery times similar to butterflies not given trefoil (mean $\pm$ SEM: final 2.2 $\pm$ 0.14, test 3.13 $\pm$ 0.73, $n = 5$).

Note that while discovery times on bellflower after experience with trefoil are not significantly different from initial discovery times, they are lower, suggesting that the butterflies may have retained something from their earlier exposure. Possibly with increased exposure butterflies could remember the handling of two species. This is even more plausible in insects with greater memory capacity. Remembering two flowers would allow the so-called majoring and minoring behavior described in bumble bees by Heinrich (1976, 1979), discussed below.

More Complex Flowers Are Usually Harder to Learn

Darwin's learning-centered mechanism for visit preference cannot be linked to flower morphology as a selective force unless flower morphology influences the learning demands placed on pollinators. Is there such an influence? Heinrich (1979) and Laverty (1980, 1985) tested learning by bumble bees on flowers varying in morphological complexity. They used a qualitative measurement of complexity: simple, bowl-shaped flowers had

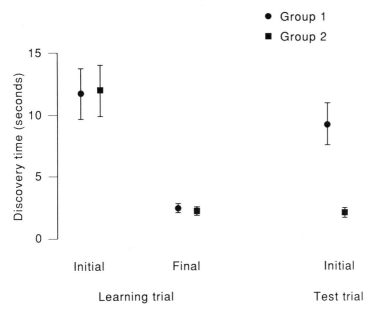

Figure 9.3. Interference test. Group 1 butterflies ($n = 17$) were given *Lotus corniculatus* between learning and test trials with *Campanula rotundifolia*. Group 2 butterflies ($n = 20$) were given no flowers during a similar period. Differences between final and initial discovery times for group 1, $p < 0.01$, paired *t* test; group 2, NS (from Lewis, A.C. 1986. *Science* 232:863–865.)

centrally placed rewards while complex flowers had hidden rewards, access being down long corolla tubes or behind overlapping petals. Learning times for bumble bees did increase as apparent floral complexity increased. In one case, about half of the bees tested on the most complex flowers gave up on their first encounter without contacting nectar (Laverty, 1985).

I compared learning by butterflies on flowers of 11 species (Fig. 9.4). The results are arranged by decreasing initial discovery times. As with the results on learning discussed earlier (Fig. 9.2), butterflies showed an improvement in performance with experience on most species. There is some correlation between learning time and morphology, with the simplest flowers requiring the least initial time for nectar discovery. Consistent with this, planned comparisons of flowers with similar complexity (to the human eye) revealed no significant differences in times within four of five pairs of similar flowers: *Oxalis stricta* and *Barbarea vulgaris*, *Nepeta cataria* and *Prunella vulgaris*, *Trifolium pratense* and *Lotus corniculatus*, Vicia cracca and *Lotus corniculatus*. The exception is the pair, *Centaurea maculosa* and *Cirsium vulgare* (Mann Whitney U = 281.5, $p < 0.05$). *C. vulgare* flowers

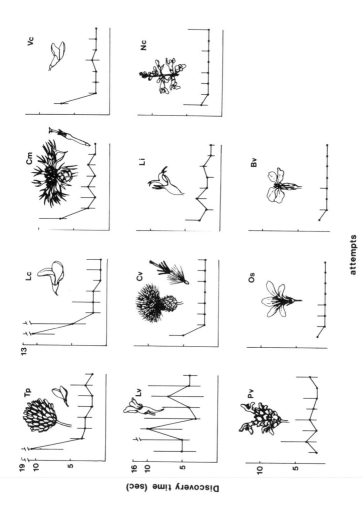

Figure 9.4. Discovery times (medians with first and third quartiles) for the first eight attempts to locate nectar by naive butterflies. Species abbreviations and number of butterflies tested are: Tp, *Trifolium pratense* (20), Lc, *Lotus corniculatus* (20), Cm, *Centaurea maculosa* (22), Vc, *Vicia cracca* (8), Lv, *Linaria vulgaris* (15), Cv, *Cirsium vulgare* (18), Li, *Lobelia inflata* (8), Nc, *Nepeta cataria* (20), Pv, *Prunella vulgaris* (18), Os, *Oxalis stricta* (14), Bv, *Barbarea vulgaris* (15) (from Lewis, A.C. and Lipani, G. 1990. *Insect-Plant Interactions* 2:95–110.)

are wider and more loosely packed on the inflorescence, allowing easier access and inspection by the butterflies.

These results also make clear the difficulty in making good a priori assessments of handling difficulty: seemingly simple flowers such as *T. pratense* present initial difficulties in nectar discovery. Corolla length alone does not explain increased probing times for butterflies as it does in some cases for bees (Harder, 1983) as corollas of *T. pratense* are relatively short. The difficulty for butterflies on *T. pratense* as well as on two species with relatively long tubes seems to be in finding the opening of the individual florets. These are often tightly packed on the inflorescence. The structure of the inflorescence may also be important: on rounded ones such as *T. pratense* and *C. maculosa*, the butterflies walked between florets, while on vertically arranged ones, they flew between florets, possibly orienting more easily from the air.

Morphology alone cannot explain differences in discovery times. In a separate experiment, butterflies were given two flowers very similar in morphology, *V. cracca* and *Coronilla varia*. The latter species did not occur at the study site and so was not included in the original series. Naive butterflies given these two species had significant differences in initial and final discovery times (U = 198.5, $p < 0.01$; U = 134, $p < 0.05$). These results suggest that these species may differ in providing cues such as markings or scent that are important once the insect has landed (Waser and Price 1983). Bumble bees tested on two similar species of *Aconitum* and two other similar species, *Chelone* and *Gentiana*, also showed differences in learning times for no readily apparent reason (Laverty, 1985).

Butterflies on *Linaria vulgaris* and *P. vulgaris* did not improve over eight attempts. *L. vulgaris* is usually classified as a bee flower although *P. rapae* will occasionally visit it in the field. The nectar spur is too long for the butterfly unless entered at one point, away from the visibly marked center. The lack of cues may account for the high variance on most attempts as the importance of chance encounters increases. Lack of improvement in handling by butterflies on *P. vulgaris* is due to characteristics of the inflorescence, which contains many empty calyces from which senescent corollas have been shed. Butterflies continue to probe these calyces, increasing their overall discovery times. They do show some tendency to explore fewer of these with time: fewer are contacted in the second half of the observation period than in the first ($p < 0.05$, sign test). Inflorescences of *N. cataria* also contain senescent flowers which presumably contain little, if any, traces of nectar. Butterflies do probe these but spend less time exploring them with successive encounters. In the case of both *P. vulgaris* and *N. cataria*, butterflies may be learning to avoid nectarless flowers. Such learning has been suggested for bees (Little, 1983; Dukas, 1987) but all cases require more rigorous testing to verify that learning is occurring.

The details of these cases show the difficulty, with present knowledge, of determining precisely how flower features influence learning time. But the results do suggest that flower morphology does influence learning time, with human judgments of morphological complexity having some predictive value.

Flower Morphology Can Be Seen as Shaped in Part by the Advantages Conferred When Pollinators Must Make a Substantial Investment in Learning to Handle a Flower

In Darwin's argument, plants benefit from the cost that insects must pay in learning to handle flowers. But in order for this cost to influence the morphology of flowers, there must be some way in which the morphology of the flowers of an individual plant can affect the visit consistency of its pollinators. There are two ways in which this can occur. First, by exacting a high initial learning cost, plants can increase the chance that pollinators that bring pollen from another species will leave without transferring pollen, reducing receipt of stigma-clogging heterospecific pollen (Galen and Gregory, 1989). Second, in the presence of memory limitations or other interference effects, plants can increase the likelihood that subsequent successful visits by a pollinator will be to plants of the same, rather than a different, species, since the pollinator will lose some or all of its ability to handle flowers of other species. In this way the plant can reduce pollen wastage.

Consistent with this analysis, Laverty (1985) found that bumble bees that learn to handle the complex *Aconitum* flowers were more consistent in their visits than ones visiting relatively simple flowers. He also analyzed data of Macior (1974) showing that pollen loads taken from bees that visited complex or long-tongued flowers contained less foreign pollen than those from bees visiting relatively open flowers.

The flower should not be so complex or difficult to handle that the insect gives up its search and switches to another individual or species. The observations of failed visits reported above show that this can happen. The complexity of the flower should reflect the overall availability of pollinators in the habitat, complexity decreasing as pollinator availability decreases (Richards, 1986).

The magnitude of rewards may also enter the balance here. It could be that the interfering effect of a visit to a new species is influenced by the magnitude of the reward obtained. At the extreme, an unrewarded visit to a new species might not interfere at all with knowledge of how to handle previously learned species, while a heavily rewarded visit might be highly interfering. At the same time, rewards can compensate the insect for the learning costs imposed by the plant. Because long initial handling times present costs to the insect in energy and risk, insects should adapt so as to avoid visits to plants which did not offer sufficient reward to balance these costs. Thus one expects to see a positive relationship between handling time and reward (reviewed by Harder and Cruzan, 1990).

One would also expect differences in selection for floral complexity when plants differ in the importance of pollinator constancy, as Laverty (1985) argues. Plants that grow in large monospecific patches do not require pollinators to be constant, because even inconstant pollinators will tend to make many visits within the patch and hence be effective in transferring pollen. Laverty (1985) reviews evidence that plants with complex flowers tend to be widely dispersed, and hence more reliant on pollinator constancy.

For visit consistency to act as a selective force on plants, via learning of handling, there must have been variation for the relevant morphological traits, although this variation need not be detectable today (Rausher, 1992). Evolutionary competence (see Stebbins, 1988) for the morphological changes necessary for pollinator shifts exists (see also Raven, 1989). Population data from various flower species reveal genetic variation for various traits of the corolla such as length, width, flare, and depth (Gottlieb, 1984; Richards, 1986; Kenrick et al., 1987; Bromer et al., 1990; Murcia, 1990; Herrera, 1990). Some specialist pollinators have been shown to exert selective pressure on floral traits of their hosts (reviewed by Galen, 1989).

Though the argument based on Darwin's hypothesis is suggestive, it is difficult to cite direct evidence that insect learning and its limitations act as selective forces on flower morphology. Crepet and Friis (1987) observe that the fossil record indicates that the second major diversification of angiosperms coincided with the appearance and diversification of so-called faithful pollinators, bees and lepidopterans capable of learning. This is consistent with Darwin's argument, since visit consistency should promote speciation of plants, and (by his account) visit consistency is enhanced in pollinators who learn. But it does not illuminate the connection between pollinator learning and flower morphology.

Finding evidence in support of the hypothesis will be complicated by other forces selecting for complex morphology and hidden rewards. Initially, adaptations to protect the carpels from herbivory or adapt it for dispersal may have been of overriding importance (Doyle and Donoghue, 1987). Morphological effects on the microclimate of the flower may also be important: for instance, elevated temperatures in closed flowers can lead to reduced nectar viscosity, which is preferred by pollinators (Corbet, 1990). Hidden rewards may lead to longer pollinator visits, thereby increasing pollen transfer in some cases (Harder and Thompson, 1989; Galen and Stanton, 1989). On the other hand, removal of more pollen by an individual pollinator can lead to greater proportional loss of pollen, and flower morphology can influence this by limiting how much pollen is removed on a single visit (Harder and Thomson, 1989).

Open Questions

Is Flower Morphology a Selective Force on Insect Learning?

Learning is assumed to be adaptive if it increases an animal's fitness. Stephens (this volume) suggests that learning should not be advantageous in cases of extreme environmental predictability or unpredictability but only in cases of intermediate predictability. This is the case for many flower-visiting insects: flower handling is similar between flower species but each requires a particular motor pattern for extraction. Stephens proposes in particular that learning is useful when environments change unpredictably between generations but are predictable within generations, a situation quite likely for the floral environment of many pollinators. Flower morphology thus appears to be a possible selective force on insect learning.

Insect learning, like flower morphology, has the underlying variability necessary as raw material for selection (reviewed by Papaj and Prokopy, 1989). However, insects perform multiple tasks, many of which may be learned. It may not be possible to separate out which task is selecting for learning, particularly as the seemingly different sorts of learning may have underlying similarities (Papaj, 1986). Selection may have favored learning ability which can be applied to several different tasks such as mate location, host plant recognition, or nest construction as well as flower handling (Darwin, 1895, p. 430).

How Is Flower Handling Learned?

A more detailed analysis of the relationships among floral structure, handling time, learning, and interference requires more insight into what pollinators learn and how. Papaj and Prokopy (1989) suggest that there are spontaneous and random movements which may come to be associated with successful nectar extraction, leading to instrumental conditioning. Meanwhile, motor patterns formerly triggered by non-neutral stimuli are elicited by previously neutral stimuli by classical conditioning. Through trial and error, the component motor patterns are assembled into an order that effectively and consistently gives nectar extraction.

Alternatively, Gould and Towne (1988) and Gould (this volume) suggest that flower handling resembles bird song learning in that individuals have an innate and limited set of motor subunits which are slightly modifiable and then are arranged in the appropriate order by trial and error. This second view is supported by Laverty (1980) and Laverty and Plowright (1988), who found that initial probing by inexperienced bumble bees on a variety of flowers was not random but directed to certain areas of the corolla. However, nectar-robbing insects show remarkable behavioral flexibility (Darwin, 1895, p. 425ff.) as do insects that apparently learn to use the holes made by nectar robbers (Lewis, unpublished observations). These observations tend to support the suggestions made by Papaj and Prokopy (1989).

Different accounts are possible of the way in which later learning interferes with earlier learning. One account posits a capacity limitation: information about flower handling can be maintained only up to some quantitative limit, and hence interference would be greater in attempting to learn a diverse collection of flowers than a collection within which certain information was true in common. This is the position argued by Waser (1983, 1986) as noted earlier. On the other hand, if flower-handling skill is represented by connections between relevant stimuli, such as colors, scents and forms, and appropriate actions, then learning to handle a group of similar-appearing flowers that required different actions would be more difficult than learning to handle a group of flowers of diverse appearance, because different behaviors would have to be associated with similar stimuli. Studies of the details of handling episodes on sequences of actual or artificial flowers differing in appearance and handling methods might clarify this matter.

Would Flowers Benefit From Insect Limitations on Learning to Recognize Flowers?

While flowers should be easily recognizable as potentially rewarding food sources, there may be limitations on the ability of insects to learn to associate information concerning reward or handling time with the appearance of a given species and particularly of several species simultaneously (see the earlier section Interference from Later Learning). As with limitations on learning handling, such recognition limitations might make visit consistency advantageous (Papaj, personal communication), an idea that awaits testing.

How Does Learning Time Influence Pollinator Flower Choice?

In addition to learning to handle flowers, pollinators are learning which ones to visit. While there is some evidence that learning influences flower choice in bumble bees (Laverty, 1985), it is unclear exactly how visit choices are made and how they are influenced by handling time and learning. The framework for most thinking about visit choice is that the insect learns the appearance and handling of rewarding flowers, using what it has learned in choosing among flowers. Bees have been shown to have this ability in laboratory studies (reviewed by Gould and Towne, 1988) but field studies of bumble bees raise questions about how leaning actually affects choice. Based on observations of bumble bees foraging in natural habitats, Heinrich (1976, 1979) proposed that bees sample the available flowers and then specialize on a so-called major, the most rewarding species at the time. The bees intersperse visits to the major with visits to a minor in a pattern that is irregular but reflects their first and second preferences. He speculated that this behavior allows them to switch species when relative rewards change. But as he points out (1984), it is unclear how the insect first narrows down its choice to one or two species, particularly if learning is insensitive to differences in quantity of reward, as suggested by data of Menzel et al. (1974) from honey bees rewarded by different concentrations of sucrose solution. It is also not known what determines the length of runs at the two flower species, or how insects switch since laboratory studies indicate that more learning should make them reluctant to switch. A further problem arises from laboratory studies showing a lag time in learning the cues associated with rewards while field choices are established relatively quickly.

How might experience in handling a flower influence the visit choice process? In one view, visit choices would be directly influenced by learning to handle a flower. Flowers of the same species encountered later would be assigned a greater estimated net reward than they were assigned before learning took place. The adjustment in estimated reward occasioned by a visit could reflect an assessment of how much was learned on the visit, or, since learning rates do not appear to vary much among flower species, the adjustment could be based only on the handling time and reward for the visit.

In another view, adjustments would be based only on reward, ignoring handling time, and visit consistency can still result. Lewis (1989) outlines a model of choice in which a priori visit probabilities for different flower species are adjusted based on the receipt of rewards, so that species which have been successfully visited have their visit probabilities increased. If these adjustments are large relative to the size of initial visit probabilities, strong visit consistency would be produced. The size of the adjustments would reflect the average cost of learning a new flower, relative to the cost of limiting visits to known flowers. Because visit consistency in this model results from large adjustments, and large adjustments reflect big savings in staying with known flowers, visit consistency is still attributable to learning and its effect in reducing handling time for known flowers. This is true even though the model does not require individual insects to monitor handling time or learning.

Refining either of these accounts would require dealing with a number of complications. Choice is influenced by the value of the rewards in other available species (Real et al., 1983; Waddington, 1983b, 1987; Heinrich, 1983, 1984; Schmid-Hempel et al., 1985; Harder, 1988) and the presence of competing insects (Roubik, 1989). Innate color preferences may override choices made for purely energetic reasons (Harder and Real, 1987).

Conclusion

The data presented here provide support for Darwin's hypothesis, strengthened to take note of the limitations on pollinator learning. Several insect species are apparently consistent in their visits to flowers; choice is influenced by experience with a flower; certain insects learn to recognize and handle flowers; their memory is constrained so that the importance of visit consistency is increased. These facts, together with the importance of pollinator visit consistency to flowering plants, suggest that adaptations in flower morphology to increase pollinator learning demands would be advantageous. Flowers should evolve to turn generalist visitors into specialists through influences on learning.

The relationship outlined here between plant morphology and insect learning may be widespread. For instance, Papaj and Prokopy (1989) suggest that plants may evolve defenses geared to the memory limitations of their insect enemies. The particular example they give is that of leaf shape, used by ovipositing butterflies as a cue in host finding, but any physical or chemical trait may be affected, in mutualistic or antagonistic relationships.

Acknowledgments

I thank D. Papaj, C. Lewis, Y. Linhart, M. Lane, C. Galen, and J. Cresswell for comments and M. Rausher and D. Campbell for papers. For assistance of various kinds I thank C. Lewis, M. Rothschild, C. Jones, J. Simms, and the staff of the IES. This paper is a contribution to the program of the Institute of Ecosystem Studies of the New York Botanical Garden. Funds were provided by the NSF (BSR 85-06072; BSR 89-08446), the Mary Flagler Cary Trust, and the American Philosophical Society.

References

Armbruster, W.S. 1990. Estimating and testing the shapes of adaptive surfaces: the morphology and pollination of *Dalechampia* blossoms. Am. Nat. **135**:14–31.

Baker, H.G. 1983. An outline of the history of anthecology, or pollination biology. In L. Real (ed.), Pollination Biology. Academic Press, Orlando, FL, pp. 7–28.

Baker, H.G., and Baker, I. 1983. Floral nectar sugar constituents in relation to pollinator type. In C.E. Jones and R.J. Little (eds.), Handbook of Experimental Pollination Biology. Van Nostrand Reinhold, New York, pp. 117–141.

Bogdany, F.J. 1978. Linking of learning signals in honey bee orientation. Behav. Ecol. Sociobiol. **3**:323–336.

Bromer, W., Barnette, J., Green, L.D., and Ervin, V. 1990. Consequences of breeding systems and pollination. Bull. Ecol. Soc. Am. **71**:103.

Buchman, S.L. 1983. Buzz pollination in angiosperms. In C.E. Jones and R.J. Little (eds.), Handbook of Experimental Pollination Biology. Van Nostrand Reinhold, New York, pp. 73–113.

Campbell, D.R. 1985. Pollen and gene dispersal: The influences of competition for pollinators. Evolution **39**:418–431.

Corbet, S.A. 1990. Pollination and the weather. Israel J. Bot. **39**:13–30.

Crepet, W.L., and Friis, E.M. 1987. The evolution of insect pollination in angiosperms. In E.M. Friis, W.G. Chaloner, and P.R. Crane (eds.) The Origins of Angiosperms and Their Biological Consequences. Cambridge University Press, New York, pp. 181–201.

Cresswell, J.A., and Galen, C. 1991. Frequency-dependent selection and adaptive surfaces for floral character combinations: The pollination of *Polemonium viscosum*. Submitted.

Darwin C. 1895. On the Effects of Cross- and Self-Fertilization in the Vegetable Kingdom. Appleton, New York.

Doyle, J.A., and Donoghue, M.J. 1987. The origin of the angiosperms: A cladistic approach. In E.M. Friis, W.G. Chaloner, and P.R. Crane (eds.), The Origins of Angiosperms and Their Biological Consequences. Cambridge University Press, New York, pp. 17–34.

Dukas, R. 1987. Foraging behavior of three bee species in a natural mimicry system: Female flowers which mimic male flowers in *Ecballium elaterium* (Cucurbitaceae). Oecologia **74**:256–263.

Feinsinger, P. 1983. Coevolution and pollination. In D. Futuyma and M. Slatkin (eds.), Coevolution. Sinauer, Sunderland, MA, pp. 282–310.

Galen, C. 1989. Measuring pollinator-mediated selection on morphometric floral traits: Bumble bees and the alpine sky pilot, *Polemonium viscosum*. Evolution **40**:882–890.

Galen, C., and Blau, S. 1988. Caste-specific patterns of flower visitation in bumble bees (*Bombus kirbyellus*) collecting nectar from *Polemonium viscosum*. Ecol. Entomol. **13**:11–17.

Galen, C., and Gregory, T. 1989. Interspecific pollen transfer as a mechanism of competition: Consequences of foreign pollen contamination for seed set in the alpine wildflower, *Polemonium viscosum*. Oecologia **81**:120–123.

Galen, C., and Stanton, M.L. 1989. Bumble bee pollination and floral morphology: Factors influencing pollen dispersal in the alpine sky pilot, *Polemonium viscosum* (Polemoniaceae). Am. J. Bot. **76**(3):419–426.

Gottlieb, L.D. 1984. Genetics and morphological evolution in plants. Am. Nat. **123**:681–709.

Gould, J.L. 1987. Honey bees store learned flower-landing behavior according to time of day. Anim. Behav. **35**:1579–1581.

Gould, J.L., and Towne, W.F. 1988. Honey bee learning. Adv. Insect Physiol. **20**:54–86.

Harder, L.D. 1983. Flower handling efficiency of bumble bees: Morphological aspects of probing time. Oecologia **57**:274–280.

Harder, L.D. 1985. Morphology as a predictor of flower choice by bumble bees. Ecology **66**:198–210.

Harder, L.D. 1988. Choice of individual flowers by bumble bees: Interactions of morphology, time and energy. Behaviour **104**:60–77.

Harder, L.D., and Cruzan, M.B. 1990. An evaluation of the physiological and evolutionary influences of inflorescence size and flower depth on nectar production. Func. Ecol. **4**:559–572.

Harder, L.D., and Real, L.A. 1987. Why are bumble bees risk averse? Ecology **68**:1104–1108.

Harder, L.D., and Thomson, J.D. 1989. Evolutionary options for maximizing pollen dispersal of animal-pollinated plants. Am. Nat. **133**:323–344.

Heinrich, B. 1976. The foraging specializations of individual bumble bees. Ecol. Monogr. **46**:105–128.

Heinrich, B. 1979. "Majoring" and "minoring" by foraging bumble bees, *Bombus vagans*: An experimental analysis. Ecology **60**:245–255.

Heinrich, B. 1983. Insect foraging energetics. In C.E. Jones and R.J. Little (eds.), Handbook of Experimental Pollination Biology. Van Nostrand Reinhold, New York, pp. 187–214.

Heinrich, B. 1984. Learning in invertebrates. In P. Marler and H.S. Terrace (eds.), The Biology of Learning. Springer-Verlag, New York, pp. 135–147.

Herrera, C.M. 1987. Components of pollinator "quality": Comparative analysis of a diverse insect assemblage. Oikos **50**:79–90.

Herrera, C.M. 1988. Pollination relationships in southern Spanish Mediterranean shrublands. J. Ecol. **76**:274–287.

Herrera, C.M. 1990. The adaptedness of the floral phenotype in a relict endemic hawkmoth-pollinated violet: 2. Patterns of variation among disjunct populations. Biol. J. Linnean Soc. **40**:275–292.

Horvitz, C.C., and Schemske, D.W. 1990. Spatiotemporal variation in insect mutualists of a neotropical herb. Ecology **71**:1085–1097.

Howe, H.F. 1984. Constraints on the evolution of mutualisms. Am. Nat. **123**:764–777.

Johnson, R.A. 1986. Intraspecific resource partitioning in the bumble bees *Bombus ternarius* and *B. pennsylvanicus*. Ecology **67**:133–138.

Kenrick, J., Bernhardt, P., Marginson, R., Bernesford, G., Knox, R.B., Baker, I., and Baker H.G. 1987. Pollination-related characteristics in the mimisoid legume *Acacia terminalis* (Leguminosae). Plant Syst. Evol. **157**:49–62.

Kevan, P.G., and Lane, M.A. 1985. Flower petal microtexture is a tactile cue for bees. Proc. Nat. Acad. Sci. USA **82**:4750–4752.

Laverty, T.M. 1980. The flower-visiting behaviour of bumble bees: Floral complexity and learning. Can. J. Zool. **58**:1324–1334.

Laverty, T.M. 1985. On the ecological significance of floral complexity and its effect on the foraging behavior of bumble bees. Ph.D. thesis, University of Toronto, Toronto, Ontario, Canada.

Laverty, T.M., and Plowright, R.C. 1988. Flower handling by bumble bees: A comparison of specialists and generalists. Anim. Behav. **36**:733.

Lewis, A.C. 1986. Memory constraints and flower choice in *Pieris rapae*. Science **232**:863–865.

Lewis, A.C. 1989. Flower visit consistency in *Pieris rapae*, the cabbage butterfly. J. Anim. Ecol. **58**:1–13.

Lewis, A.C., and Lipani, G.A. 1990. Learning and flower use in butterflies: Hypotheses from honey bees. In E.A. Bernays (ed.), Insect–Plant Interactions, Vol. II. CRC Press, Boca Raton, FL, pp. 95–110.

Little, J.R. 1983. A review of floral food deception mimicries with comments on floral mutualism. In C. Jones and R.J. Little (eds.), Handbook of Experimental Pollination Biology. Van Nostrand Reinhold Company, New York, pp. 294–309.

Macior, L.W. 1974. Pollination ecology of the front range of the Colorado Rocky Mountains. Melanderia **15**:1–59.

Menzel, R. 1985. Learning in honey bees in an ecological and behavioral context. In B. Holldobler and M. Lindauer (eds.), Experimental Behavioral Ecology and Sociobiology. Sinauer, Sunderland, MA, pp. 55–74.

Menzel, R., Erber, J., and Mashur, J. 1974. Learning and memory in the honey bee. In L.B. Browne (ed.), Experimental Analysis of Insect Behavior. Springer-Verlag, New York, pp. 195–217.

Müller, H. 1883. The fertilisation of flowers. Macmillan, London.

Murawski, D.A., and Gilbert, L.E. 1986. Pollen flow in *Psiguria warscewiczii:* A comparison of *Heliconius* butterflies and hummingbirds. Oecologia **68**:161–167.

Murcia, C. 1990. Effect of floral morphology and temperature on pollen receipt and removal in *Ipomoea trichocarpa*. Ecology **71**:1098–1109.

Newell, A., and Rosenblum, P.S. 1981. In J.R. Anderson (ed.), Cognitive Skills and Their Acquisition. Lawrence Erlbaum Associates, Hillsdale, NJ, pp. 1–56.

Nilsson, L.A. 1988. The evolution of flowers with deep corolla tubes. Nature **334**:147–149.

Nilsson, L.A., Jonsson, L., Ralison, L., and Randrianjohany, E. 1987. Angraecoid orchids and hawkmoths in central Madagascar: Specialized pollination systems and generalist foragers: Biotropica **19**:310–318.

Papaj, D.R. 1986. Interpopulation differences in host preference and the evolution of learning in the butterfly, *Battus philenor*. Evolution **40**:518–530.

Papaj, D.R. and Prokopy, R.J. 1989. Ecological and evolutionary aspects of learning in phytophagous insects. Annu. Rev. Entomol. **34**:315–350.

Rathcke, B. 1988. Interactions for pollination among coflowering shrubs. Ecology **69**:446–457.

Rausher, M.D. 1992. Natural selection and the evolution of plant-insect interactions. In B.D. Roitberg and M.B. Isman (eds.), Evolutionary Perspectives in Insect Chemical Ecology. Chapman and Hall, New York, pp. 20–38.

Raven, P.H. 1989. Onagraceae as a model of plant evolution. In L.D. Gottlieb and S.K. Jain (eds.), Plant Evolutionary Biology. Chapman and Hall, New York, pp. 85–108.

Real, L., Ellner, S., and Harder, L.D. 1990. Short-term energy maximization and risk-aversion in bumble bees: A reply to Possingham et al. Ecology **71**:1625–1628.

Real, L., Otte, J., and Silverfine, E. 1983. On the tradeoff between the mean and variance in foraging: An experimental analysis of bumble bees. Ecology **63**:1617–1623.

Richards, A.J. 1986. Plant Breeding Systems. George Allen & Unwin Ltd., London, England.

Roubik, D.W. 1989. Ecology and Natural History of Tropical Bees. Cambridge University Press, New York.

Roubik, D.W., and Buchmann, S.L. 1984. Nectar selection by *Melipona* and *Apis mellifera* (Hymenoptera: Apidae) and the ecology of nectar intake by bee colonies in a tropical forest. Oecologia **61**:1–10.

Schemske, D.W. 1983. Limits to specialization and coevolution in plant–animal mutualisms. In M.H. Nitecki (ed.), Coevolution. University of Chicago Press, Chicago, pp. 67–110.

Schmid-Hempl, P., Kacelnik, A., and Houston, A.J. 1985. Honey bees maximize efficiency by not filling their crops. Behav. Ecol. Sociobiol. **17**:61.

Shoen, D.J., 1983. Relative fitness of selfed and outcrossed progeny in *Gilia achilleifolia* (Polemoniaceae). Evolution 37:292–301.

Staddon, J.E.R. 1983. Adaptive Behavior and Learning. Cambridge University Press, New York.

Stanton, M.L. 1984. Searching in a patchy environment: foodplant selection by *Colias eriphyle* butterflies. Ecology **63**:839–853.

Stebbins, G.L. 1988. An overview of evolutionary biology. In L.D. Gottlieb and S.K. Jain (eds.), Plant Evolutionary Biology. Chapman and Hall, New York, pp. 1–20.

Waddington, K.D. 1983a. Floral-visitation-sequences by bees: Models and experiments. In C.E. Jones and R. Little (eds.), Handbook of Experimental Pollination Biology. Scientific & Academic Editors, New York, pp. 461–473.

Waddington, K.D. 1983b. Foraging behavior of pollinators. in L. Real (ed.), Pollination Biology. Academic Press, Orlando, FL, pp. 213–239.

Waddington, K.D. 1987. Perception of foraging costs and intakes and foraging decisions. In R. Menzel and A. Mercer (eds.), Neurobiology and Behavior of Honey Bees. Springer-Verlag, New York, pp. 66–83.

Waser, N.M. 1983. The adaptive nature of floral traits: Ideas and evidence. In L. Real (ed.), Pollination Biology. Academic Press, Orlando, FL, pp. 241–285.

Waser, N.M. 1986. Flower constancy: Definition, cause, and measurement. Am. Nat. **127**:593–603.

Waser, N.M., and Price. M.V. 1983. Optimal and actual outcrossing in plants, and the nature of plant-pollinator interaction. In C. Jones and R.J. Little, Handbook of Experimental Pollination Biology. Van Nostrand Reinhold, New York, pp. 341–359.

10

Automatic Behavior and the Evolution of Instinct: Lessons From Learning in Parasitoids

Daniel R. Papaj

> By instinct is meant the fixed tendencies displayed by animals in their actions; and many people have held that these tendencies are the produce of a reasoned choice, and therefore the fruit of experience.
>
> —Lamarck (1809), p. 350

The Evolution of Instinct—an Historical Perspective

Among contemporary biologists, learning is often presumed to be an evolutionarily derived trait (Mayr, 1974; Dethier, 1978; Shepherd, 1983; but see Tierney, 1986). Yet, as the opening quote from Lamarck's *Zoological Philosophy* illustrates, early naturalists believed just the opposite. Lamarckians contended that instincts were derived from learned behavior. Their perspective was based on two basic observations made of learned and instinctive behavior. First, behavior regarded as instinctive was observed to be "automatic"; i.e., in a given individual, an instinctive behavior varied little in form from one time to the next. Second, behavior which was learned was observed to be rather variable at first but became increasingly automatic as the animal gained experience. It was commonly remarked that learned behavior eventually became as automatic as behavior termed instinctive.

These dual observations led early naturalists to propose that instincts evolved from behavior that was originally learned. Stephen Jay Gould (1977) summarized their argument succinctly: "Instincts are the unconscious remembrance of things learned so strongly, impressed so indelibly into memory, that the germ cells themselves are affected and pass the trait to future generations." So convinced were Lamarckians of the correctness of their views that memory (or some unknown but exactly analogous process) was invoked as a possible mechanism for the inheritance of physiological and morphological as well as behavioral characters. This proposed mechanism was embraced by morphologists in particular because it dove-tailed neatly with the concept that ontogeny recapitulated phylogeny: an animal's ontogeny was a kind of transcript or remembrance of its phylo-genetic history. It is not an exaggeration to say that one of the most controversial principles in the history of biology turned in part on anecdotal observations made of automatism in behavior.

The idea that instincts evolved from behavior that was originally learned did not die with the appearance of Darwin's theory of natural selection and the fall from grace of the Lamarckian mode of evolution (cf. Spalding, 1872). Indeed, Darwin himself did not dismiss the possibility (Darwin, 1868) and Romanes argued as late as 1884 that, while most instincts (termed primary instincts) arose de novo through natural selection, some secondary instincts evolved through an essentially Lamarckian mechanism of "lapsing intelligence." With the advent of Mendelian genetics and the acceptance of Weismann's views on the germ tissue's invulnerability to environmental influences, Lamarckian explanations were largely abandoned.

Arguments about the evolution of instincts through natural selection emphasized their spontaneous origin. There were exceptions. The French entomologist Bouvier (in a 1908 volume with the endearing title, *The Psychic Lives of Insects*) attempted to reconcile the notion that instincts were derived from learned behavior with Darwinian principles of adaptive evolution. Bouvier's arguments were not convincing. More successful were Baldwin (1896), Morgan (1896), and Osborn (1896), who collectively con-structed a theory of "organic selection" in which phenotypic "accommo-dation" (= "plasticity" in the modern vernacular) facilitated the evolution of congenitally expressed traits through natural selection. In their view, learning permitted a population to persist in a new environment long enough for latent genetic variation to be acted upon by selection and for congenital responses (i.e., instincts) to arise.

These efforts notwithstanding, the original issue of such importance to Lamarck and Darwin, that of the evolution of instinct, gradually faded from prominence [see Klopfer and Hailman (1967) and Wcislo (1989) for more thorough histories of the subject]. With it, observations of early naturalists on patterns of automatism in learned and instinctive behavior were virtually forgotten. In this chapter, I critically evaluate the obser-vations made by the early naturalists. In doing so, I review work by myself and Louise Vet on parasitoid learning which demonstrates how experience affects the automatism or, to use modern parlance (Boake, 1989), the consistency of insect behavior. I then use a simple model to argue that behavioral consistency can be of functional significance to an animal. Fi-nally I appeal to another model to illustrate how learning to be consistent (or any learning of a functional sort) can influence the evolution of instinc-tive behavior.

What is Instinct?

Discourse on the evolution of instinct suffers from problems in semantics. Over time, the terms "instinct" and "innateness" acquired multiple and often conflicting meanings (Bateson, 1983) and, at least partly for this reason, they are little used today. The confusion has deep historical roots. Like Lamarck, many early naturalists regarded instincts as fixed behavior, i.e., behavior expressed in complete form the first time it is expressed and relatively insensitive to experience. However, that meaning was not universally accepted. Another meaning for the term may date back as far as the early 17th century. In Shakespeare's play *Coriolanus* is found the remark, "I'll never be such a gosling to obey instinct, but stand as if a man were author of himself and knew no other kin." Other interpretations are possible, but Shakespeare may have been inferring from resemblances in behavior among relatives the existence of some factor which prevented an animal from exercising control over its own behavior (i.e., prevented the gosling from being author of itself). Instinctive behavior was predetermined in some way, a predetermination manifested in resemblances among kin.

Similarities in genetic constitution are now known to account, at least in part, for consistent resemblances among individuals. Should instinct be defined as behavior for which variation among individuals within a population is primarily genetically based? Such a characterization would clash with one of instinct as fixed behavior as there is no a priori reason to think that fixed behavior should be any more heritable than behavior influenced by the environment (e.g., learned behavior). Indeed, it is entirely possible that natural selection has minimized all sources of variation, both environmental and genetic, in a behavior that is observed to be fixed. In that case, fixed behavior would not be heritable at all and could not be altered further by the action of natural selection. This point notwithstanding, portraying instincts as highly heritable behavior seems to be exactly what some authors intended when they referred to "genetically-determined," "genetically-programmed," or "genetically-controlled" behavior (Bateson, 1983).

In this chapter, an instinct will be defined as a behavior which is expressed in complete form the very first time performed and which is relatively insensitive to experience of a specified kind. Under this definition, instinctive behavior need be fixed only with respect to the specified experience. It is not completely fixed if, by that, is meant "completely insensitive to any environmental condition." No behavior is completely fixed in that sense. Of course, neither are many behaviors regarded as instincts likely to be absolutely insensitive to a specific kind of experience; attempting to pigeonhole all behavior as learned or instinctive is ultimately an exercise in futility. Behavior which is expressed congenitally is often modified by experience (Hailman, 1967; Smith and Menzel, 1989a) and learning itself is frequently biased in some way (Marler and Peters, 1981; Menzel, 1985; Smith and Menzel, 1989b; see Gould, this volume; see Menzel et al., this volume). Few behavioral programs (*sensu* Mayr, 1974) are completely "closed" or completely "open" to effects of experience and it is with good reason that the concept of a strict learning/instinct dichotomy was condemned (Hinde, 1970; Lehrman, 1970; Bateson, 1983). At the same time, few biologists deny that behavior ranges from that which is not influenced by specific experiences to that which is wholly dependent on them (Bateson, 1983; Huntingford, 1984). What this chapter aims to do is examine how behavior evolves along the continuum between those extremes.

Evaluating the Observations of Early Naturalists

Observation 1: Instinctive behavior is more consistent than learned behavior

The notion that instinctive behavior is associated with lower moment-to-moment variability than learned behavior is strongly implied in the concept of the *fixed-action pattern* (FAP). FAPs are behavioral sequences which are released by simple yet highly specific stimuli (termed releasers). The waggle dance of honey bees, brood provisioning and nest construction in wasps and aspects of oviposition behavior in many insects (cf. Mowry et al., 1989) are examples of FAPs. FAPs are typically characterized as both instinctive and automatic in form: (1) they are expressed in complete form the first time performed; (2) they are relatively insensitive to experience; and (3) they are predictable and stereotyped (i.e., consistent) (Dewsbury, 1978).

Barlow (1977) challenged the concept of a fixed-action pattern, arguing that behavior considered to be extremely invariant and predictable under one set of conditions was actually quite variable over a range of conditions. He scrutinized a number of traditional examples of FAPs in a variety of animals (none of them insects) and noted that any behavior, no matter how stereotyped, was associated with some variance greater than zero. To characterize the degree of stereotypy of a behavior more precisely than had been done previously, Barlow calculated its coefficient of variation (C.V.). Simply the standard deviation expressed as a percentage of the mean with an adjustment for sample size, the C.V. is a measure of variation that is statistically independent of the mean (Sokal and Rohlf, 1981). Barlow found that the C.V. of supposedly fixed behavior was often significantly

greater than zero.[1] Such results prompted Barlow to recommend that behavior formerly termed fixed-action patterns would better be described as
modal-action patterns, or MAPs. In Barlow's terminology, a MAP would
be characterized by both a modal value and a coefficient of variation.

Smith and Menzel (1989a) found support for Barlow's position in the
feeding behavior of honey bees. An oft-cited example of a FAP in insects,
the feeding motor pattern in honey bees varied considerably according to
the kind of stimuli used to release it. Smith and Menzel also noted that
different portions of the honey bees' feeding motor pattern differed in
degree of stereotypy. Some components (e.g., the duration of a licking
movement with the glossa) were more or less completely fixed, while others
(e.g., time between proboscis extension and retraction) were associated
with considerable variability over the same range of stimulus conditions.
This kind of observation has been made with respect to egg-laying behavior
in insects, too. Some components of egg deposition tend to be extremely
consistent (e.g., the onion fly's egg-deposition motor program; Mowry et
al., 1989), while others are notoriously variable from one time to the next
(e.g., the onion fly's finding and examination of an oviposition site or even
the tendency for egg deposition to be initiated; Harris and Miller, 1991).

Even if examples of extreme stereotypy in instinctive behavior were
found that survived all of these qualifications (e.g., aspects of courtship in
insects; Boake, 1989), they would not by themselves constitute evidence
that instinctive behavior is significantly more consistent than learned behavior. What is needed are studies in which moment-to-moment variability
in a species that learns a particular behavior is compared with variability
in a closely related species that expresses virtually the same behavior without benefit of experience. The opinion of the early naturalists would be
supported if the behavior of naive individuals of the former species was
more variable than that of individuals of the latter species. Such comparisons have not been made to date.

Despite the lack of crucial comparative evidence, most behavioral biologists probably agree that certain patterns of behavior are associated
with an extreme degree of stereotypy and that such patterns, whether
termed FAPs or MAPs, tend to be expressed even in the absence of experience. In that respect, our perspective is not unlike that of the early
naturalists: instinctive behavior is automatic behavior. Unfortunately, our

---

[1]It is important to note that Barlow generally assessed variability in behavior *among* individuals, whereas here we are mainly concerned with variability in behavior *within* individuals. As discussed below, within-individual variability can generate among-individual variability. However, the two kinds of variability are not equivalent. They have, for example,
very different implications for a population's response to selection on behavior. See Machlis
et al. (1985) and Boake (1989) for a more complete discussion of this issue.

perspective is little better supported by hard data today than it was over a century ago.

Observation 2: Behavior Becomes More Consistent With Experience

Does learning reduce moment-to-moment variability in behavior? Once more, little hard evidence can be brought to bear on the pattern observed by early naturalists. However, older work in vertebrate ethology is certainly suggestive. Variability in food-begging by gull chicks, for instance, declines with experience (Hailman, 1967). Squirrels opening nuts become more regular in their movements as time goes on (Eibl-Eibesfeldt, 1951). A cursory review of the literature turned up just a few anecdotal examples in insects (Bouvier, 1908; McDougall, 1923). However, recent work by Louise Vet and myself indicates that experience at least sometimes reduces moment-to-moment variability in behavior in insects. The insect of interest is the parasitoid, *Leptopilina heterotoma*, and the behavior of interest is upwind movement in a plume of odor from a microhabitat containing the parasitoid's host.

Leptopilina heterotoma is a generalist parasitoid (Order Hymenoptera; Family Eucoilidae) which attacks larvae of a variety of *Drosophila* species (family Drosophilidae; order Diptera) that inhabit fermenting fruits and sap fluxes as well as decaying mushrooms and plant material. Oviposition experience with hosts infesting a particular microhabitat (such as a mixture of fermenting apple-yeast or a mass of decaying mushroom) enhances responses of female *L. heterotoma* to the odor of that microhabitat in olfactometer (Vet, 1988, and references within), greenhouse (Papaj and Vet, unpublished), and field assays (Papaj and Vet, 1990). Contact with uninfested microhabitat is not itself sufficient to learn the odor of the microhabitat, suggesting that learning is associative (Vet, 1988; Vet and Groenewold, 1990).

In an effort to quantify precisely how learning affects movement in an odor plume, Vet and Papaj (1992) employed a Kramer-type locomotion compensator or "servosphere" positioned at the outlet of a wind tunnel. This device (which is conceptually nothing more than a two-dimensional treadmill; Fig. 10.1) permitted us to make very precise records of the walking tracks of parasitoids in odor plumes (Fig. 10.2). In an experiment described in detail in Vet and Papaj (1992), some females were given a brief 2-hour opportunity to lay eggs in *D. melanogaster* larvae in either apple-yeast or mushroom medium. Other females (so-called naïve females) were given no opportunity to lay eggs at all. Oviposition experience had a number of effects on walking movement in odor from different host "microhabitats." Females walked faster and straighter, made narrower turns, and spent more time in upwind movement (i.e., in movement toward

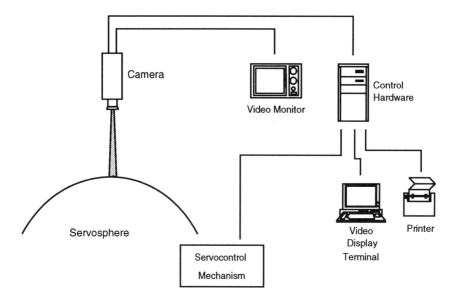

Figure 10.1. Diagram of main components of locomotion compensator apparatus. Wasp bearing reflective tag is placed on servosphere (50-cm diameter) beneath camera which projects beam of visible light onto the wasp. Light reflected from the tag permits position of wasp to be monitored by camera. As the wasp walks, two motors (servocontrol mechanism) rotate the sphere in the opposite direction and at the same speed, thus keeping the wasp in a fixed position. Pulse generators in combination with a computer (control hardware) record walking speed and direction. The servosphere is positioned at the outlet of a wind tunnel, which is not shown.

the source) in odor from a particular microhabitat with which they had experience than in odor from another microhabitat. Naive females, in contrast to experienced ones, showed little difference in responses to alternative odors (Vet and Papaj, 1992).

In addition to affecting mean walking speed, mean straightness, and mean turning angle, did experience affect the moment-to-moment variability around those means? To answer this question, we arbitrarily divided each 120-second observation period into twelve 10-second segments. We then calculated the coefficient of variation (C.V.; see above) among segments for various movement parameters. Oviposition experience with a substrate had a dramatic effect on variability in walking movement by individual females in odor from that substrate. Females experienced with a particular substrate (either apple-yeast or mushroom) showed signifi-

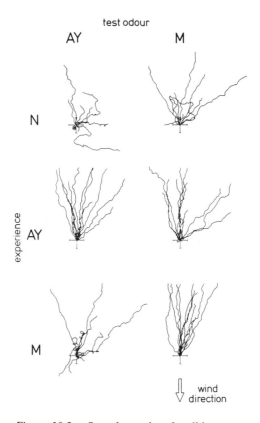

Figure 10.2. Sample tracks of walking parasitoids. Insects were given oviposition experience on either apple-yeast (AY) or mushroom (M) medium or no experience at all (N). They were then tested in presence of odor from either apple-yeast (AY) or mushroom (M) media. Figure taken from Vet and Papaj (1992).

cantly less variability among path segments in walking speed in a plume of odor from that substrate than did naïve females (Fig 10.3A).

A similar pattern was observed with respect to turning angles in apple-yeast odor: females experienced with apple-yeast substrate showed significantly less variability among path segments in turning angle than did naive females (Fig. 10.3B). With respect to mushroom test odor, however, the pattern was weak: females experienced with mushroom substrate showed slightly less variability in turning angle than did naïve females, but the difference was not significant. Patterns in path straightness (measured as

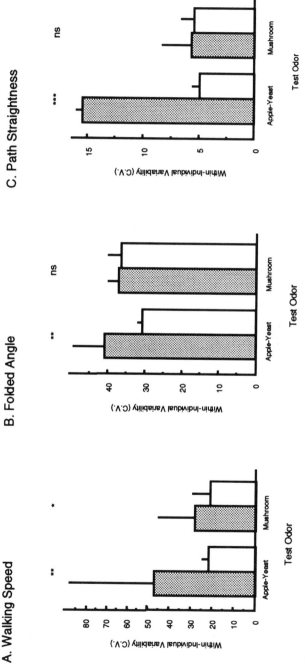

Figure 10.3. Pattern of variability among segments of an individual *Leptopilina heterotoma*'s walking path. Gray bars indicate naive females; empty bars indicate females given oviposition experience with a particular substrate and tested in the odor of the *same* substrate. Also shown are standard error bars. Sample size varies from 16 to 23. A. Walking speed. B. Folded angle. C. Path straightness. Asterisks indicate significant differences between experienced and naive individuals according to a Mann-Whitney U-test. A lack of significant differences is designated "ns."

the net distance moved divided by the total length of the walking path) resembled those in turning angle (Fig. 10.3C). Females experienced with apple-yeast substrate showed significantly less variability among path segments in path straightness in apple-yeast odor than naïve females. With respect to mushroom test odor, however, that pattern was once again weak: females experienced with mushroom substrate showed slightly less variability in path straightness than did naïve females, but the difference was not significant.

An important assumption in the interpretation of these results is that any differences between experienced and naïve individuals were due to differences in learning and not to differences in egg load, which can also influence parasitoid foraging behavior (see Rosenheim, this volume). That this assumption may be valid here is suggested by data for females experienced with one odor but tested in the other odor. If the difference in variability between experienced and naïve females was due only to egg load, variability in experienced females tested in a novel odor should also differ from that of naïve females tested in the same odor. In fact, variability for experienced females tested in a novel odor was never significantly different from that of naïve females tested in that odor, regardless of any differences in egg load (Papaj and Vet, unpublished data). The observed difference in consistency between experienced and naïve females thus seems to be a difference due to learning.

Finally, it is worth noting that everywhere we found a difference in variability within individual tracks (Fig. 10.3), we found a corresponding difference in variability among individual tracks (Fig. 10.4). Movement parameters for a cohort experienced with a particular substrate (either apple-yeast or mushroom) were generally less variable in odor from that substrate than was a cohort of naïve individuals. This result is noteworthy because, while there is little evidence for changes in individual consistency as a consequence of experience, there is growing evidence, for a number of parasitoid species, that cohorts of experienced individuals are less variable in behavior than cohorts of naïve individuals (Vet et al., 1990). We might anticipate that patterns in variability *among* individuals in these species will be associated with patterns in variability *within* individuals, as they are in *L. heterotoma*.

Individual Consistency and Reproductive Success

With experience, wasps seem to become more consistent in their behavior or, to use the terminology of the early naturalists, more automatic in their movements. How might consistency profit the parasitoid, or insects in general? We formalized one advantage of consistent behavior in a simple

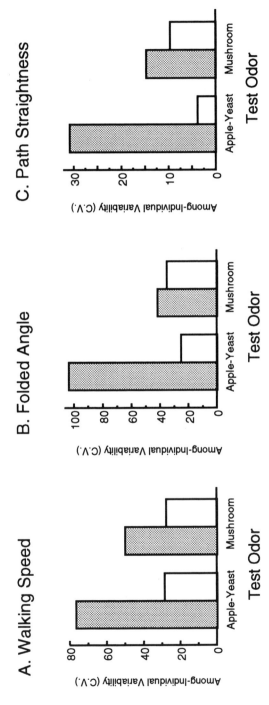

Figure 10.4. Pattern of variability among walking paths of individual *Leptopilina heterotoma*. Gray bars indicate naïve females; empty bars indicate females given oviposition experience with a particular substrate and tested in the odor of that substrate. A. Walking speed. B. Folded angle. C. Path straightness.

simulation model of upwind movement by parasitoids in odor plumes. The model assumes that a parasitoid orients to an odor source simply by alternating right and left turns over consecutive time intervals (Fig. 10.5). If successive turns were always of the same magnitude (i.e., if an error in heading during one step was always exactly compensated by counterturning in the next step) and if movement between turns was constant in velocity, the insect would maintain a heading directly upwind and would reach the source. The smaller the angles, the straighter would be the parasitoid's path and the sooner it would reach the odor source. In fact, simulated insects neither turn at constant angle nor walk at constant speed. Rather, the model is *stochastic* in the sense that the angle at which the insect turns

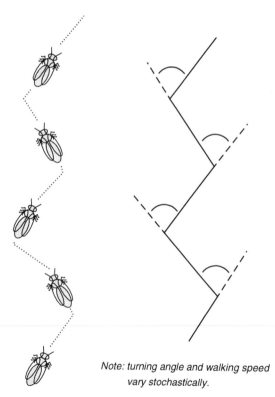

Note: turning angle and walking speed
vary stochastically.

Figure 10.5. Diagram of counterturning assumed in stochastic simulation of wasp movement. Consistency was varied by changing variability associated with turning angle and move length (walking speed).

as well as its instantaneous speed varies over time in an unpredictable manner.

A walking path is simulated as follows. An insect starts out with a heading directly toward the source. In the following second, its heading deviates from straight ahead by some variable amount. In the next second, it turns to compensate for that error. Were compensation perfect, the insect would always head directly at the source. In fact, compensatory turns are also associated with some error. Between turns, the insect walks at some speed which is constant within the 1-second interval but variable across intervals. Both angle and speed are determined by drawing randomly from normal distributions approximating actual data (cf. Vet and Papaj, 1992). In successive simulation runs, mean turning angle and mean walking speed among path segments were held constant at 15° and 4.5 mm/second, respectively (cf. Vet and Papaj, 1992), while the standard deviation of values around those means was doubled progressively.

Mean walking speed and turning angle for a cohort of simulated insects hovered around 4.5 mm/second and 15° respectively, regardless of the variance around those values. Mean path straightness, by contrast, declined exponentially as variability around mean speed and angle increased (Fig. 10.6A). This effect is not due to changes in mean speed and angle; means were more or less constant across simulation runs. Rather, the effect is due primarily to variability in turning angle: because successive right and left turns are progressively less likely to be of similar magnitude, the insect is less and less able to maintain a constant heading and its path becomes more and more tortuous.

One might reasonably expect that differences in path straightness will be associated with differences in how effectively individuals find an odor source. One measure of effectiveness derived easily from these simulations is the net distance moved by the insect in the direction of the source. While probably an incomplete measure of foraging success, an insect that has moved further toward an occupied microhabitat over a certain period of time must often be a more effective forager. Not surprisingly, mean net distance moved toward the source decreased steadily as variability in walking speed and turning angle among path segments increased (Fig. 10.6B). On this basis alone, a reduction in moment-to-moment variability in orientation toward a microhabitat should yield improvements in the insect's foraging success. In addition, a reduction in mean path straightness corresponds to a reduction in the tendency for the insect to be heading directly toward the source (not shown). Errant headings could conceivably cause the insect to lose the odor plume and, for this reason, fail to find the source.

In short, *L. heterotoma* females appear to be learning to counterturn in an odor associated with an oviposition reward. By increasing the rate at

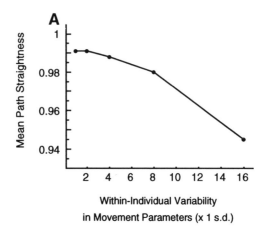

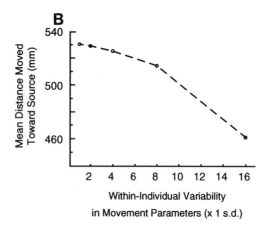

Figure 10.6. Effect of simulated within-individual variability in walking speed and turning angle on (A) mean path straightness and (B) net distance moved toward the plume.

which the insect moves toward the source and the probability that the insect is heading directly toward the source, adoption of consistent behavior ought to facilitate discovery of a host microhabitat on at least two counts. First, it should increase the tendency to find a microhabitat given detection of its odor. Second, it should decrease the time required to arrive at the microhabitat once odor is detected (i.e., travel time). It is worth noting that each of these effects was observed in field releases of *L. heterotoma* (Papaj and Vet, 1990).

Learning and the Evolution of Instinct

The advantage of consistent behavior posed here should be applicable to other kinds of foraging behavior in other insects, including flower-foraging by bees, host-plant–finding by butterflies, and the like. In fact, any behavior pattern which involves compensatory motor movements (which is to say, almost all behavior) might benefit from consistency in the manner outlined here. Such patterns include not only walking, swimming, or flight movements, but also the movement of groups of appendages (as when a preying mantis manipulates its prey) or the movement of a single-jointed appendage (as when an insect's proboscis is extended to drink). The same arguments might also apply to the coordinated actions of pairs of individuals (as in elaborate courtship sequences in butterflies and other insects). These arguments may even be applicable to coordinated movements among members of large groups of insects (as in ant or termite colonies).

Whereas early naturalists inferred from patterns in automatism that instincts evolved from behavior that was originally learned, nothing in the arguments presented above suggests that instincts are less likely to arise de novo than through the acquisition of habit. At best, patterns in automatism constitute a metaphor for the evolution of instinct from learned behavior. This is not to say that instincts do not evolve through the acquisition of habit, only that patterns of automatism in learned and instinctive behavior do not bear directly on the issue. What bears directly on the issue is the tendency for learning to be consistent (and learning in general) to be functional. Insects, like other animals, learn to do the right thing. They learn to orient toward stimuli that are rewarding and away from stimuli that are punishing. Learning, especially associative learning, tends to adapt the insect to its current environment.

The adapting properties of learning can be used to frame a plausible Darwinian scenario in which instincts evolve through the acquisition of habit. While the following argument is set forth in terms of a parasitoid foraging for host microhabitats, it should apply equally well to other insects and other types of behavior. It is worth noting at the outset that behavioral consistency is not an explicit element of this model.

Suppose a population of parasitoids initially exploits a host species that occupies a number of microhabitats, each of which is rather unpredictable in abundance over time or space and/or degree of infestation. Selection has presumably favored females which have a congenital predisposition (albeit a weak one) to respond to each potential microhabitat odor as well as an ability to adopt (through learning) strong responses to odor or odors that yield the largest number of reproducing progeny. Suppose the environment suddenly changes such that one and only one host microhabitat is available and potentially occupied by hosts. Assuming learning is costly in predictable environments, there will be immediate selection for the immediate expression of fully functional behavior, i.e., selection for instinctive behavior of a functional sort. The response to selection may conceivably take either of two forms. First, selection might favor a stronger congenital response to the odor of the only available microhabitat, i.e., a stronger predisposition to orient toward the odor even in the absence of oviposition experience (Fig. 10.7A). This is obvious. The second kind of response to selection may be less obvious: selection might favor individuals which learn faster to exhibit a strong response (Fig. 10.7B).

Here then is a paradox of sorts: assuming genetic variation in each trait (learning ability and congenital response), selection in predictable environments should simultaneously favor both congenital responses and faster

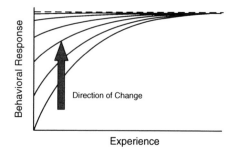

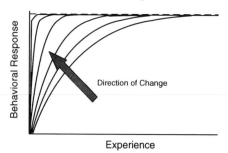

Figure 10.7. Alternative routes to evolution of instinct. A, Evolution of congenital response; B, evolution of faster learning. In each figure, dashed line indicates value of optimal behavioral response in predictable environment.

learning. An instinct could be generated either when insects become so congenitally predisposed to respond to the odor that responses cannot be improved through learning (as seems intuitively reasonable; Fig. 10.7A) or when insects come to learn so quickly that only a single experience is required in order to express the appropriate behavior (Fig. 10.7B). In the latter case, one could imagine that a smaller and smaller "reward" becomes necessary to evoke the appropriate response until eventually no reward at all is required. A formerly acquired response to odor will have become genetically assimilated (*sensu* Waddington, 1953). This scenario is wholly consistent with the Lamarckian view that instincts represent the acquisition of habits and yet wholly consistent with the Darwinian model of evolutionary change through natural selection.[2]

These alternative responses to selection, however paradoxical, are not mutually exclusive. A parasitoid might conceivably evolve both congenital responses and faster learning. Might one or the other effect predominate? The answer from ongoing simulation studies is a tentative yes. The basic algorithm used in these simulations is presented in Figure 10.8 and is a modification of one constructed by Jaenike (Jaenike and Papaj, 1992; see Dukas and Real, 1991, for a similar formula). A behavioral response (R) has an instinctual component (denoted by I, the instinct coefficient) and a learned component. The learned component is a function of I, but also of a learning coefficient (L) and the number of experiences (N) that modify the response. This algorithm generates a learning curve (Fig. 10.8). All individuals with L greater than zero eventually adopt a response of value equal to one. Because the model assumes that this value is optimal in the sense of yielding highest future fitness, this learning is a sort of "adapting" learning. More precisely, it is an "optimizing" learning.

At the outset of a simulation run, either I or L or both varied genetically. Where either coefficient was genetically variable, heritability was always set at a value of 0.50 (i.e., half of the variation in either trait has a genetic basis). Population sizes were set at 1,000. Once the simulation begins, selection acts generation by generation on mean response and molds whatever genetic variability in instinct and learning coefficients is available. The fitness function on which selection is based is an inverted parabola truncated at zero (Fig. 10.9; see also Stephens, this volume). Individuals

---

[2]The notion that instincts might evolve through assimilation of learned behavior was addressed to varying degrees by, among others, Haldane and Spurway (1954, 1956), Spurway (1955), Ewer (1956), and Tierney (1986). Not all of these authors were favorable toward the idea. Haldane and Spurway (1956) commented that, "[genetic assimilation of learned behavior] would require special selection pressures; for example, adaptedness would have to be made to confer more fitness than adaptability during postnatal life." This would not seem to be a problem under the simulation conditions as any learning, no matter how rapid, has some cost in terms of the time required to achieve an optimal response.

$$R = I + \left[(1 - I)\left(1 - e^{-NL}\right) \right]$$

R = Behavioral Response

I = Instinct Coefficient

L = Learning Coefficient

N = Number of Trials

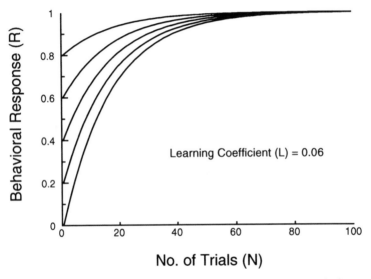

Figure 10.8. Algorithm assumed in simulation model of evolution of instinct. Also shown are learning curves generated by algorithm, for various values of *I* with *L* = 0.06.

contribute progeny in direct proportion to the fitness value associated with their particular behavioral response profile.

While genetic in nature, the model flauts "genetic reality" in many respects. For example, individuals reproduce asexually, yet genetic variation in particular traits (if any) is maintained at a constant level even in the face of strong selection. This amounts to supposing that mutation compensates perfectly each generation for the loss of genetic variation due to selection, although this process is not considered explicitly in the model.

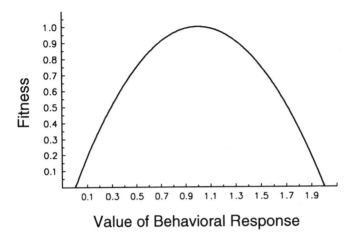

Value of Behavioral Response

Figure 10.9. Fitness function assumed in simulations of instinct evolution: Fitness $= F_{max} - [c(1-R)^2]$ where $F_{max} =$ maximum fitness $= 1$, c is a flatness parameter also equal to 1, and R is the behavioral response. Individuals in each generation contribute progeny to the next generation in direct proportion to their overall fitness.

Despite its obvious limitations, the model yields some interesting results that should hold up under more realistic genetic conditions.

In the first set of simulations, I asked whether learning influenced the evolution of a congenital response to an odor in a predictable environment where such a response would be favored (cf. Fig. 10.7A). In this simulation, only the congenital response was genetically variable. The ability to learn, if any, was identical for all individuals. The results are shown in Figure 10.10A. In the absence of learning, congenital responses (expressed in terms of I, the instinct coefficient) evolve in a negatively accelerating fashion until individuals express the optimal response congenitally. In the presence of learning, congenital responses evolve more slowly. Exactly how much more slowly they evolve depends on how fast individuals learn. Fast learning (denoted by high values of L, the learning coefficient) slows the rate of evolution of instinctive behavior more than slow learning. When learning is relatively rapid ($L = .0.5$), evolution of a congenital response (i.e., the de novo evolution of instinct) virtually grinds to a halt. Though rapid (Figure 10.10B), learning at this point is far from single-trial learning. In fact, rates of learning and this magnitude are considerably lower than those reported for odor learning in honey bees (Gould, this volume; Menzel

A. Evolution of instincts when animals learn

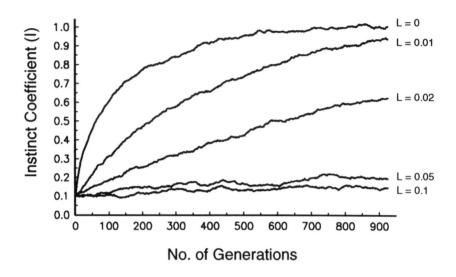

B. Learning curves assumed in simulations

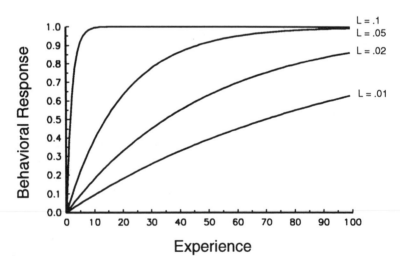

Figure 10.10. Results of simulation in which only *I* is genetically variable. *L* is varied in successive simulation runs between 0 and 0.1. A, Evolutionary trajectory of congenital response, denoted by instinct coefficient, *I*; B, corresponding learning curves associated with values of *L* assumed in simulation.

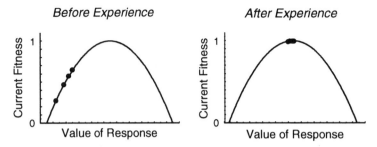

Figure 10.11. Diagram showing how optimizing learning moves individuals to optimal responses in terms of current (and future) fitness. Individuals are represented as points on a current fitness function. Experience causes individuals to converge on the optimal response, regardless of initial congenital differences.

et al., this volume) and consistent with those reported for bumble bees (Dukas and Real, 1991).

The tendency for learning to suppress the evolution of instinct is explained easily in intuitive terms. Optimizing learning like that assumed here permits any individual regardless of genetic constitution to eventually adopt the optimal behavior in the current environment (in this case, an optimal response to the microhabitat odor). Such learning makes all individuals more similar in overall fitness than if they did not learn (Fig. 10.11). The faster that individuals learn, the smaller are the effects of congenital differences on overall fitness. When individuals learn rapidly, even relatively large genetic differences in congenital response are associated with almost negligible differences in overall fitness. Under selection, evolutionary change is correspondingly negligible.[3] In this way, optimizing learning may preserve genetic variation in behavior for long periods of time even in the face of apparently strong selection. This may explain why we often find learned responses when we expect instinctive ones (e.g., in feeding or egg-laying behavior by specialist insects; Papaj, 1986; Papaj and Prokopy, 1989; but see Dukas and Real, this volume, for an alternative explanation).

The notion that learning might inhibit the evolution of instincts is seemingly at odds with the conclusion of Morgan (1896) and his contemporaries that learning facilitates it (see above). The crux of the conflict lies in assumptions about initial conditions. In my model, at least some genotypes in the population have nonzero fitness even in the absence of learning. In

---

[3]The same result is obtained in a model of genetic transmission of behavior put forth by Boyd and Richerson (1985). In their words (p. 121), "as the organism's skill at moving towards the optimum increases, selection for a good a priori guess about the environment decreases."

Morgan's (1896) formulation, individuals that do not learn (or do not learn well enough) have no fitness in the new environment. Genetic variation in congenital response in the population is absent (or latent), having been removed (or suppressed) under past selective regimes. If at least some individuals cannot learn the appropriate responses, the entire population goes extinct. Learning in the situation envisaged by Morgan permits the population time enough to generate (through mutation, recombination, and migration) individuals that would have nonzero fitness even in the absence of learning.

My next simulation was like the first in that individuals learned and there was genetic variation in congenital response. In addition, there was genetic variation in learning ability. Here I was essentially asking whether, in a predictable environment where an instinctive response to the odor of a host microhabitat was favored, such a response evolved de novo (i.e., through the evolution of a congenital predisposition to the odor) or through the acquisition of habit (i.e., through faster and faster learning). Results shown in Figure 10.12A indicate that both congenital responses and faster learning evolve rapidly at first. However, learning soon evolves to a rate at which evolution of the congenital response is essentially arrested. At this point, the mean congenital response in the population is rather weak. Learning, by contrast, is relatively rapid (Fig. 10.12B). Moreover, learning at this point is reminiscent of α-conditioning in which a preexisting response is heightened by experience (see Menzel et al., this volume, for a discussion of α-conditioning). Eventually, individuals would presumably learn so fast that virtually no experience would be needed to elicit the optimal behavioral response. Under the conditions of this simulation, instincts seem to evolve primarily (though not completely) as the early naturalists believed, i.e., through the acquisition of habit.

In this simulation, learned behavior was presumed to be an initial or ancestral condition, as supposed by early naturalists. One might argue that, in the absence of convincing phylogenetic evidence for that condition, the deck was unfairly stacked against the de novo evolution of instinct. In a final simulation, I therefore considered a population in which individuals originally had almost no learning ability and only a very weak congenital response, but for which there was genetic variation in both traits. As before, I asked whether instinctive behavior would evolve through the generation of congenital responses or faster learning. The results are shown in Figure 10.13. Once again, both congenital response (I) and learning ability (L) increase in value at first (Fig. 10.13A). Once again, learning quickly evolved to a rate at which evolution of the congenital response was essentially arrested. The congenital response at this point is far less than one, the value at which the individual is behaving wholly instinctively. By contrast, learning at this point is rather rapid (Fig. 10.13B) and increasing in strength.

A. Evolution of Instinct and Learning Coefficients

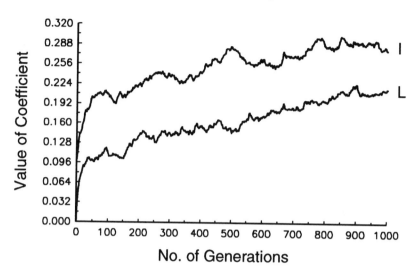

B. Evolution of Behavioral Response Function

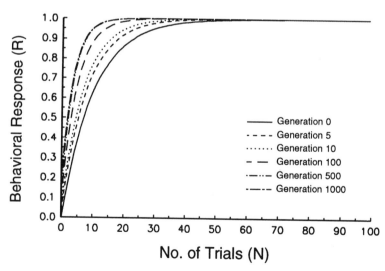

Figure 10.12. Results of simulation in which both *I* and *L* are genetically variable. *L* in generation 0 was equal to ca. 0.06. A. Evolutionary trajectory of mean values of *I* and *L*; B. change in mean behavioral response function, composed of congenital responses and learning curve. Selection causes only a slight change in mean congenital response, while causing a conspicuous change in the mean rate at which individuals learn.

A. Evolution of Learning and Instinct Coefficients

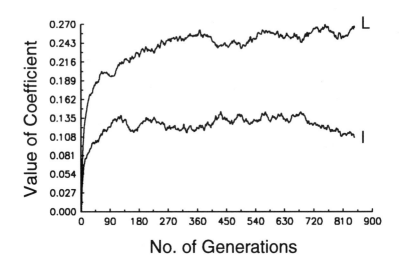

B. Evolution of Response Function

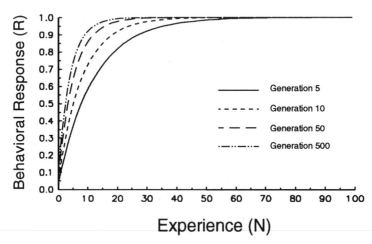

Figure 10.13. Results of second simulation in which both *I* and *L* are genetically variable. In this simulation *L* in generation 0 was vanishingly small and $I = 0.1$. A. Evolutionary trajectory of mean values of *I* and *L*; *B*. change in mean behavioral response function, composed of congenital responses and learning curve. Selection again causes only a slight change in mean congenital response, while causing a conspicuous change in the mean rate at which individuals learn.

As before, instincts are evolving primarily through the acquisition of habit. More precisely, instincts are evolving first through the formation of habits and then through their acquisition.

The reason that evolution of fast learning dominates that of congenital responses is not particularly profound: in my algorithm (Fig. 10.8), L has an exponential effect on the value of R while I has only an additive one. Thus, a change of a given magnitude in the value of L generally causes a much greater change in R than does a change of the same magnitude in I. Even though both coefficients have the same heritability, initial changes in L have larger impact on behavior and fitness than similar changes in I and so learning ability appears to evolve faster than congenital response. Whether this is biologically realistic—whether changes at loci affecting learning have larger effects on behavior than changes at loci affecting congenital responses—is anybody's guess. At this point, one can only note that the outcome of the model is sensitive to the algorithm used to specify learning (which is itself a subject of debate in the psychological literature, Shaw and Ailey, 1985). One should also note that nothing in the model requires that instinctive behavior be automatic or that learned behavior become more automatic with experience. If early naturalists were right that instincts evolve through the acquisition of learned behavior, they were right for the wrong reasons.

A Prediction

> Those insects which possess the most wonderful instincts are certainly the most intelligent.
>
> Darwin (1871, p. 37)

The model presented here lends itself to a prediction. If instincts commonly arise through the acquisition of habit, we might expect to see the most well-developed instincts in groups of insects that exhibit the most remarkable learning abilities. It would be in exactly those groups that learned behavior might be most readily assimilated into instinctive behavior. If instincts generally arise de novo, by contrast, we might expect exactly the opposite pattern: instincts might be best developed in groups where learning was not fast enough to suppress the evolution of congenital responses. As evidenced in the above quotation from *The Descent of Man*, Darwin himself noted a positive association between complexity of instincts and learning ability in insects. Darwin was almost certainly referring to members of the order Hymenoptera, for whom there also exists some limited evidence that behavioral plasticity is an ancestral trait (Evans, 1966; Wcislo, 1987). The tendency for members of the Hymenoptera, among all insect orders, to exhibit both the most remarkable learning abilities and the most remarkable fixed-action patterns (even in the same species in the case with the honey bee; see Gould, this volume) certainly invites speculation that the path favored by the early naturalists has been taken in at least one group of insects.

Closing Remarks

The Hymenoptera would seem to be ideal candidates for phylogenetic studies (cf. Rosenheim, this volume) that might establish the degree to which learned behavior is ancestral to fixed behavior. However, all manner of information on all manner of taxa must be acquired before the ideas presented here can be evaluated fully. Field studies, for example, are needed to assess costs associated with learning in predictable environments. Also needed are data on the relative heritabilities of learning and congenital responses. Detailed information on the physiological basis of learning and instinct will permit insight into possible mechanisms through which learned behavior could be assimilated. Finally, additional models of learning and evolutionary change should be constructed and existing ones refined. Interestingly, such models are fast emerging not in the field of evolutionary biology but in the areas of robotics and artificial intelligence (Hinton and Nowlan, 1987; Maynard Smith, 1987; Parisi, et al., 1990; Nolfi, et al., 1990; Cecconi and Parisi, 1991; Meyer and Guillot, 1991). In one notable model (Nolfi, et al, 1990), adaptive change in a neural network is mediated by changes in both the network's congenital response and its ability to learn. Given the significance of automatism in early discussions of the evolution of instinct, it would be curious if the path to understanding how learning guides evolution runs through discourse on robots, devices which evoke images of extremes in automatic behavior.

Aknowledgments

Portions of the research described here were supported by a visitor's grant from the Netherlands Organization for the Advancement of Research (N.W.O.) and by an Agricultural University research grant and an NSF traineeship at the Center for Insect Science in Tucson. I am grateful to John Hildebrand, Tom Tobin, and the Division of Neurobiology at the Arizona Research Laboratories for providing an environment conducive to the development of many of the ideas expressed here. I am grateful to Ann Hedrick, John Jaenike, Alcinda Lewis, Jay Rosenheim, Ted Turlings, Louise Vet, and Bill Wcislo for comments on an earlier draft.

References

Alcock J. 1989. Animal Behavior: An Evolutionary Approach. Sinauer, Sunderland, MA.

Baldwin, J.M. 1896. A new factor in evolution. Am. Nat. **30**:441–553.

Barlow, G.W. 1977. Modal action patterns. In Sebeok, T.A. (ed.), In How Animals Communicate. Indiana University Press, Bloomington, pp. 98–134.

Bateson, P. 1983. Genes, environment and the development of behaviour. In Animal Behaviour. Vol. 3: Genes, Development and Learning. Blackwell Scientific Publications, London, pp. 52–81.

Boake, C.R.B. 1989. Repeatability: Its role in evolutionary studies of mating behavior. Evol. Ecol. **3**:173–182.

Bouvier, E.L. 1908. The Psychic Life of Insects. Century Co., New York. 1922 English translation.

Boyd, R., and Richerson, P.J. 1985. Culture and the Evolutionary Process. University of Chicago Press, Chicago, IL.

Cecconi, F., and D. Parisi. 1991. Evolving organisms that can reach for objects. In J.A. Meyer and S.W. Wilson (eds.) From Animals to Animals. MIT Press, Cambridge, MA.

Darwin, C. 1868. Variation of Animals and Plants Under Domestication. D. Appleton, New York.

Darwin, C. 1871. The Descent of Man, and Selection in Relation to Sex. Princeton University Press, Princeton, NJ.

Dethier, V.G. 1978. The Hungry Fly: A Physiological Study of the Behavior Associated with Feeding. Harvard University Press, Cambridge, MA.

Dewsbury, D.A. 1978. Comparative Animal Behavior. McGraw-Hill, New York.

Dukas, R., and L.A. Real. 1991. Learning foraging tasks by bees: A comparison between social and solitary species. Anim. Behav. **42**:269–276.

Eibl-Eibesfeldt, I. 1951. Beobachtungen zur Fortpflanzungsbiologie und Jungendentwicklung des Eichhörnchens (*Sciuris vulgaris*). Z. Tierpsychol. **8**:370–400.

Evans, H.E. 1966. The Comparative Ethology and Evolution of the Sand Wasps. Harvard University Press, Cambridge, MA.

Ewer, R.F. 1956. Imprinting in animal behaviour. Nature **177**:227–228.

Gould, S.J. 1977. Ontogeny and Phylogeny. Belknap Press, Cambridge, MA.

Hailman, J.P. 1967. The ontogeny of an instinct: the pecking response in chicks of the laughing gull (*Larus atricilla* L.) and related species. Behav. Suppl. **15**:1–196.

Haldane, J.B.S., and Spurway, H. 1954. A statistical analysis of communication in "Apis mellifera" and a comparison with communication in other animals. Insectes Sociaux **1**:247–283.

Haldane, J.B.S., and H. Spurway. 1956. Imprinting and the evolution of instincts. Nature **178**:85–86.

Hinde, R. A. 1970. Animal Behavior: A Synthesis of Ethology and Comparative Psychology, 2nd ed. McGraw-Hill, New York.

Hinton, H., and Nowlan, S.J. 1987. How learning guides evolution. Compl. Syst. **1**:495.

Huntingford, F. 1984. The Study of Animal Behaviour. Chapman and Hall, New York, 411 pp.

Jaenike, J. and Papaj, D.R. 1992. Learning and patterns of host use in insects. In B.D. Roitberg, and M. Isman (eds.). Chemical Ecology: Ecological and Evolutionary Perspectives. Chapman and Hall, New York.

Klopfer, P.H., and Hailman, J.P. 1967. An Introduction to Animal Behavior: Ethology's First Century. Prentice-Hall, Englewood Cliffs, NJ.

Lamarck, J.B. 1809. Zoological Philosophy. 1984 ed. University of Chicago Press, Chicago.

Lehrman, D.S. 1970. Semantic and conceptual issues in the nature-nurture problem. pp. 17–52. In Development and Evolution of Behavior: Essays in Memory of T.C. Schneirla. Aronson, L.R., Tobach, E., Lehrman, D.S., and Rosenblatt, J.S. (eds). Freeman and Co., New York.

Machlis, L., Dodd, P.W.D., and Fentress, J.C. 1985. The pooling fallacy: Problems arising when individuals contribute more than one observation to the data set. Z. Tierpsychol. **68**:201–204.

Marler, P. and Peters, S. 1981. Sparrows learn song and more from memory. Science **213**:780–782.

Maynard Smith, J. 1987. When learning guides evolution. Nature **329**:761–762.

Mayr, E. 1974. Behavior programs and evolutionary strategies. Am. Sci. **62**:650–659.

McDougall, W. 1923. An Outline of Psychology. 2nd ed. Methuen, London.

Menzel, R. 1985. Learning in honey bees in an ecological and behavioral context. In B. Hölldobler and M. Lindauer, (eds.), Experimental Behavioral Ecology. Gustav Fischer Verlag, Stuttgart, pp. 55–74.

Meyer, J.A., and Guillot, A. 1991. Simulation of Adaptive Behavior in Animals: Review and prospect, In From Animals to Animals. J.A. Meyer and S.W. Wilson (eds.), MIT Press, Cambridge, MA.

Mowry, T.M., Spencer, J.L., Keller, J.E., and Miller, J.R. 1989. Onion fly (*Delia antique*) egg depositional behaviour: Pinpointing host acceptance by an insect herbivore. J. Insect Physiol. **35**:331–339.

Morgan, L. 1896. Habit and Instinct. Arnold, London.

Nolfi, S., Elman, J.L., and Parisi, D. 1990. Learning and evolution in neural networks. CRL. Technical Report 9019.

Osborn, H.F. 1896. Ontogenic and phylogenic variation. Science **4**:786–789.

Papaj, D.R. 1986. Interpopulation differences in host preference and the evolution of learning in the butterfly *Battus philenor*. Evolution **40**:518–530.

Papaj, D.R., and Prokopy, R.J. 1989. Ecological and evolutionary aspects of learning in phytophagous insects. Annu. Rev. Entomol. **34**:315–350.

Papaj, D.R., and Vet, L.E.M. 1990. Odor learning and foraging success in the parasitoid, *Leptopilina heterotoma*. J. Chem. Ecol. **16**:3137–3150.

Parisi, D., Cecconi, F. and Nolfi, S. 1990. Econets: Neural networks that learn in an environment. Network **1**:149–168.

Romanes, G.J. 1884. Mental Evolution in Animals. Keegan, Paul, Trench & Co., London.

Shaw, R.E., and Ailey, T.R. 1985. How to draw learning curves: Their use and justification. In A.T. Pietrewicz and T.D. Johnston, (eds.), Issues in the Ecological Study of Learning. Lawrence Erlbaum Associates, Hillsdale, NJ, pp. 275–304.

Shepherd, G.M. 1983. Neurobiology. Oxford University Press, Oxford, England.

Smith, B.H., and Menzel, R. 1989a. An analysis of variability in the feeding motor program of the honey bee: The role of learning in releasing a modal action pattern. Ethology **82**:68–81.

Smith, B.H., and Menzel, R. 1989b. The use of electromyogram recordings to quantify odourant discrimination in the honey bee. *Apis mellifera*. J. Insect Physiol. **35**:369–375.

Sokal, R.R., and Rohlf, F.J. 1981. Biometry. W.H. Freeman, San Francisco.

Spalding, D.A. 1872. On instinct. Nature **6**:485–486.

Spurway, H. 1955. The causes of domestication: An attempt to integrate some ideas of Konrad Lorenz with evolution theory. Genetics **53**:325–362.

Stephens, D.W. 1987. On economically tracking a variable environment. Theor. Popul. Biol. **32**:15–25.

Tierney, A.J. 1986. The evolution of learned and innate behavior: Contributions from genetics and neurobiology in a theory of behavioral evolution. Anim. Learn. Behav. **14**:339–348.

Vet, L.E.M. 1988. The influence of learning on habitat location and acceptance by parasitoids. Les Colloques de l'INRA **48**:29–34.

Vet, L.E.M., and Groenewold, A.W. 1990. Semiochemicals and learning in parasitoids. J. Chem. Ecol. **16**:3119–3136.

Vet, L.E.M., Lewis, W.J., Papaj, D.R., and van Lenteren, J.C., 1990. A variable-response model for parasitoid foraging behavior. J. Insect Behav. **3**:471–490.

Vet, L.E.M., and Papaj, D.R. 1991. Effects of experience on parasitoid movement in odour plumes. Physiol. Entomol., **17**:90–96.

Waddington, C.H. 1953. Genetic assimilation of an acquired character. Evolution 7:118–126.

Wcislo, W.T. 1987. The roles of seasonality, host synchrony, and behavior in the evolutions and distributions of nest parasites in Hymenoptera (Insecta) with special reference to bees (Apoidea). Biol. Rev. **62**:515–543.

Wcislo, W.T. 1989. Behavioral environments and evolutionary change. Annu. Rev. Ecol. Syst. **20**:137–169.

11

Comparative and Experimental Approaches to Understanding Insect Learning

Jay A. Rosenheim

Introduction

The comparative approach has repeatedly been advocated as an important tool for investigating learning in animals. This suggestion has emanated primarily from comparative psychologists studying learning in vertebrates (Bitterman, 1965; Johnston, 1982; Domjan and Galef, 1983; Kalat, 1985; Kamil and Clements, 1990) but also more recently from evolutionary biologists studying insects (Menzel, 1985; Papaj and Prokopy, 1989; Lewis and Lipani, 1990). Nevertheless, there have been few attempts to apply a comparative approach to testing hypotheses concerning the physiology, ontogeny, function, or evolution of learning in insects, and, as will be reviewed below, no comparative study reported to date is sufficiently rigorous statistically to provide a strong insight into insect learning. Why?

This review addresses this question and, in doing so, attempts to identify fruitful means of applying the comparative approach to current questions in insect learning. The discussion is organized so as to

1. Define the comparative approach and describe in general terms the role of the comparative approach in evolutionary biology.

2. Review the substantial recent progress in developing statistically sound analyses for comparative data and suggest two analyses that may be particularly appropriate for studies of insect learning.

3. Catalogue those primary hypotheses advanced regarding insect learning that are amenable to comparative tests.

4. Outline the key procedural and interpretational limitations of comparative tests of learning.

5. Provide a brief overview of some published comparative studies of insect learning.

I then broaden the discussion to apply the same considerations of interpretational limitations to experimental approaches to the study of insect learning. My purpose in addressing experimental approaches is twofold: first, to establish some sense of the relative robustness of the comparative approach, and second, to consider directly some important problems experienced in the experimental study of insect learning. The latter purpose is important, because the experimental approach is generally considered to be the most powerful investigative tool available to scientists and as such is less frequently subject to critical scrutiny. This is in stark contrast to comparative analyses, the criticism of which has recently become something akin to a sport among biologists.

Finally, I suggest some future directions for comparative and experimental investigations.

Definition and Functions of the Comparative Approach

The comparative approach refers to any empirical, nonmanipulative analysis employing comparisons among populations, species, or higher-ranked taxa. It is, therefore, a class of "natural experiments" observed at levels higher than the individual organism (Thornhill and Alcock, 1983; Endler, 1986). For example, Sessions and Larson (1987) employed the comparative approach to investigate the relationship between genome size (the mass of DNA in a haploid nucleus) and the rate of limb regeneration in plethodontid salamanders. An analysis of 23 species demonstrated a significant negative correlation between genome size and limb differentiation rate, suggesting that the proliferation of nongenic sequences of DNA within the genome has produced an intrinsic constraint on development.

The comparative approach is one of the most widely used means of generating and testing hypotheses in biology and complements other investigative approaches, including (1) nonmanipulative, observational experiments conducted at the individual level (also termed "natural experiments"), (2) manipulative experiments, and (3) theoretical investigations, including the construction of verbal and mathematical models. Because a complete understanding of animal behavior can best be developed through investigations from the multiple standpoints of mechanism, development, function, and evolution (Tinbergen, 1963; Thornhill and Alcock, 1983; Stamp-Dawkins, 1989), each of these investigative tools can make important contributions.

The comparative approach has traditionally been viewed as having two primary applications in biology: first, comparative analyses can be used to study the phylogeny of traits by focusing on the progressive modification of homologous traits or associations of traits within evolutionary lineages;

and second, comparative analyses can be used to study the function of traits by focusing on convergent evolution of analogous traits across a number of independent evolutionary lineages (Hodos and Campbell, 1969; Lauder, 1981; Kamil and Yoerg, 1982; Ridley, 1983; Barlow, 1989; Gittleman, 1989; Donoghue, 1989). Comparative analyses of trait function rely on the premise that independent evolutionary lineages experiencing similar selection pressures will sometimes converge on the same adaptation. Recently, however, the distinction between the dual functions of comparative analyses has become blurred in practice by the incorporation of explicit phylogenies into comparative analyses of trait function (see below).

This chapter focuses primarily on using the comparative approach to investigate the function of learning in insects. Although in many cases the function of learning is self-evident within a given life-history framework [Kamil and Yoerg, 1982; e.g., the learning of nest location by sand wasps and their cleptoparasites (van Iersel, 1975; Rosenheim, 1987)], in other cases the adaptive significance of learning is not at all obvious [e.g., induced feeding preferences or induced preferences for host acceptance by ovipositing insect parasites (Prokopy et al., 1986; Cooley et al., 1986; Papaj and Prokopy, 1989; Ward et al., 1990; but see Karowe, 1989) and food aversion learning (Papaj and Prokopy, 1989; Bernays, this volume)]. In these latter cases the comparative approach may be able to provide insights into the functions of learning by identifying ecological correlates of learning traits.

Statistical Analyses for Comparative Data

Any statistical analysis for comparative data must cope first and foremost with the problem of nonindependence of different observations (Lauder, 1981; Ridley, 1983; Dobson, 1985; Felsenstein, 1985; Huey, 1987; Pagel and Harvey, 1988, 1989; Donoghue, 1989; Grafen, 1989; Maddison, 1990). This problem, first recognized by Darwin (1859, p. 185), must be understood to appreciate the strengths and weaknesses of different statistical treatments of comparative data. I will attempt to illustrate with a hypothetical example. Consider the evolution of learning ability within a family of insect herbivores comprising 20 species divided equally into two genera, A and B. It has been suggested that herbivore diet breadth may be an important factor in the evolution of learning (see discussion below). Assume that the family contains ten species that may be considered to be specialists in their use of host plants and ten species that are more generalized, and that each of the 20 species has been tested to determine whether or not it is able to learn new host plant preferences. One way to test the hypothesis that learning ability is associated with generalist feeding might be to cross-tabulate species according to the specialist/generalist and learner/nonlearner dichotomies; one potential outcome of such a cross-tabulation is shown in Figure 11.1A. The data seem to support a clear association between the two variables (generalist feeding is perfectly associated with learning ability), and a standard test of independence would strongly support this conclusion (e.g., G-test, $p < 0.0001$). However, standard tests of independence rely on the assumption that different observations are independent; is this the case for our 20 species? To answer this question we must consider the phylogenetic relationships within the family.

Two hypothetical phylogenies for the 20 species are shown in Figure 11.1B and C. Assume for both phylogenies that cladistic analyses incorporating outgroups (not shown) have revealed that the common ancestor for the family was a specialist feeder without an ability to learn. In the first phylogeny, all ten species in genus B are generalists and learners; the evolution of these traits may be inferred using the principle of parsimony to have occurred only once along the branch leading to genus B. Clearly, the assumption of species independence that we made when cross-tabulating the 20 species is invalid under this phylogenetic hypothesis; the species in genus B are generalists and learners because they have inherited these two traits from a common ancestor. A more appropriate cross-tabulation of the data would count only the single independent evolutionary step (Fig. 11.1B); this reconstruction is consistent with, but does not provide strong evidence for, an association of learning and generalized feeding. If, however, the phylogeny of the family is as shown in Figure 11.1C, we can infer the occurrence of eight independent instances of the simultaneous evolution of generalist feeding and learning; these data are therefore more compelling [we could analyze these data with Maddison's (1990) method; see below]. Note that for each of the three scenarios considered in Figure 11.1, we have the same final outcome of a perfect association of learning and generalist feeding among the 20 terminal species. Only by considering the phylogeny of the group are we able to assess the number of independent evolutionary events.

Statistical analyses that cope with nonindependence of observations have been developed along three lines. One group of analyses attempts to resolve statistically significant relationships in comparative data sets without an explicit use of a phylogeny; this group of analyses includes the nested ANOVA technique of Clutton-Brock and Harvey (1984), the phylogenetic autocorrelation method of Cheverud et al. (1985), the nested analysis of covariance technique of Bell (1989), and a variety of closely related approaches (see the excellent review by Pagel and Harvey, 1988). These techniques all attempt to mitigate the nonindependence problem by combining various statistical procedures with "pseudophylogenies" provided

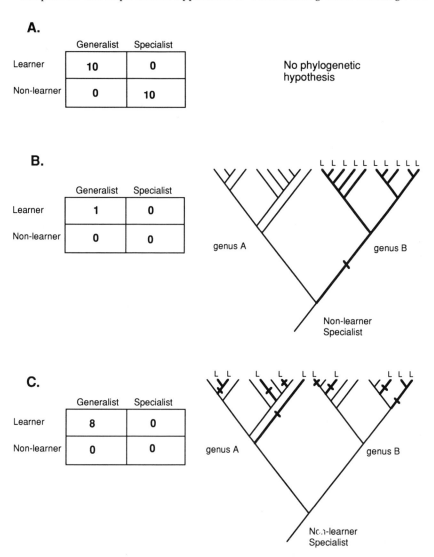

A.

	Generalist	Specialist
Learner	10	0
Non-learner	0	10

No phylogenetic hypothesis

B.

	Generalist	Specialist
Learner	1	0
Non-learner	0	0

genus A

genus B

Non-learner Specialist

C.

	Generalist	Specialist
Learner	8	0
Non-learner	0	0

genus A

genus B

Non-learner Specialist

Figure 11.1. Influence of different phylogenetic hypotheses on the number of independent evolutionary events inferred. A. In the absence of a phylogenetic hypothesis, each of the 20 species is treated as an independent observation. B. With the specified phylogenetic hypothesis, parsimony suggests the occurrence of only a single evolutionary transition to generalist feeding and learning. Thick lines, generalist feeders; thin lines, specialist feeders; L, species able to learn; black crossbars indicate the origin of learning abilities. C. An alternate phylogenetic hypothesis suggests the occurrence of eight independent origins of generalist feeding and learning.

by current taxonomic classifications (i.e., by controlling for membership in the conventional taxonomic ranks).

These analyses have been criticized on three grounds. First, none of the techniques does more than modestly reduce the severity of the non-independence problem (Ridley, 1983; Huey, 1987; Donoghue, 1989; Grafen, 1989). Second, these analyses fail to consider information on the sequence of evolutionary change (Donoghue, 1989); sequence information can be extremely helpful in interpreting any observed comparative relationships, for instance in allowing us to infer the direction of causation and distinguish between adaptation and exaptation (Lauder, 1981; Ridley, 1983; Huey, 1987; Huey and Bennett, 1987; Sillén-Tullberg, 1988; Donoghue, 1989). Third, because taxonomic classifications are at best incomplete phylogenetic hypotheses and at worst are not phylogenetic at all, analyses may generate a variety of spurious results. Taxonomic classifications that are constructed using noncladistic techniques may not reflect evolutionary relationships between taxa, and resulting supraspecific taxa may therefore be para- or polyphyletic (Ridley, 1983; Cheverud et al., 1985; Dobson, 1985; Felsenstein, 1985). Even cladistic taxonomies are incomplete representations of phylogenies; taxonomies do not address the branching events that occur between taxonomic ranks (Dobson, 1985; Felsenstein, 1985). For instance, standard taxonomies do not provide information on the relationships between species within a genus, but as demonstrated in Figure 11.1, these relationships can be crucial to evaluating comparative hypotheses. In essence, taxonomies are at best skeletal phylogenetic hypotheses that represent phylogenies as a hierarchically arranged series of unresolved polychotomies (= star phylogenies). Finally, different taxa within a given supraspecific taxonomic rank (genera, families, etc.) are in no sense equivalent; they do not represent clades of a given range of evolutionary ages (Cheverud et al., 1985; Huey, 1987). Rather, the designation of higher-level taxonomic ranks reflects subjective decisions made by taxonomists, generally following taxonomic convention for a given specialized group of organisms. For all these reasons, a taxonomy is not an adequate foundation for this first group of comparative analyses.

A second group of analyses is based on the explicit use of complete phylogenies to identify independent instances of the evolution of the trait(s) in question. Techniques have been developed for both continuous variables (Felsenstein, 1985; Huey, 1987; Huey and Bennett, 1987) and discrete variables (Lauder, 1981; Ridley, 1983; Sillén-Tullberg, 1988; Donoghue, 1989; Madisson, 1990). For discrete variables, the analysis developed by Maddison (1990) and demonstrated by Donoghue (1989) and Maddison (1990) appears to represent an important advance in testing hypotheses related to the correlated evolution of two traits; the analysis permits assessment of whether gains or losses of one trait are significantly concentrated on branches of the phylogeny where the second trait has a specified state. As usual, although associations between traits can potentially be demonstrated, causality remains elusive.

Analyses based upon explicit phylogenies are substantially superior to ahistorical analyses; however, because phylogeny reconstruction is based upon cladistic techniques and the underlying principle of parsimony, these analyses are associated with a set of unique statistical problems (Felsenstein, 1985; Maddison, 1990). Because parsimony is also used to map the traits of interest onto the phylogeny, phylogenetic reconstruction may fail if the behavioral or ecological trait being studied is so evolutionarily labile that multiple character-state transitions occur on individual phylogeny branch segments. Perhaps most importantly, the application of these analyses is currently constrained by the requirement for an exact phylogeny of the group being studied. Detailed phylogenies are often not available, and many existing phylogenies are associated with a large degree of uncertainty (Huey, 1987; Donoghue, 1989; Maddison, 1990). The increasing availability of molecular phylogenies, which can provide information on both cladogenesis and estimated divergence times of different clades, promises to partially alleviate this problem in the future.

Are there then no comparative analyses that solve the problem of nonindependence without a detailed knowledge of phylogenetic relationships? A third group of analyses, reviewed by Pagel and Harvey (1989), provides us with just such techniques, including those developed by Felsenstein (1985) and Grafen (1989). Felsenstein's (1985) analysis is based on replicated paired comparisons of closely related species or sister groups of higher-ranked taxa. To test the previously considered hypothesis that learning ability is associated with generalist feeding, for example, one might identify pairs of closely related species (e.g., congeners) that differed markedly in their degree of diet breadth. Each pair of species can then provide one independent datum either supporting the hypothesis (if the more generalized species performs better in a learning assay) or arguing against the hypothesis (if the more specialized species performs better). Felsenstein (1985) suggested replicating the study (in this example, never using more than one pair of species from a given genus to maintain the independence of the contrasts) and assessing statistical significance with a nonparametric test, such as a sign test.

This technique does suffer from two drawbacks. First, like other techniques that do not incorporate explicit phylogenies, we lose information on the sequence of evolutionary change (i.e., does a generalist feeding habit evolve before or after the evolution of learning?). Second, the technique has relatively low power: a study of n species provides only $n/2$ independent observations, compared to the $n-1$ independent observations that would be obtained from Felsenstein's (1985) phylogenetic analysis

(Huey, 1987). However, the comparison of closely related species has been strongly advocated by Kamil and Yoerg (1982) and Huey (1987) for reasons unrelated to what I am here considering its primary advantage of providing independent observations. These authors note that because closely related species will tend to be similar, there is a decreased probability that the relationship being studied (in our example, the relationship between diet breadth and learning ability) will be confounded or obscured by other differences between the two species (for example, differences in sensory acuity or general feeding behavior). The increased similarity of closely related species should also facilitate the design of a standardized learning assay that can be applied to both species (see below). Because this paired-species experimental design provides independent observations without requiring extensive knowledge of phylogeny, and because previous studies have demonstrated the feasibility of comparing the learning abilities of closely related species (or populations) (e.g., Hanson, 1976; Chew, 1980; Daly et al., 1982; Jaenike, 1982; Vet and van Opzeeland, 1984; Greenberg, 1985; Papaj, 1986; Papaj et al., 1987; Thomson and Stinner, 1990), the paired-species experimental design and analysis appears to be particularly appropriate for comparative studies of insect learning. The paired-species analysis is especially suitable when the data set for the analysis is generated in a single experimental study rather than extracted piecemeal from the literature.

Comparative analyses may, however, be conducted with literature-based data sets (see below), and for these cases, where appropriate pairs of taxa may not be available or where more than two closely related species contribute observations, the "phylogenetic regression" technique developed by Grafen (1989) or the closely related technique discussed by Pagel and Harvey (1989) may be most appropriate. These analyses apply multiple regression techniques to comparative data and require knowledge of only a skeletal phylogeny (i.e., unresolved polychotomies are treated successfully). Unlike the other techniques based on incomplete knowledge of phylogeny, but similar to Felsenstein's (1985) technique, these regression analyses produce a set of independent observations by distilling data from all species within a higher-level taxon into a single observation.

Some Hypotheses Testable With the Comparative Approach

The development of sound statistical analyses for comparative data provides an opportunity to critically assess a variety of current hypotheses in the insect learning literature that are framed in explicitly comparative terms. Here I describe some prominent hypotheses; the list is far from complete.

Empirical evidence pertinent to some of these hypotheses is considered later in the overview of published comparative studies.

Environmental Unpredictability

The literature on animal learning is perfused with the notion that the adaptive value of learning is directly related to unpredictable environmental variability that occurs over a time frame short enough to prevent natural selection from maintaining populations at or near adaptive peaks (i.e., either within an individual's lifetime or over a small number of generations; Baker, 1978; Dethier and Yost, 1979; Dethier, 1980; Turner, 1981; Johnston, 1982; Daly et al., 1982; de Boer and Hanson, 1984; Gould and Marler, 1984; Greenberg, 1985; Papaj, 1986; Bernays and Lee, 1988; Papaj and Prokopy, 1989; Sheehan and Shelton, 1989; Sivinski, 1989; Vet et al., 1990; Lewis and Lipani, 1990; Gould, this volume; Stephens, this volume). This basic hypothesis has been formulated in more specific terms to reflect the particular aspect of the environment that demonstrates heterogeneity. First, heterogeneity of host association for insect herbivores (i.e., diet breadth) or parasitoids (i.e., host range) may influence the ability (1) to learn new host preferences (Hanson, 1976; de Boer and Hanson, 1984; Gould and Marler, 1984; Sheehan and Shelton, 1989; Vet et al., 1990), (2) to habituate to feeding deterrents (Papaj and Prokopy, 1989), or (3) to learn aversions to toxic foods (Dethier and Yost, 1979; Dethier, 1980; Bernays and Lee, 1988; Papaj and Prokopy, 1989; Bernays, this volume). Second, heterogeneity of the spatial location of key resources, such as foraging, oviposition, or nesting sites, may influence the development of locality learning (Baker, 1978; Turner, 1981; Rosenheim, 1987; Sivinski, 1989; Lewis and Lipani, 1990). And third, heterogeneity in the timing of resource availability (e.g., nectar availability in flowers) may influence the ability to learn the time at which to forage for different resources (Gould, 1984).

Rate of Encounter With Different Resources

Menzel (1983, 1985; Menzel et al., this volume) has hypothesized that the temporal dynamics of short- to long-term memory transfer reflect the rate of encounter with different types of resources under natural foraging conditions.

Feeding Style

Dethier (1980), Bernays and Lee (1988), and Lee and Bernays (1988, 1990) have suggested that insects that take discrete meals with long inter-meal times are more likely to possess an ability to learn aversions to toxic foods than insects that feed more or less continuously as grazers. See Bernays (this volume) for a detailed consideration of this hypothesis and a discussion of the roles of resource fidelity and herbivore mobility in food aversion learning.

Cues Associated With Key Resources

Cues or motor response patterns that are intimately associated with obtaining key resources may be incorporated into learned responses more readily than less pertinent cues or responses (see review by Menzel, 1985; Papaj and Prokopy, 1989; Menzel et al., this volume). Essentially the same hypothesis of "preparedness" or "predisposition" to learn certain stimuli has been advanced in the comparative psychology literature (e.g., Domjan and Galef, 1983; Pietrewicz and Richards, 1985; Timberlake, 1990).

Brain Morphology

Sivinski (1989) has hypothesized that the ability to learn is positively correlated with the size of the corpora pedunculata ("mushroom bodies") of the insect protocerebrum. Again, there is a corresponding brain size-learning ability hypothesis in the vertebrate literature (reviewed by Johnston, 1982).

Procedural and Interpretational Limitations

I have thus far described the functions of the comparative approach, the statistical analysis of comparative data, and a group of hypotheses that are amenable to comparative tests. The next step is to consider some of the procedural issues associated with constructing a sound data set with which an analysis can be performed. The most common approach is to extract scattered results from the literature to compile a data set for analysis; this approach relies, however, on the underlying quality and homogeneity of the primary studies (Gittleman, 1989). Another less commonly applied approach is to generate the needed data with experiments (e.g., Brattsten and Metcalf, 1970; Bernays, 1988). In this section I review some of the prerequisites for basing a comparative analysis on a data set extracted from the literature; I conclude that it will be difficult to meet these requirements with the literature on insect learning. I then consider some of the more important interpretational limitations of comparative studies, which are applicable to studies based on data sets that are extracted from the literature or generated experimentally.

Literature-Based Data Sets

The last decade has witnessed a surge of interest in insect learning, with a resulting proliferation of published studies. These studies have substantially increased our understanding of learning in insects (as attested to by advances described in this volume), but can they also be used to test comparative hypotheses? What follows is a summary of problems associated with use of this literature.

Biases in Data Reported

One potentially important, but difficult to evaluate, source of bias in the literature is the failure to report negative results. Although negative results are sometimes published (e.g., Hanson, 1976; Dethier and Yost, 1979; Chew, 1980; Jaenike, 1982; de Boer and Hanson, 1984; Hedrick et al., 1990), it seems likely that many studies producing exclusively negative results fail to leave the laboratory. For example, in a review of induced feeding preferences in insect herbivores, Jermy (1987) listed two species (*Heliothis zea* and *Manduca sexta*) for which he and his colleagues had obtained positive results, and which were published (Jermy et al., 1968), and two species (*Hyphantria cunea* and *Mamestra brassicae*) for which negative results had been obtained, and which had remained unpublished. A heavily biased literature will not provide meaningful insights into insect learning.

Variable Methodologies

One of the problems most frequently encountered when attempting to compare learning studies is how to cope with the highly variable methodologies used by different investigators (Bitterman, 1965; Jaenike, 1982; Johnston, 1982; Kamil and Yoerg, 1982; Domjan and Galef, 1983; Hoffmann, 1985; Menzel, 1985; McGuire et al., 1990). There are two tiers of related issues here. The first issue is the obvious one: comparisons between studies are difficult to make when studies use different experimental protocols. A second potential problem exists, however, if different species tested with a standardized protocol perceive the experimental treatments differently. Daly et al. (1982) explored this problem in their comparative study of food aversion learning in two congeneric species of kangaroo rats. Food aversion learning was closely tied to the initial level of a food's acceptability, and their two species differed in initial preferences for the experimental foods. Thus, what initially appeared to be differences in ability to acquire aversions to foods was later reinterpreted as a potential artifact related to initial differences in food acceptability. Parallels may exist in food aversion learning by insects (Lee and Bernays, 1990). Vet et al. (1990) proposed a general model for associative learning in insect parasitoids that postulates that learning will generate larger changes in response to stimuli that initially elicit low-level responses; if this postulate is verified (e.g., Sheehan and Shelton, 1989), it will mean that testing protocols for comparative studies will need to be standardized with respect to the initial level of responsiveness to the test stimuli.

Quantifying Key Variables

Most comparative hypotheses of insect learning involve relationships among continuous variables. Learning ability, measured as a change in behavior with experience, is generally considered to be a continuous variable, and most of the key factors whose influence on learning ability we wish to study (e.g., habitat heterogeneity, rate of encounter with different resources, intermeal times, etc.) also vary continuously. Unfortunately for the researcher interested in compiling a data base on insect learning, many learning studies present results in a binary form (i.e., insects do, or do not, demonstrate learning), while studies that quantify learning performance use an array of different, non-interchangeable indices (e.g., de Boer and Hanson, 1984; Greenberg, 1985; Hoffmann, 1985; Menzel, 1985; Papaj et al., 1987; McGuire et al., 1990). Furthermore, quantifying environmental parameters that might influence learning may be challenging. For example, to test whether a generalist feeding habit favors the evolution of learning, we need to quantify diet breadth. This is not a trivial task, in part because the degree of specialization observed may depend on the spatial scale at which species are observed (Fox and Morrow, 1981). Furthermore, the most meaningful measures of diet breadth for insect herbivores might be based on diversity of phytochemistry rather than taxonomic diversity of the host plants.

Intraspecific Variation

Further complicating the quantification of key variables is the observation of significant intraspecific variation. For instance, learning may be age-specific (e.g., Jaisson, 1980; Wardle and Borden, 1985), strain-specific (Papaj et al., 1987; McGuire et al., 1990), or cue-specific (e.g., Jaenike, 1982; de Boer and Hanson, 1984; Lee and Bernays, 1990; see above). Herbivorous insects may recognize and respond behaviorally to variation occurring within a single host plant species (e.g., Prokopy and Papaj, 1988; Papaj and Prokopy, 1989; Greenfield et al., 1989), making it difficult to identify the level of biologically relevant environmental heterogeneity.

Biased reporting of learning studies, variable methodologies, and difficulties in quantifying key variables and coping with intraspecific variation combine to create a formidable obstacle to extracting a useful data set on insect learning from the literature. Some of these problems can, however, be circumvented by experimentally generating the appropriate comparative data.

Interpretational Issues

As discussed previously, the key statistical issue in the analysis of comparative data is maintaining the independence of different observations. In the same way the key interpretational issue is assessment of causality; comparative studies are, by definition, correlative studies, and correlation cannot be extended with assurance to infer causation (Hodos and Campbell, 1969; Ridley, 1983; Clutton-Brock and Harvey, 1984; De Boer and Hanson, 1984; Endler, 1986; Donoghue, 1989; Maddison, 1990). An observed correlation of two variables may reflect the fact that both covary with a third, unmeasured variable, which may be the key causal factor in the system. Furthermore, even if an observed correlation between two variables does reflect a causal link, it may be difficult to determine the direction of causality. Although incorporating potentially influential third variables into statistical analyses and using explicit phylogenies can help to minimize these problems (Ridley, 1983; Clutton-Brock and Harvey, 1984; Huey and Bennett, 1987; Bell, 1989; Donoghue, 1989), difficulty in assessing causality will remain a limitation of comparative studies (and indeed all nonmanipulative studies).

A second class of interpretational problems stems from our incomplete understanding of the physiological basis of learning. If different manifestations of learning share a common physiological basis, then attempts to identify correlations between specific learning abilities and specific environmental characteristics may be confounded by selection for learning ability in different contexts; different manifestations of learning may not evolve independently (Johnston, 1982; Kamil and Yoerg, 1982; Domjan and Galef, 1983; Papaj, 1986; Papaj and Prokopy, 1989). This problem may be important (1) if different types of learning (e.g., locality learning, associative learning, food aversion learning, induced preferences, etc.) share some or all of the same physiological bases in the insect nervous system or (2) if a given type of learning is important in a variety of contexts. As an example of the second factor, Turner (1981) and Mallet et al. (1987) have suggested that locality learning in *Heliconius* butterflies may function in the location of (1) oviposition sites (*Passiflora* meristems), (2) adult food sites (*Anguria* flowers), (3) mating sites, (4) overnight communal roosting sites, and (5) sites where predators were previously encountered (for subsequent avoidance). It might be fruitless to search within the genus *Heliconius* for a correlation between the degree of spatial heterogeneity of one factor alone (e.g., adult food sites) and the degree to which locality learning has been developed. A more complete understanding of the neuronal basis for learning in insects will help to evaluate the degree to which different forms of learning may evolve independently.

I close this discussion of interpretational limitations with one warning regarding sample size requirements. The problem here is the facility with which verbal arguments can replace statistical ones in the evaluation of comparative data (e.g., compare Sillén-Tullberg [1988] and Maddison [1990]). For example, a single comparison of two taxa cannot satisfactorily reject any null hypothesis; p levels less than 0.5 are unobtainable. Sample size requirements for a comparative study can be calculated in the same way as for manipulative experiments, given estimates of the magnitude of the effect to be identified, an α error rate, and levels of variability of key parameters. For comparative studies to make substantial contributions to our understanding of learning, comparative data must be subjected to formal statistical analysis.

Published Comparative Studies

As noted in the Introduction, few explicitly comparative analyses of insect learning have been attempted, and none has been evaluated in a statistically rigorous manner. These studies have, however, helped to define key issues, identify procedural and interpretational difficulties, and suggest avenues for enhancing future investigations.

Most comparative studies have attempted to evaluate some facet of the environmental unpredictability hypothesis; these studies were reviewed recently by Papaj and Prokopy (1989), Prokopy et al. (1989a), and Vet et al. (1990), who concluded that there is no strong evidence supporting the hypothesis that species experiencing greater environmental heterogeneity have more highly developed learning abilities. This conclusion seems well-founded and, with one exception, I will not reevaluate the individual studies here.

The exception is the study reported by de Boer and Hanson (1984), which is unique in that sufficient data were compiled to permit a statistical analysis. De Boer and Hanson (1984) created a literature-based data set on induced preferences in 14 species of lepidopteran larvae; for all 14 species a common "induction index" was used to quantify changes in feeding preferences following conditioning treatments to each of a pair of potential host plants. Using this data base, they confirmed a result that they had observed experimentally in extensive studies of preference induction in *Manduca sexta*: induced preferences were stronger when the paired host plants were in more distantly related taxonomic groups (e.g., different genera or families; see also Wasserman, 1982). Noting that this pattern could produce an apparent trend for generalist herbivores to show stronger preference induction than specialists if generalists are often tested with more distantly related pairs of plants than are specialists, they then ranked the 14 species by approximate degree of dietary specialization. The correlation of dietary specialization and preference induction was, however, then assessed without the use of formal statistics. I therefore reanalyzed their data using nonparametric rank correlation tests. When the paired host plants tested were confamilial but not congeneric, there was a significant direct relationship between dietary specialization and strength of induced preference (Fig. 11.2A; $r_s = -0.70, N = 9, p < 0.05$); generalist feeders showed significantly *weaker* induction of preference. When the host plants tested were in different families, there was no significant relationship between dietary specialization and induction index scores (Fig. 11.2B; $r_s = 0.05, N = 9, p > 0.50$). Although this study and the just-completed statistical analysis suffer from many of the shortcomings discussed above (e.g., different training and testing protocols used in different studies, nonindependence of different observations, no explicit procedure presented for quantifying dietary specialization), the results nevertheless provide some of the clearer evidence arguing against a positive relationship between diet breadth and learning ability in phytophagous insects.

Confounding Variables in Experimental Analyses of Insect Learning

The foregoing exposition of the problems and limitations of the comparative approach might suggest to some readers that the whole undertaking should be jettisoned in favor of a complete reliance on manipulative experiments. Aside from the fact that some questions simply cannot be addressed with experiments (e.g., what is the phylogenetic history of learning traits?), would it be safe to conclude that the key problems outlined above would be avoided by using a strictly experimental approach?

I think the answer is no. Although comparative and experimental approaches are associated with somewhat different sets of statistical and interpretational problems, these problems are not as dissimilar as might initially be suspected. I will focus the discussion on the relevance to experimental studies of what may be the single most significant limitation of the comparative approach: interpretational problems due to unseen correlations with third variables. However, a brief mention of the second key problem discussed above, the nonindependence of observations, is perhaps also in order.

Nonindependence of Observations

Nonindependence of replicate observations is a surprisingly pervasive problem in experimental ecology and behavior (Hurlbert, 1984; Machlis et al., 1985). Although errors of "pseudoreplication" of treatments or the "pooling fallacy" (reflecting non-independent repeated measurements of

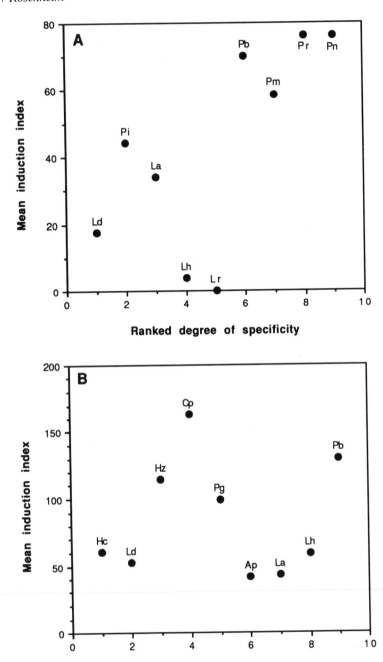

the same individual) are now perhaps more clearly recognized than they were 10 years ago, many studies, and especially critical field studies, continue to succumb to these statistical pitfalls. For example, any field study employing natural populations of insects that does not include explicit recognition of individuals (either by uniquely marking individuals or by capturing individuals after observations) runs the risk of unwittingly making multiple, nonindependent observations on the same individual; some of my own work may suffer from this problem (e.g., Rosenheim et al., 1989). Non-independence of observations is an issue that demands the attention of both the experimental and comparative biologist; solutions are available for both.

Confounding Variables

Learning is a proximate explanation for behavioral variability within and among individuals of an insect population. There are, however, many other possible proximate bases for such variability (Papaj and Rausher, 1983; Parker, 1984; Roitberg, 1990; Lewis et al., 1990; Vet et al., 1990; Rosenheim and Rosen, 1991), including

1. Genetic effects
2. Exogenous environmental effects

   a. Host-contact
   b. Availability of key resources
   c. Abiotic factors
   d. Density of conspecifics

3. Endogenous environmental effects

   a. Age

Figure 11.2. Relationship between the degree of dietary specialization and the strength of induced feeding preferences in lepidopteran caterpillars; data from de Boer and Hanson (1984; see their Table 4 for references to the original studies). Each point represents the mean induction index and ranked degree of dietary specialization (1 = least specialized) for a single species. A, Paired host plants tested were confamilial but not congeneric ($r_s = -0.70, p < 0.05$). B, Paired host plants were in different families ($r_s = 0.05, p > 0.50$). Ap, *Antheraea polyphemus*; Cp, *Callosamia promethea*; Hc, *Hypantria cunea*; Hz, *Heliothis zea*; La, *Limenitis astyanax*; Ld, *Lymantria dispar*; Lh, *Limenitis hybrid rubidus*; Lr, *Limenitis archippus*; Pb, *Pieris brassicae*; Pg, *Papilio glaucus*; Pi, *Polygonia interrogationis*; Pm, *Papilio machaon*; Pn, *Pieris napi macdunnoughii*; Pr, *Pieris rapae*.

    b. Egg load

    c. Size

    d. Maternal effects

    e. Migration and diapause

    f. Nutritional status

    g. Other physiological factors

4. "Mixed" evolutionarily stable strategies

Fortunately, by randomly assigning treatments to experimental units in manipulative experiments, we control for variation in many of these factors and thereby prevent them from confounding our analysis. *Vigilance is required, however, whenever the experimental treatment by which insects are given an opportunity to learn also has an impact on other factors that may influence subsequent behavior.* In such cases it may be inappropriate to ascribe resulting shifts in behavior to the influence of learning.

Hints that confounding factors may be important in the experimental study of learning have been provided by recent advances in our understanding of search images (Guilford and Dawkins, 1987) and superparasitism (van Alphen and Visser, 1990). In each case, what was initially viewed as learning (to form search images or to discriminate between parasitized and unparasitized hosts) was subsequently reinterpreted as a conditional strategy based on the perceived density of a key resource (cryptic prey or unparasitized hosts). Thus, it may have been perceived resource availability and not learning that generated the changes in behavior observed experimentally.

I suggest that similar instances of the confounding of learning effects with other bases for variable behavior are widespread in current experimentation. Examples concerning two broad classes of experimental designs in insect learning are presented in support of this thesis.

Example 1: Learning vs. Genetic Effects in Studies of
Preimaginal "Conditioning"

Preimaginal conditioning in the current context refers to the influence of the host experienced by an immature insect herbivore or parasitoid on the subsequent oviposition site preference expressed by the resulting adult insect. Although preimaginal conditioning has received considerable theoretical and empirical attention because of its possible role in promoting genetic subdivision of populations (Jaenike 1982, 1988), it has been acknowledged that demonstration of true larval conditioning is difficult (Jermy et al., 1968; Jaenike, 1982, 1983; Papaj and Rausher, 1983; Corbet, 1985; Vet et al., 1990). Most attention has focused on the difficulty of distinguishing between larval conditioning and effects of early adult experience with chemical cues remaining from the host or the host's microhabitat.

Genetic effects may be equally important, however. Many researchers recognize the inadvisability of testing for preimaginal conditioning by contrasting the oviposition behaviors of populations reared for many generations on different hosts because of the possibility that the populations may be genetically differentiated (e.g., Papaj and Rausher, 1983; Vet, 1983; Debolt, 1989). Less frequently considered, however, is the possibility that a single generation of rearing on different hosts (the minimum sufficient to generate the conditioning treatments) may also change resulting adult preferences through strictly genetic effects. In the typical experiment in which a group of larvae is reared on two different hosts, this may happen in either of two ways. First, if ovipositing females are allowed to actively choose between the two hosts to create the larval populations used in the conditioning treatments, selection can act directly on within-population genetic variation in host preference (Jaenike, 1982, 1989; Singer et al., 1988; Prevost and Lewis, 1990). Second, even if the experimenter intervenes to randomly allocate eggs to the two host treatments (e.g., Jaenike, 1982, 1983; Rausher, 1983; Taylor, 1986), selection can still act indirectly on adult preference if there exists within-population genetic covariation of larval performance (i.e., the ability of larvae to develop successfully on the host) and adult preference (Thompson, 1988; Singer et al., 1988).

How then are experiments to be designed and analyzed to differentiate conditioning from genetic effects? Experiments that combine quantitative genetic designs with conditioning treatments appear to achieve a partitioning of observed behavioral variation into environmental (i.e., learning-based) and genetic components (e.g., Rausher, 1983; Taylor, 1986). A closer inspection reveals, however, that this partitioning may be incomplete; genetic effects can still contribute to the environmental component in these experimental designs if genetic variation for preference or performance-preference genetic covariation exists. [Within-family or within-strain genetic covariation of performance and preference may have generated this effect in the studies of Rausher (1983) and Taylor (1986), respectively.]

To exclude a genetic response to selection pressure during the conditioning treatments, we need to demonstrate that either (1) there is no within-population genetic variation for preference or performance or (2) there is no selection pressure. Genetic variation can be shown experimentally to be absent (or, at least, not expressed), or can be inferred to be absent if we (1) use the clonal offspring of a parthenogenetically reproducing individual in the experiment (e.g., Via 1989) or (2) remove genetic variation from a sexually reproducing population through a program of inbreeding. (In some cases, however, inbreeding will only reduce genetic variability rather than completely eliminate it.) Alternatively, if naturally laid eggs are harvested manually and then randomly reallocated to hosts by the experimenter, and there is no mortality during the larval growth period, then the selection pressure is zero and genetic factors can be ruled out. Only by clearly removing the opportunity for genetic factors to operate can the putative effects of larval conditioning be isolated.

Example 2: Learning vs. Changes in Perceived Host Availability and/or Egg Load in Studies of Induction of Preference for Oviposition Site

A substantial research effort has recently been focused on determining how acceptance of hosts for oviposition by herbivorous "parasites" (e.g., fruit flies, bruchid beetles) or parasitoids is influenced by learning. Most studies have contrasted the behavior of insects with different ovipositional experiences, for example comparing insects that (1) are "naïve" (i.e., no previous host contact), (2) have had a period of ovipositional activity on host A, and (3) have had a period of ovipositional activity on host B. Conditioning on host A or B might typically comprise a period of hours or days of foraging and oviposition in a small arena with excess hosts. Statistically significant differences among groups in posttreatment behavior are then generally interpreted as manifestations of learning ("induced preferences").

Insects with different ovipositional histories are likely to differ, however, in at least three ways (1) learning-based changes to the nervous system, (2) perception of the density and quality of available hosts, and (3) egg load, the number of mature oocytes present in the ovaries (e.g., Mangel, 1989). There is little agreement among researchers as to whether changes in perceived host availability following foraging experience should be considered a form of learning. Here I sidestep this issue and simply treat induction of preference, perceived host availability, and egg load as three potentially important influences on insect behavior whose roles we may wish to evaluate independently.

Because the ability of perceived host availability and egg load to influence oviposition decisions has until recently remained obscure, few researchers have explicitly considered the roles of these variables in their learning experiments (for noteworthy exceptions see Prokopy et al., 1986; Papaj et al., 1989; Drost and Cardé, 1990). Recent studies have demonstrated, however, that perceived host availability and egg load may have profound influences on virtually all aspects of insect foraging and oviposition decisions. The bulk of these studies has documented differences following some sort of experimentally enforced partial or complete host deprivation. Deprivation treatments generally have simultaneous influences on perceived host availability and egg load; thus, some (largely unknown) combination of the effects of perceived host availability and egg load has been observed to influence host acceptance (i.e., host range or diet breadth; Ikawa and Suzuki, 1982; Singer, 1982; Roitberg and Prokopy, 1983; Root and Kareiva, 1984; Fitt, 1986; Reeve, 1987; Simbolotti et al., 1987, Harris and Rose, 1989; Odendaal and Rausher, 1990; Vökl and Mackauer, 1990), clutch size (Podoler et al., 1978; van Lenteren and DeBach, 1981; Ikawa and Suzuki, 1982; Ikawa and Okabe, 1985; Pilson and Rausher, 1988; Strand and Godfray, 1989; Fitt, 1990; Odendaal and Rausher, 1990; Tatar, 1991), and searching intensity (Jones, 1977; Pak et al., 1985; Collins and Dixon, 1986; Donaldson and Walter, 1988; Odendaal, 1989; Odendaal and Rausher, 1990).

For an example, consider the isolated influence of egg load on foraging and oviposition by the gregarious parasitoid *Aphytis lingnanensis* (Hymenoptera: Aphelinidae) (Rosenheim and Rosen, 1991). Egg load was manipulated by exploiting size-related variation in parasitoid fecundity and by holding parasitoids at different temperatures to modulate the rate of egg production; thus, egg load was manipulated without concurrent changes in perceived host availability or opportunity to learn. Egg load strongly influenced parasitoid searching intensity, clutch sizes allocated to large hosts, and total host handling time (Fig. 11.3); no effect of egg load on acceptance of a small host was observed (data not shown). Egg load did not appear to operate as a threshold trait; parasitoid behavior changed progressively as egg load increased from 3 to 23. These results combined with the studies of host deprivation cited above suggest that egg load and perceived host availability must be considered along with learning when interpreting any experiment that uses treatments of differential access to hosts.

How severe a problem is the confounding of learning experiments by correlated variation in egg load and perceived host availability? First, the magnitude of the problem is likely to depend on the degree to which different conditioning treatments generate different levels of ovipositional activity. Thus, comparisons between naïve females and females given access to hosts will frequently be seriously confounded. Comparisons between treatments comprising access to different hosts types will also generally confound learning, perceived host availability, and egg load effects. Insects rarely respond equally to different host types; the relatively few studies that have quantified ovipositional activity during conditioning treatments on different host species have uniformly documented large differences in the number of eggs deposited (Papaj et al., 1987; Prokopy and Fletcher, 1987; Prokopy et al., 1990; Thomson and Stinner, 1990). Conditioning treatments that involve contact with host-related semiochemicals without actual host encounters and opportunities to oviposit, for example, allowing parasitoids to contact chemical cues associated with host frass, webbing, or host-generated feeding damage (Lewis and Tumlinson, 1988; Prevost

A

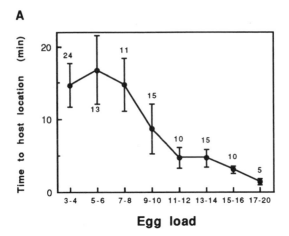

B

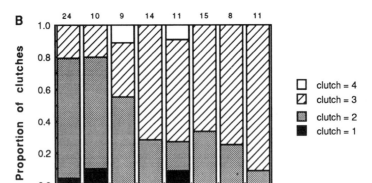

clutch = 4
clutch = 3
clutch = 2
clutch = 1

C

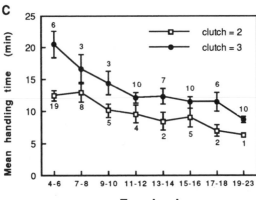

and Lewis, 1990; Turlings et al., 1990), control for egg-load effects but may still influence perceived host availability.

The degree to which learning effects are confounded by egg load and perceived host availability is also likely to vary greatly with the type of experimental design employed. Papaj and Rausher (1983) have suggested that learning may influence host preferences in two ways: first, by generating changes in the rank order of preferences and second, by generating changes in the degree of preference. Because egg load and perceived host availability are likely to generate changes in the degree of preference but unlikely to generate changes in the rank order of preference, studies that observe changes in rank order of preference after exposure to different host types may still reliably infer the operation of learning. The magnitude of the learning effect will still be difficult or even impossible to assess, however, without accounting for the effects of egg load and perceived host availability.

A worst-case scenario will perhaps make the point most clearly. Considerable attention has focused on the question of whether induction of preference involves primarily (1) learning to increase acceptance of familiar hosts, (2) learning to decrease acceptance of novel hosts, or (3) some combination of (1) and (2) (Cooley et al., 1986; Prokopy et al., 1986, 1989a,b; Prokopy and Fletcher, 1987; Papaj et al., 1989; Prokopy and Papaj, 1988; Lewis et al., 1990; Vet et al., 1990). Figure 11.4 shows how a hypothetical experiment designed to distinguish between alternative hypotheses (1) and (2) by comparing naive parasitoids with parasitoids experienced with oviposition on one of two experimental hosts could be rendered uninterpretable by the simultaneous action of egg load and learning effects. Assume that a conditioning treatment of oviposition on host A generates a substantial change in egg load and/or perceived host availability; perceived host availability will henceforth be ignored in this example, but will generally tend to amplify the effects of egg load. The influences on host acceptance patterns of six possible combinations of egg

---

Figure 11.3. Influence of egg load on foraging and oviposition by the parasitoid *Aphytis lingnanensis* (from Rosenheim and Rosen, 1991). A, Influence of egg load on the time required to discover a host (a third-instar armored scale, *Aonidiella aurantii*) in a small foraging arena. Shown are the mean search times ± 1 SE; numbers above or below the SE bars are sample sizes. B, Influence of egg load on clutch-size decisions. Numbers above columns are sample sizes. C, Influence of egg load on total host handling time when clutches of two or three eggs were deposited. Shown are the mean handling times ± 1 SE; numbers above or below the SE bars are sample sizes. For A, B, and C the effect of egg load is significant, $p < 0.001$.

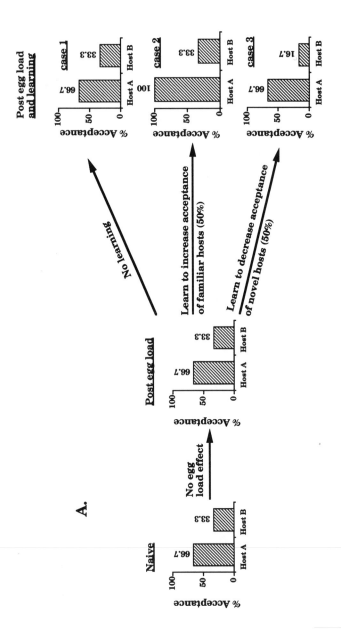

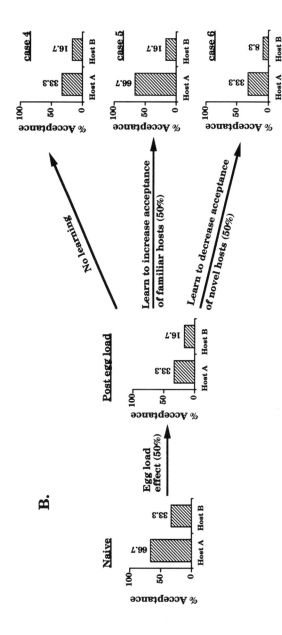

Figure 11.4. Schematic depiction of a hypothetical experiment that confounds the effects of learning and egg load. Naïve individuals are compared with individuals experiencing a period of ovipositional activity on host A. Cases 1–3, egg load has no effect; cases 4–6, decreased egg load in experienced females generates a 50% decrease in acceptance of hosts A and B. Cases 1 and 4, learning has no effect on host acceptance; cases 2 and 5, learning generates a 50% increase in acceptance of the familiar host (host A); cases 3 and 6, learning generates a 50% decrease in acceptance of the novel host (host B). For ease of presentation only, egg load and learning effects are depicted as if they occurred sequentially. Note that the outcomes of cases 3 and 5 are identical.

load effect (absent v. present) and learning effect [(1) absent, (2) learning causes increased acceptance of the familiar host, host A, and (3) learning causes decreased acceptance of the novel host, host B] are depicted. The key point is that two of the predicted outcomes are identical; oviposition by parasitoids that exhibit no egg-load effect and learn to decrease acceptance of novel hosts (case 3) is identical to oviposition by parasitoids that exhibit an egg-load effect and learn to increase acceptance of familiar hosts (case 5). This experiment cannot, therefore, reliably distinguish between these two alternative explanations. If egg load were to influence acceptance levels for hosts A and B differentially (which theory suggests will be common; e.g., Iwasa et al., 1984), a large range of additional outcomes could be generated in our hypothetical experiment, further complicating attempts to disentangle the effects of learning and egg load.

Designing experiments to distinguish the effects of learning, perceived host availability, and egg load will not always be easy; for two explicit attempts to deal with this problem, see Prokopy et al. (1986) and Rosenheim and Rosen (1991). Effects of egg load can generally be removed statistically by simply dissecting insects after behavioral assays and including egg load along with learning treatments as independent variables in multivariate statistical analyses. Reliably distinguishing the influences of preference induction and perceived host availability will be less easy, because neither can be measured directly in the way that egg load can.

The primary purpose of this discussion has been to highlight the complexity of reliably isolating learning effects from the many other possible bases for variable insect behavior. The two examples are, I feel, representative of a broader class of problems (e.g., Hoffmann, 1985; Hoffmann and Turelli, 1985) that must be addressed to improve our understanding of insect learning.

Future Directions

Comparative studies can complement manipulative experiments in addressing questions of the function and phylogeny of learning. We have an opportunity to capitalize on the very substantial recent advances in analytical techniques for comparative data to produce valid statistical inferences concerning comparative hypotheses; the increasing availability of molecular phylogenies should catalyze these investigations. Because of the many difficulties encountered when attempting to compile a comparative data set from the literature, experimental studies that test a number of species with a single experimental protocol may hold the greatest promise. The choice of meaningful ecological settings for these comparative studies will also be important. Although improperly performed comparative studies are associated with a number of potentially serious procedural and interpretational limitations, solutions to these problems are being developed, and these limitations need not, therefore, be crippling.

The recent surge of interest in insect learning has resulted in a wave of studies documenting within- and between-individual variability in behavior, most of which has been attributed to learning. Other bases for variable behavior, including most notably various physiological (endogenous) factors and genetic variability, have received proportionately little attention. Furthermore, many studies, both comparative and experimental, potentially confound multiple bases for variable behavior. This discussion has attempted to highlight these problems and thereby encourage experimentation that will distinguish between alternative proximate causes of variable behavior.

Acknowledgments

I thank H. Dingle, O.P.J.M. Minkenberg, D.R. Papaj, P.S. Ward, and an anonymous reviewer for their very helpful critical reviews of the manuscript and B.C. Murphy for assistance with creating the figures. I gratefully acknowledge the support of the Fulbright Foundation and the Hebrew University of Jerusalem, where initial work on this review was conducted.

References

Alphen, J.J.M. van, and Visser, M.E. 1990. Superparasitism as an adaptive strategy for insect parasitoids. Annu. Rev. Entomol. **35**:59–79.

Baker, R.R. 1978. The Evolutionary Ecology of Animal Migration. Holmes & Meier, New York.

Barlow, G.W. 1989. Has sociobiology killed ethology or revitalized it? In P.P.G. Bateson and P.H. Klopfer (eds.), Perspectives in Ethology, Vol. 8. Plenum Press, New York, pp. 1–45.

Bell, G. 1989. A comparative method. Am. Nat. **133**:553–571.

Bernays, E.A. 1988. Host specificity in phytophagous insects: Selection pressure from generalist predators. Entomol. Exp. Appl. **49**:131–140.

Bernays, E.A., and Lee, J.C. 1988. Food aversion learning in the polyphagous grasshopper *Schistocerca americana*. Physiol. Entomol. **13**:131–137.

Bitterman, M.E. 1965. Phyletic differences in learning. Am. Psychol. **20**:396–410.

Brattsten, L.B., and Metcalf, R.L. 1970. The synergistic ratio of carbaryl with piperonyl butoxide as an indicator of the distribution of multifunction oxidases in the Insecta. J. Econ. Entomol. **63**:101–104.

Cheverud, J.M., Dow, M.M., and Leutenegger, W. 1985. The quantitative assessment of phylogenetic constraints in comparative analyses: Sexual dimorphism in body weight among primates. Evolution **39**:1335–1351.

Chew, F.S. 1980. Foodplant preferences of *Pieris* caterpillars (Lepidoptera). Oecologia **46**:347–353.

Clutton-Brock, T.H., and Harvey, P.H. 1984. Comparative approaches to investigating adaptation. In J.R. Krebs and N.B. Davies (eds.), Behavioural Ecology, an Evolutionary Approach. Blackwell Scientific, Oxford, England, pp. 7–29.

Collins, M.D., and Dixon, A.F.G. 1986. The effect of egg depletion on the foraging behaviour of an aphid parasitoid. J. Appl. Entomol. **102**:342–352.

Cooley, S.S., Prokopy, R.J., McDonald, P.T. and Wong, T.T.Y. 1986. Learning in oviposition site selection by *Ceratitis capitata* flies. Entomol. Exp. Appl. **40**:47–51.

Corbet, S.A. 1985. Insect chemosensory responses: A chemical legacy hypothesis. Ecol. Entomol. **10**:143–153.

Daly, M., Rauschenberger, J., and Behrends, P. 1982. Food aversion learning in kangaroo rats: A specialist-generalist comparison. Anim. Learn. Behav. **10**:314–320.

Darwin, C. 1859. The Origin of Species. Penguin, New York.

de Boer, G., and Hanson, F.E. 1984. Foodplant selection and induction of feeding preference among host and non-host plants in larvae of the tobacco hornworm *Manduca sexta*. Entomol. Exp. Appl. **35**:177–193.

Debolt, J.W. 1989. Host preference and acceptance by *Leiophron uniformis* (Hymenoptera: Braconidae): Effects of rearing on alternate *Lygus* (Heteroptera: Miridae) species. Ann. Entomol. Soc. **82**:399–402.

Dethier, V.G. 1980. Food-aversion learning in two polyphagous caterpillars, *Diacrisia virginica* and *Estigmene congrua*. Physiol. Entomol. **5**:321–325.

Dethier, V.G., and Yost, M.T. 1979. Oligophagy and absence of food-aversion learning in tobacco hornworms, *Manduca sexta*. Physiol. Entomol. **4**:125–130.

Dobson, F.S. 1985. The use of phylogeny in behavior and ecology. Evolution **39**:1384–1388.

Domjan, M., and Galef, B.G., Jr. 1983. Biological constraints on instrumental and classical conditioning: retrospect and prospect. Anim. Learn. Behav. **11**:151–161.

Donaldson, J.S., and Walter, G.H. 1988. Effects of egg availability and egg maturity on the ovipositional activity of the parasitic wasp, *Coccophagus atratus*. Physiol. Entomol. **13**:407–417.

Donoghue, M.J. 1989. Phylogenies and the analysis of evolutionary sequences, with examples from seed plants. Evolution **43**:1137–1156.

Drost, Y.C., and Cardé, R.T. 1990. Influence of experience on the sequential and temporal organization of host-acceptance behavior in *Brachymeria intermedia* (Chalcididae), an endoparasitoid of gypsy moth. J. Insect Behav. **3**:647–661.

Endler, J.A. 1986. Natural Selection in the Wild. Princeton University Press, Princeton, NJ.

Felsenstein, J. 1985. Phylogenies and the comparative method. Am. Nat. **125**:1–15.

Fitt, G.P. 1986. The influence of a shortage of hosts on the specificity of oviposition behaviour in species of *Dacus* (Diptera, Tephritidae). Physiol. Entomol. **11**:133–143.

Fitt, G.P. 1990. Comparative fecundity, clutch size, ovariole number and egg size of *Dacus tryoni* and *D. jarvisi*, and their relationship to body size. Entomol. Exp. Appl. **55**:11–21.

Fox, L.R., and Morrow, P.A. 1981. Specialization: Species property or local phenomenon? Science **211**:887–893.

Gittleman, J.L. 1989. The comparative approach in ethology: Aims and limitations. In P.P.G. Bateson and P.H. Klopfer (eds.), Perspectives in Ethology, Vol. 8. Plenum Press, New York, pp. 55–83.

Gould, J.L. 1984. Natural history of honey bee learning. In P. Marler and H.S. Terrace (eds.), The Biology of Learning. Springer-Verlag, Berlin, pp. 149–180.

Gould, J.L., and Marler, P. 1984. Ethology and the natural history of learning. In P. Marler and H.S. Terrace (eds.), The Biology of Learning. Springer-Verlag, Berlin, pp. 47–74.

Grafen, A. 1989. The phylogenetic regression. Phil. Trans. R. Soc. Lond. B **326**:119–157.

Greenberg, R. 1985. A comparison of foliage discrimination learning in a specialist and a generalist species of migrant wood warbler (Aves: Parulidae). Can. J. Zool. **63**:773–776.

Greenfield, M.D., Alkaslassy, E., Wang, G.-Y., and Shelly, T.E. 1989. Long-term memory in territorial grasshoppers. Experientia **45**:775–777.

Guilford, T., and Stamp Dawkins, M. 1987. Search images not proven: A reappraisal of recent evidence. Anim. Behav. **35**:1838–1845.

Hanson, F.E. 1976. Comparative studies on induction of food choice preferences in lepidopterous larvae. In T. Jermy (ed.), The Host-Plant in Relation to Insect Behaviour and Reproduction. Plenum Press, New York, pp. 71–77.

Harris, M.O., and Rose, S. 1989. Temporal changes in the egglaying behaviour of the Hessian fly. Entomol. Exp. Appl. **53**:17–29.

Hedrick, P.W., Barker, J.S.F., and Armstrong, T. 1990. Effect of adult experience on oviposition choice and short-distance attraction in *Drosophila buzzatii*. J. Insect Behav. **3**:689–697.

Hodos, W., and Campbell, C.B.G. 1969. *Scala naturae*: Why there is no theory in comparative psychology. Psychol. Rev. **76**:337–350.

Hoffmann, A.A. 1985. Effects of experience on oviposition and attraction in *Drosophila*: comparing apples and oranges. Am. Nat. **126**:41–51.

Hoffmann, A.A., and Turelli, M. 1985. Distribution of *Drosophila melanogaster* on alternative resources: Effects of experiences and starvation. Am. Nat. **126**:662–679.

Huey, R.B. 1987. Phylogeny, history, and the comparative method. In M.E. Feder, A.F. Bennett, W.W. Burggren, and R.B. Huey (eds.), New Directions in Ecological Physiology. Cambridge University Press, Cambridge, England, pp. 76–98.

Huey, R.B., and Bennett, A.F. 1987. Phylogenetic studies of coadaptation: Preferred temperatures versus optimal performance temperatures of lizards. Evolution **41**:1098–1115.

Hurlbert, S.H. 1984. Pseudoreplication and the design of ecological field experiments. Ecol. Monogr. **54**:187–211.

Iersel, J.J.A. van. 1975. The extension of the orientation system of *Bembix rostrata* as used in the vicinity of its nest. In G. Baerends, C. Beer, and A. Manning (eds.), Function and Evolution in Behaviour. Clarendon, Oxford, England, pp. 142–168.

Ikawa, T., and Okabe, H. 1985. Regulation of egg number per host to maximize the reproductive success in the gregarious parasitoid, *Apanteles glomeratus* L. (Hymenoptera: Braconidae). Appl. Entomol. Zool. **20**:331–339.

Ikawa, T., and Suzuki, Y. 1982. Ovipositional experience of the gregarious parasitoid, *Apanteles glomeratus* (Hymenoptera: Braconidae), influencing her discrimination of the host larvae, *Pieris rapae crucivora*. Appl. Entomol. Zool. **17**:119–126.

Iwasa, Y., Suzuki, Y., and Matsuda, H. 1984. Theory of oviposition strategy of parasitoids. I. Effect of mortality and limited egg number. Theor. Popul. Biol. **26**:205–227.

Jaenike, J. 1982. Environmental modification of oviposition behavior in *Drosophila*. Amer. Nat. **119**:784–802.

Jaenike, J. 1983. Induction of host preference in *Drosophila melanogaster*. Oecologia **58**:320–325.

Jaenike, J. 1988. Effects of early adult experience on host selection in insects: Some experimental and theoretical results. J. Insect Behav. **1**:3–15.

Jaenike, J. 1989. Genetic population structure of *Drosophila tripunctata*: Patterns of variation and covariation of traits affecting resource use. Evolution **43**:1467–1482.

Jaisson, P. 1980. Environmental preference induced experimentally in ants (Hymenoptera: Formicidae). Nature **286**:388–389.

Jermy, T. 1987. The role of experience in the host selection of phytophagous insects. In E.A. Bernays and J.G. Stoffolano, Jr. (eds.), Perspectives in Chemoreception and Behavior. Springer-Verlag, New York, pp. 143–157.

Jermy, T., Hanson, F.E., and Dethier, V.G. 1968. Induction of specific food preference in lepidopterous larvae. Entomol. Exp. Appl. **11**:211–230.

Johnston, T.D. 1982. Selective costs and benefits in the evolution of learning. Adv. Study Behav. **12**:65–106.

Jones, R.E. 1977. Movement patterns and egg distribution in cabbage butterflies. J. Anim. Ecol. **46**:195–212.

Kalat, J.W. 1985. Taste-aversion learning in ecological perspective. In T.D. Johnston and A.T. Pietrewicz (eds.), Issues in the Ecological Study of Learning. Lawrence Erlbaum Associates, Hillsdale, NJ, pp. 119–141.

Kamil, A.C., and Yoerg, S.I. 1982. Learning and foraging behavior. In P.P.G. Bateson and P.H. Klopfer (eds.), Perspectives in Ethology, Vol. 5. Plenum Press, New York, pp. 325–364.

Kamil, A.C., and Clements, K.C. 1990. Learning, memory, and foraging behavior. In D.A. Dewsbury (ed.), Contemporary Issues in Comparative Psychology. Sinauer Associates, Sunderland, MA, pp. 7–30.

Karowe, D.N. 1989. Facultative monophagy as a consequence of prior feeding experience: Behavioral and physiological specialization in *Colias philodice* larvae. Oecologia **78**:106–111.

Lauder, G.V. 1981. Form and function: Structural analysis in evolutionary morphology. Paleobiology **7**:430–442.

Lee, J.C., and Bernays, E.A. 1988. Declining acceptability of a food plant for the polyphagous grasshopper *Schistocerca americana*: The role of food aversion learning. Physiol. Entomol. **13**:291–301.

Lee, J.C., and Bernays, E.A. 1990. Food tastes and toxic effects: Associative learning by the polyphagous grasshopper *Schistocerca americana* (Drury) (Orthoptera: Acrididae). Anim. Behav. **39**:163–173.

Lenteren, J.C. van, and DeBach, P. 1981. Host discrimination in three ectoparasites (*Aphytis coheni*, *A. lingnanensis* and *A. melinus*) on the oleander scale (*Aspidiotus nerii*). Netherlands J. Zool. **31**:504–532.

Lewis, A.C., and Lipani, G.A. 1990. Learning and flower use in butterflies: Hypotheses from honey bees. Focus on Insect–Plant Interactions, E.A. Bernays (ed.), Vol. II. CRC Press, Boca Raton, FL, pp. 95–110.

Lewis, W.J., and Tumlinson, J.H. 1988. Host detection by chemically mediated associative learning in a parasitic wasp. Nature **331**:257–259.

Lewis, W.J., Vet, L.E.M., Tumlinson, J.H., Lenteren, J.C. van, and Papaj, D.R. 1990. Variations in parasitoid foraging behavior: Essential element of a sound biological control theory. Environ. Entomol. **19**:1183–1193.

Machlis, L., Dodd, P.W.D., and Fentress, J.C. 1985. The pooling fallacy: Problems arising when individuals contribute more than one observation to the data set. Z. Tierpsychol. **68**:201–214.

Maddison, W.P. 1990. A method for testing the correlated evolution of two binary characters: Are gains or losses concentrated on certain branches of a phylogenetic tree? Evolution **44**:539–557.

Mallet, J., Longino, J.T., Murawski, D., Murawski, A., and De Gamboa, A.S. 1987. Handling effects in *Heliconius*: Where do all the butterflies go? J. Anim. Ecol. **56**:377–386.

Mangel, M. 1989. An evolutionary interpretation of the "motivation to oviposit." J. Evol. Biol. **2**:157–172.

McGuire, T.R., Tully, T., and Gelperin, A. 1990. Conditioning odor-shock associations in the black blowfly, *Phormia regina*. J. Insect Behav. **3**:49–59.

Menzel, R. 1983. Neurobiology of learning and memory: The honeybee as a model system. Naturwissenschaften **70**:504–511.

Menzel, R. 1985. Learning in honey bees in an ecological and behavioral context. In B. Hölldobler and M. Lindauer (eds.), Experimental Behavioral Ecology and Sociobiology. Sinauer, Sunderland, MA. pp. 55–74.

Odendaal, F.J. 1989. Mature egg number influences the behavior of female *Battus philenor* butterflies. J. Insect Behav. **2**:15–25.

Odendaal, F.J., and Rausher, M.D. 1990. Egg load influences search intensity, host selectivity, and clutch size in *Battus philenor* butterflies. J. Insect Behav. **3**:183–193.

Pagel, M.D., and Harvey, P.H. 1988. Recent developments in the analysis of comparative data. Q. Rev. Biol. **63**:413–440.

Pagel, M.D., and Harvey, P.H. 1989. Comparative methods for examining adaptation depend on evolutionary models. Folia Primatol. **53**:203–220.

Pak, G.A., Halder, I. van, Lindeboom, R., and Stroet, J.J.G. 1985. Ovarian egg supply, female age and plant spacing as factors influencing searching activity in the egg parasite *Trichogramma* sp. Mededeling. Fac. Landbouww. Rijksuniv. Gent. **50**:369–378.

Papaj, D.R. 1986. Interpopulation differences in host preference and the evolution of learning in the butterfly, *Battus philenor*. Evolution **40**:518–530.

Papaj, D.R., Opp, S.B., Prokopy, R.J., and Wong, T.T.Y. 1989. Cross-induction of fruit acceptance by the medfly *Ceratitis capitata*: The role of fruit size and chemistry. J. Insect Behav. **2**:241–254.

Papaj, D.R., and Prokopy, R.J. 1989. Ecological and evolutionary aspects of learning in phytophagous insects. Annu. Rev. Entomol. **34**:315–350.

Papaj, D.R., Prokopy, R.J., McDonald, P.T., and Wong, T.T.Y. 1987. Differences in learning between wild and laboratory *Ceratitis capitata* flies. Entomol. Exp. Appl. **45**:65–72.

Papaj, D.R., and Rausher, M.D. 1983. Individual variation in host location by phytophagous insects. In S. Ahmad (ed.), Herbivorous Insects, Host-Seeking Behavior and Mechanisms. Academic Press, New York, pp. 77–124.

Parker, G.A. 1984. Evolutionarily stable strategies. In J.R. Krebs and N.B. Davies (eds.), Behavioural Ecology, an Evolutionary Approach. Blackwell Scientific, Oxford, England, pp. 30–61.

Pietrewicz, A.T., and Richards, J.B. 1985. Learning to forage: An ecological perspective. In T.D. Johnston and A.T. Pietrewicz (eds.), Issues in the Ecological Study of Learning. Lawrence Erlbaum Associates, Hillsdale, NJ, pp. 99–117.

Pilson, D., and Rausher, M.D. 1988. Clutch size adjustment by a swallowtail butterfly. Nature **333**:361–363.

Podoler, H., Rosen, D., and Sharoni, M. 1978. Ovipositional responses to host density in *Aphytis holoxanthus* (Hymenoptera: Aphelinidae), an efficient gregarious parasite. Ecol. Entomol. **3**:305–311.

Prevost, G., and Lewis, W.J. 1990. Heritable differences in the response of the braconid wasp *Microplitis croceipes* to volatile allelochemicals. J. Insect Behav. **3**:277–287.

Prokopy, R.J., Cooley, S.S., and Opp, S.B. 1989b. Prior experience influences the fruit residence of male apple maggot flies, *Rhagoletis pomonella*. J. Insect Behav. **2**:39–49.

Prokopy, R.J., and Fletcher, B.S. 1987. The role of adult learning in the acceptance of host fruit for egglaying by the Queensland fruit fly, *Dacus tryoni*. Entomol. Exp. Appl. **45**:259–263.

Prokopy, R.J., Green, T.A., and Wong, T.T.Y. 1989a. Learning to find fruit in *Ceratitis capitata* flies. Entomol. Exp. Appl. **53**:65–72.

Prokopy, R.J., Green, T.A., and Vargas, R.I. 1990. *Dacus dorsalis* flies can learn to find and accept host fruit. J. Insect Behav. **3**:663–673.

Prokopy, R.J., and Papaj, D.R. 1988. Learning of apple fruit biotypes by apple maggot flies. J. Insect Behav. **1**:67–74.

Prokopy, R.J., Papaj, D.R., Cooley, S.S. and Kallet, C. 1986. On the nature of learning in oviposition site acceptance by apple maggot flies. Anim. Behav. **34**:98–107.

Rausher, M.D. 1983. Conditioning and genetic variation as causes of individual variation in the oviposition behaviour of the tortoise beetle, *Deloyala guttata*. Anim. Behav. **31**:743–747.

Reeve, J.D. 1987. Foraging behavior of *Aphytis melinus*: Effects of patch density and host size. Ecology **68**:530–538.

Ridley, M. 1983. The Explanation of Organic Diversity: The Comparative Method and Adaptations for Mating. Clarendon Press, Oxford, England.

Roitberg, B.D. 1990. Variation in behaviour of individual parasitic insects: Bane or boon? In M. Mackauer, L.E. Ehler, and J. Roland (eds.), Critical Issues in Biological Control. Intercept, Andover, England, pp. 25–39.

Roitberg, B.D., and Prokopy, R.J. 1983. Host deprivation influence on response of *Rhagoletis pomonella* to its oviposition deterring pheromone. Physiol. Entomol. **8**:69–72.

Root, R.B., and Kareiva, P.M. 1984. The search for resources by cabbage butterflies (*Pieris rapae*): Ecological consequences and adaptive significance of Markovian movements in a patchy environment. Ecology **65**:147–165.

Rosenheim, J.A. 1987. Host location and exploitation by the cleptoparasitic wasp *Argochrysis armilla*: The role of learning (Hymenoptera: Chrysidae). Behav. Ecol. Sociobiol. **21**:401–406.

Rosenheim, J.A., Meade, T., Powch, I.G., and Schoenig, S. 1989. Aggregation by foraging insect parasitoids in response to local variations in host density: Determining the dimensions of a host patch. J. Anim. Ecol. **58**:101–117.

Rosenheim, J.A., and Rosen, D. 1991. Foraging and oviposition decisions in the parasitoid *Aphytis lingnanensis*: Distinguishing the influences of egg load and experience. J. Anim. Ecol. **60**:873–893.

Sessions, S.K., and Larson, A. 1987. Developmental correlates of genome size in plethodontid salamanders and their implications for genome evolution. Evolution **41**:1239–1251.

Sheehan, W., and Shelton, A.M. 1989. The role of experience in plant foraging by the aphid parasitoid *Diaeretiella rapae* (Hymenoptera: Aphidiidae). J. Insect Behav. **2**:743–759.

Sillén-Tullberg, B. 1988. Evolution of gregariousness in aposematic butterfly larvae: A phylogenetic analysis. Evolution **42**:293–305.

Simbolotti, G., Putters, F.A., and Assem, J. van den. 1987. Rates of attack and control of the offspring sex ratio in the parasitic wasp *Lariophagus distinguendus* in an environment where host quality varies. Behaviour **100**:1–32.

Singer, M.C. 1982. Quantification of host preference by manipulation of oviposition behavior in the butterfly *Euphydryas editha*. Oecologia **52**:224–229.

Singer, M.C., Ng, D., and Thomas, C.D. 1988. Heritability of oviposition preference and its relationship to offspring performance within a single insect population. Evolution **42**:977–985.

Sivinski, J. 1989. Mushroom body development in nymphalid butterflies: A correlate of learning? J. Insect Behav. **2**:277–283.

Stamp Dawkins, M. 1989. The future of ethology: How many legs are we standing on? In P.P.G. Bateson and P.H. Klopfer (eds.), Perspectives in Ethology, Vol. 8. Plenum Press, New York, pp. 47–54.

Strand, M.R., and Godfray, H.C.J. 1989. Superparasitism and ovicide in parasitic Hymenoptera: Theory and a case study of the ectoparasitoid *Bracon hebetor*. Behav. Ecol. Sociobiol. **24**:421–432.

Tatar, M. 1991. Clutch size in the swallowtail butterfly, *Battus philenor*: The role of host quality and egg load within and among seasonal flights in California. Behav. Ecol. Sociobiol. **28**:337–344.

Taylor, C.E. 1986. Habitat choice by *Drosophila pseudoobscura*: The roles of genotype and of experience. Behav. Genet. **16**:271–279.

Thompson, J.N. 1988. Evolutionary ecology of the relationship between oviposition preference and performance of offspring in phytophagous insects. Entomol. Exp. Appl. **47**:3–14.

Thomson, M.S., and Stinner, R.E. 1990. The scale response of *Trichogramma* (Hymenoptera: Trichogrammatidae): Variation among species in host specificity and the effect of conditioning. Entomophaga **35**:7–21.

Thornhill, R., and Alcock, J. 1983. The Evolution of Insect Mating Systems. Harvard University Press, Cambridge, MA.

Timberlake, W. 1990. Natural learning in laboratory paradigms. In D.A. Dewsbury (ed.), Contemporary Issues in Comparative Psychology. Sinauer Associates, Sunderland, MA, pp. 31–54.

Tinbergen, N. 1963. On aims and methods in ethology. Z. Tierpsychol. **20**:410–433.

Turlings, T.C.J., Scheepmaker, J.W.A., Vet, L.E.M., Tumlinson, J.H., and Lewis, W.J. 1990. How contact foraging experiences affect preferences for host-related odors in the larval parasitoid *Cotesia marginiventris* (Cresson) (Hymenoptera: Braconidae). J. Chem. Ecol. **16**:1577–1589.

Turner, J.R.G. 1981. Adaptation and evolution in *Heliconius*: A defense of neoDarwinism. Annu. Rev. Ecol. Syst. **12**:99–121.

Vet, L.E.M. 1983. Host-habitat location through olfactory cues by *Leptopilina clavipes* (Hartig) (Hym.: Eucoilidae), a parasitoid of fungivorous *Drosophila*: The influence of conditioning. Netherlands J. Zool. **33**:225–248.

Vet, L.E.M., Lewis, W.J., Papaj, D.R., and Lenteren, J.C. van. 1990. A variable-response model for parasitoid foraging behavior. J. Insect Behav. **3**:471–490.

Vet, L.E.M., and Opzeeland, K. van. 1984. The influence of conditioning on olfactory microhabitat and host location in *Asobara tabida* (Nees) and *A. rufescens* (Foerster) (Braconidae: Alysiinae) larval parasitoids of Drosophilidae. Oecologia **63**:171–177.

Via, S. 1989. Field estimation of variation in host plant use between local populations of pea aphids from two crops. Ecol. Entomol. **14**:357–364.

Völkl, W., and Mackauer, M. 1990. Age-specific pattern of host discrimination by the aphid parasitoid *Ephedrus californicus* Baker (Hymenoptera: Aphidiidae). Can. Entomol. **122**:349–361.

Ward, K.E., Tamaswamy, S.B., and Nebeker, T. E. 1990. Feeding preferences and their modification in early and late instar larvae of the bagworm, *Thyridopteryx ephemeraeformis* (Lepidoptera: Psychidae). J. Insect Behav. **3**:785–795.

Wardle, A.R., and Borden, J.H. 1985. Age-dependent associative learning by *Exeristes roborator* (F.) (Hymenoptera: Ichneumonidae). Can. Entomol. **117**:605–616.

Wasserman, S.S. 1982. Gypsy moth (*Lymantria dispar*): Induced feeding preferences as a bioassay for phenetic similarity among host plants. In J.H. Visser and A.K. Minks (eds.), Proceedings of the 5th International Symposium on Insect–Plant Relationships. Wageningen, Centre for Agricultural Publishing and Documentation, The Netherlands, pp. 261–267.

12

Application of Learning to Pest Management
Ronald J. Prokopy and W. Joseph Lewis

Introduction

There are several kinds of learning that may occur in pest and beneficial insects. Described more fully elsewhere (Papaj and Prokopy, 1989), these include habituation (a waning of response to stimuli with repeated exposure to the stimulus), sensitization (the counterpart of habituation, involving a gradual increase in response to a stimulus with repeated exposure, even when unpaired with any other stimulus), associative learning (establishment through experience of an association between two stimuli or between a stimulus and a response), and induction of preference (effects of experience on diet choice that cannot be assigned readily to any of the preceding kinds of learning). This chapter is concerned principally with how these various kinds of learning in harmful or beneficial insects might play a role in current and future strategies and tactics of pest management.

How widespread is learning in pest and and beneficial insects? Beginning with the classic investigation by Thorpe and Jones (1937) on the parasitoid *Venturia (Nemeritis) canescens*, there has been a blossoming of studies demonstrating learning in several other genera of insect parasitoid wasps: for example, in *Argochrysis, Asobara, Brachymeria, Bracon, Campolitis, Cotesia, Diaeretiella, Exeristes, Itoplectis, Leptopilina*, and *Trichogramma* wasps (Vet et al., 1990; Turlings et al., this volume). There are also notable cases of demonstrated learning in several genera of predaceous arthropods: for example, *Anax* dragonfly larvae (Blois and Cloarec, 1985); *Phytoseiulus* predatory mites (Dicke et al., 1990); *Pterostichus* carabid beetles (Plotkin, 1979); *Stagmatoptera* praying mantids (Maldonado et al., 1979); *Stethorus* coccinellid beetles (Houck, 1986); and *Tenthredo* sawfly adults (Pasteels and Gregoire, 1984). Among insects that are pests, current appreciation of the breadth of learning may be deceptive. This chapter focuses on pests of plants, and indeed there are insects in several genera known to feed on plants that exhibit learning behavior: for example, *Battus*, *Colias*, *Heliconius*, and *Pieris* butterfly adults; *Acrolepiopsis* leek moths; *Ceratitis*, *Dacus*, and *Rhagoletis* fruit fly adults; *Schistocerca* and *Locusta* locust and *Melanoplus* grasshopper nymphs; *Callosobruchus*, *Deloyala*, *Haltica*, and *Leptinotarsa* beetles; and many species of lepidopterous larvae (Jermy, 1987; Papaj and Prokopy, 1986, 1989; Szentesi and Jermy, 1990). We suggest, however, that herbivores in these genera (several of which represent important pests) are only a small proportion of insect phytophages that do in fact learn something about their environment. The proportion is almost certainly not as great as with hymenopterous parasitoids (where learning may be ubiquitous) or predaceous arthropods (where it may be high). But we anticipate that once appropriate and rigorous experimentation is undertaken, the proportion of insect plant pests that learn will turn out to be considerably greater than is currently recognized.

Before proceeding, it may be useful to discuss briefly the context within which we will view learning. It is a context that considers the insect as forager for essential resources such as food, mates, egg-laying sites, or refugia and that considers the sources of variation that shape behaviors associated with resource finding, examination, and acceptance. As pointed out by Papaj and Rausher (1983), Bell (1990), Rosenheim (this volume), and others, principal sources of variation that affect behavioral decisions of a foraging insect are the current state of the environment (e.g., spatial and temporal aspects of abundance, quality and distribution of resources, abiotic conditions, presence of conspecifics or enemies), the physiological state of the forager (e.g., degree of hunger or thirst, mating status, egg load, age, maternal influences), and the genetic and informational states of the forager. The degree to which learning, as a component of informational state, affects foraging cannot be discerned without careful attention to all of these potentially interacting variables.

It may likewise be helpful to provide a brief perspective on the philosophy and practice of agricultural pest management over the course of this century (Dethier, 1976; Prokopy, 1986). Up to the 1950s, the pest-control practices of most growers in developed countries contravened natural processes only to a moderate degree. Many species of beneficial insects were able to survive treatments of botanical and inorganic pesticides used to control key insect and disease pests. Predators and parasites, together with the planting of cultivars or biotypes of crops at least partly resistant to pest attack, formed the foundation for other pest-management practices. From the 1950s through much of the 1970s, coincident with massive introduction of synthetic organic pesticides, growers profoundly intervened with nature in their use (often overuse) of pesticides as a "magic bullet" that would cure their pest problems and eliminate the need to consider behavioral, ecological and evolutionary processes in pest management. Since the mid-1970s, following large outbreaks of pests due to resistance to pesticides, growers and other pest managers have come to realize that future success in pest management can be achieved best through an ecological approach to control within a framework of integrated pest management (IPM). The philosophy of IPM is holistic, emphasizing integration of several diverse approaches to maintaining pests below damaging numbers and giving attention to effects of human intervention on multiple components of the ecosystem.

To date, the practice of IPM has centered largely on use of temperature-driven models to predict rates of pest (and beneficial organism) development and the use of a variety of techniques to sample first appearance and abundance of pests and beneficials on site in commercial plantings. For control of pests, treatment is with selective, properly timed pesticides when pests are not effectively suppressed by natural enemies. Recently, there has been increased interest in integrated use of relevant ecologically sound tactics of pest management, including biological control, environmental management, genetic manipulation, and behavorial manipulation methods, with pesticides employed only as a last resort.

In this chapter, we will consider how learning in pest and beneficial insects might affect accurate sampling of populations of insect pests and beneficials in a crop as well as how learning might affect each of the various methods of pest control. Unfortunately, the number of case histories in which insect learning has been shown through experimentation to be a factor in a pest-management program already in practice is nil. For this reason, our treatment will be more in the realm of deduction and speculation than established fact. We will not deal overtly with learning in insects that are intended as biological agents of weed control, although some concerns with learning in managing pest insects through natural enemies (particularly as discussed in the section on biological control) surely apply to managing insects for weed control. Nor will we deal overtly with learning by pest-insect larvae because larvae are generally insufficiently mobile to be significantly affected by the consequences of learning in relation to pest-management tactics.

We emphasize at the outset that throughout this century, control of pests in developed countries has, for the most part, been quite effective without knowledge of insect learning on the part of pest managers, just as bee-keeping has been effective without full knowledge of bee learning behavior. We believe, however, that fuller understanding of insect learning in pest and beneficial insects and its relevance to the pest-management process will lead eventually to application of more effective and environmentally safe pest-management practices. Failure to consider the potential impact of insect learning on sampling procedures and management tactics could compromise the breadth of applicability and the durability of an apparently effective pest-management approach, much as failure to consider the evolution of pest resistance to pesticides has compromised breadth and durability of many control practices using pesticides.

Insect Learning and Sampling Populations of Pests and Natural Enemies

An accurate estimate of the population sizes of pests and beneficials in an agroecosystem comprises the foundation upon which all arthropod pest management practices are built. Without an accurate estimate, the need and timing of human intervention in the system (for example, pesticide application) and the ability to measure and predict pest population growth or decline are substantially compromised.

Southwood (1978) provides a comprehensive treatment of the numerous existing methods of sampling arthropod populations. These include methods for estimating absolute numbers of individuals per area of habitat at one time versus another time or in one habitat versus another habitat, and products or effects of individuals present in a habitat, such as frass, webs, exuvieae or evidence of damage to plants. Several commonly used absolute and relative population estimation methods involve the behavior of sampled individuals. Accuracy of such methods is therefore susceptible to change in behavior, including change as a consequence of prior experience with stimuli associated with one or more environmental-state variables. Among these stimuli might be wind, light, temperature, and humidity or moisture as components of the abiotic environment, and visual, odor, acoustical, mechanical, and tactile stimuli from the biotic environment of resources, conspecifics, and enemies.

To our knowledge, there exist no published data showing an impact of insect learning on the performance of any method of estimating abundance of an arthropod population. There do exist, however, a few lines of evidence which suggest that arthropod learning should be considered when employing certain sampling methods.

Absolute Population Density Estimation

Methods of estimating absolute population density are integral to development of life tables for determining principal causes of mortality in populations, which is an essential (though underutilized) approach in evaluating the relative impact of various predators, parasitoids, pathogens, and other agents in IPM programs. Population density estimates are also important in accurately assessing the optimum time of initiation and the progress of programs that involve mass release of sterile insects to eradicate wild populations.

One method of absolute population density estimation common to both of these purposes is the capture-recapture method (Begon, 1979). This method is rooted in the principle that if a known number of individuals were to be marked in some fashion, released, and allowed to completely mix with unmarked individuals in the population, then the ratio of marked to unmarked individuals during repeated sampling should reflect the absolute number of unmarked individuals in the population. Two approaches have been used to obtain individuals for marking: capture of members of the existing population and rearing of individuals under laboratory conditions. Under either approach, learning could modify the behavior of the marked, released individual in a way that would alter the probability of its recapture relative to that of an unmarked member of the population.

For example, a marked insect that originally was a member of the existing population may have undergone exposure to predatory birds or insects during its lifetime and become adept at escaping (Steiner, 1981; Pearson, 1985; Srygley and Chai, 1990). According to Gould (1986), recognition of potential enemies increases as a consequence of prior experience with enemies to a surprisingly great degree. A fast-approaching insect net, commonly used in capture-recapture studies (e.g., Bateman and Sonleitner, 1967; Papaj, 1986), might be perceived by a predator-experienced insect as if it were an approaching predator and elicit an avoidance or escape response. Indeed, repeated capture of the same individual over successive days could itself elicit avoidance or escape through learning. This appears to have occurred in capture-recapture studies of *Heliconius* butterflies (Mallet et al., 1987).

If marked, released individuals had not been taken initially from the wild population but had been reared under laboratory conditions, they might have learned stimuli associated with particular resources present in laboratory cages but not present in nature (Wardle and Borden, 1986, 1991; Prokopy et al., 1990a). Such learning could compromise the extent to which laboratory-reared individuals mix completely with members of the wild population, which might learn stimuli associated with naturally present resources.

Perhaps most frequently, absolute population density is estimated by sampling completely a unit area of habitat. This may involve the use of suction or rotary net traps to sample insects from a unit of air, a variety of methods to count insects upon or remove insects from a unit of vegetation, or several methods to remove and extract insects from a unit of litter or soil (Southwood, 1978). Methods used in sampling a unit of habitat are so numerous and varied that it is impractical to speculate on how the accuracy of each might be affected by learning, but a few comments may be useful.

Almost nothing is known about insects' ability to learn abiotic environmental stimuli, several of which (e.g., light, temperature, moisture) are used to induce insects to leave sampled units of plants, litter, or soil. It is known, however, that some nectivorous insects are able to time visits to flowers to coincide with periods of nectar flow (Gould, 1986; Harrison and Breed, 1987). Thus, it is conceivable that unwanted variation across sampling periods could arise if temporal learning of environmental stimuli by the target insects were not considered when designing protocols for sampling units of habitat. Some techniques for sampling a unit of vegetation (e.g., direct visual counts, jarring foliage over a framed cloth or funnel to collect fallen individuals, use of a motor-driven vacuum insect collector) involve potential disturbance of target insects by the observer or sampling device. Just as if an insect net were swung, some individuals may drop to the ground or fly away and avoid being counted or capture owing to prior experience with a predator or other disturbing agent.

As with all methods discussed here for estimating absolute population density, intensity of pressure from predators may change from one locale to the next. Such locale-associated variation ought to be considered when using methods whose accuracy is subject to avoidance of detection or capture through memory of prior experience with a predator.

Relative Population Density Estimation

Methods of estimating relative population density, being less intensive and laborious than methods of estimating absolute population density, are more widely used than the latter. Within an IPM framework, relative methods are employed extensively to determine when the density of a pest population is approaching or has reached a level requiring human intervention. This level is termed the economic injury level or action level (Horn, 1988). It is fundamental to determining when insecticide should be applied.

The most commonly used relative method, in addition to sweeping with an insect net, involves placing odor or visual traps in or nearby crop vegetation. Recently, considerable data have been published showing that a variety of herbivorous, parasitoid, and predatory insects are able to learn to respond to odor or visual stimuli associated with one or more essential resources or other environmental variables (e.g., Bernays and Wrubel, 1985; Papaj, 1986; Papaj and Prokopy, 1986; Visser and Thiery, 1986; Traynier, 1987; Prokopy et al., 1989a; Sheehan and Shelton, 1989; Turlings et al., 1989; Wardle, 1990; Dicke et al., 1990). Adjustments in both the design and placement of a trap used to monitor the abundance of the target insect ought to be made if the insect is capable of learning properties of a stimulus employed in the trap.

For example, sticky-coated plastic spheres that mimic host fruit visual characteristics are effective traps for monitoring Mediterranean fruit flies (medflies), *Ceratitis capitata* (Wiedemann) (Nakagawa et al., 1978; Katsoyannos, 1987). Following ovipositional experience in host fruit of a particular size, medfly females find inanimate fruit models of that size to a substantially greater degree than they find inanimate fruit models of other sizes (Prokopy et al., 1989a). Together, these studies suggest that the size of sphere traps used to monitor medfly females in a particular crop ought to be adjusted to be roughly equivalent to the size of individual fruit of that crop. Otherwise, captures on the sticky-sphere traps might not reflect accurately fly population size.

Similar reasoning can be applied to use of hollow prepunctured plastic spheres used in collecting eggs from wild medflies to estimate egg fertility in population-suppression programs involving involving release of sterile males (McInnis, 1989). After successive oviopositions into host fruit of a given type, medfly females learn to reject unfamiliar host fruit of different size and surface chemistry properties (Papaj et al., 1988). Thus, prepunctured plastic spheres should mimic host fruit in size and odor. Species-specific odors can be obtained by coating the sphere with surface chemical extracts of the fruit (Prokopy et al., 1990a).

Little or no attempt has been made to modify trap characteristics or trap use patterns according to stimuli learned by the insect for insects other than medfly. This could be especially important in designing odor/visual traps for estimating the abundance of parasitoid adults (Trimble et al., 1990), which are now known to be highly capable of learning to associate odor or visual stimuli with characteristics of the host habitat or host (e.g., Turlings et al., 1990, this volume).

As a final note on possible effects of learning on sampling populations of pests and beneficial insects, it should be mentioned that reliability of data obtained from any sampling method, be it an absolute or relative method, is a function of how well the sampled sites reflect the actual distribution of the population. Most insect populations in nature are clumped (Stanton, 1983). Learning to find or accept particular resources or to avoid or escape enemies could lead to an even greater-than-normal degree of clumping in a particular microhabitat patch. Learning to avoid conspecifics (Roitberg and Prokopy, 1981) could lead to a random or possible uniform dispersion pattern. In either case, the sampling scheme might require adjustment over time to reflect changes in population dispersion patterns as a consequence of learning.

Insect Learning and Tactics of Pest Management

Several tactics, alone or in combination, have been used to prevent arthropod pests of crops from reaching population levels causing economic injury or, when injury-causing levels have been reached, to reduce populations below such levels. In broad terms, these tactics consist of cultural control through habitat management; control through host plant resistance; pesticidal control; behavioral control through use of stimuli that attract, arrest, disrupt or repel pests; genetic control through release of sterile males; and biological control through beneficial natural enemies. Some of these tactics affect management of the crop environment in ways that would impact upon a pest or beneficial natural enemy. Others affect direct management of the pest or beneficial natural enemy independent of the crop environment. We are aware of no data demonstrating an effect of learning on the efficacy of any of the tactics of pest management in a real-world agricultural setting, but there is suggestive evidence.

Cultural Control Through Habitat Management

There exist a variety of cultural practices designed to reduce pest injury to acceptable levels (Herzog and Funderburk, 1986). Some of these have been in use for centuries, even millennia. Others are of more recent origin. One method is the modification of noncrop vegetation or crop vegetation through diversification, rotation, intercropping, or trap cropping. Another method is the modification of planting or harvesting date, irrigation or fertilization practices, and soil tillage systems. We will address here how pest learning might affect the outcome of modifying vegetational composition.

Population densities of specialist herbivorous insects in most (though not all) studies are lower in diverse agrohabitats than in homogeneous ones (i.e., monocultures) (Stanton, 1983; Risch et al., 1983; Andow, 1991). Three hypotheses have been put forward to explain this apparent pattern: (1) the resource concentration hypothesis, wherein herbivorous specialist insects might be more abundant in vegetationally simple habitats because, for a variety of reasons (Andow, 1991), essential resources are easier to find than in diverse habitats (Root, 1973), (2) the enemies hypothesis, wherein predators and parasitoids of herbivorous insects might be less abundant and cause less mortality in simple than in diverse habitats (Root, 1973), and (3) the plant-odor masking hypothesis, wherein nonhost plant volatiles might mask herbivore detection of host plant odor (Visser and Thiery, 1986; Nottingham, 1988). Within-field vegetational diversity may arise either through interplanting of agriculturally valuable crops or through allowing nonharvested plants (weeds) to grow within cropped areas (Altieri, 1987). In either case, the behavior of a pest insect might be modified by experience in a way that could affect foraging efficiency and plant injury under the first and third hypotheses.

Consider the cabbage butterfly, *Pieris rapae*. Australian *P. rapae* females are known to search widely for crucifer plants on which to oviposit (Jones and Ives, 1979). They also learn (through sensitization and association) visual and chemical stimuli affiliated with properties of host plants and used as cues in finding plants and for ovipositing (Traynier, 1987). Furthermore, evidence suggests that *P rapae* adults learn to find species of flowers that provide nectar (Lewis and Lipani, 1990; Lewis, this volume). However, learning to extract nectar from a novel species of flower interferes with ability to extract nectar from familiar species (Lewis, 1986). As observed frequently in vertebrates (Marler and Terrace, 1984), a new or novel experience may interfere with recall of an earlier learned experience.

Suppose periodic contact with non–host plant stimuli were to interfere with memory of host plant ovipositional stimuli in *P. rapae*, much as experience of nectar extraction from one flower type interferes with memory of nectar extraction from another flower type. One might then postulate that foraging of *P. rapae* females for oviposition sites could be less efficient in a vegetationally diverse agrohabitat than in a monoculture of host plants, despite the tendency of some populations of *P. rapae* to lay relatively few eggs within any one host patch regardless of host density (Root and Kareiva, 1984; Jones, 1987). This could apply as well to other herbivorous insects (e.g., *Colias* and *Battus* butterflies), whose memory of particular host plant types is known to diminish after frequent encounters with other host types or nonhosts (Stanton, 1984; Papaj, 1984).

To expand upon this theme, a variety of insects engage in local (area-concentrated) search as opposed to ranging after discovery and use of a rewarding resource (Bell, 1990). In addition to genetic- and physiological-state factors, learning may play a role in shaping local search behavior (Bell, 1990). Local presence of resource items other than the most recently encountered rewarding type or local presence of nonresource items could interfere with short-term memory of encounter with a rewarding resource and give rise to ranging behavior, causing the insect to bypass unused resources. This could be one cause, among others (Stanton, 1983; Kareiva, 1983), underlying the pattern of lesser damage to crop plants in vegetationally diverse agrohabitats as opposed to monocultures.

Besides the presence, *per se*, of some proportion of alternate resource items or nonresource items within a cropped area, the spatial arrangement of these items relative to recently encountered rewarding items could be decisive in affecting the degree to which memory of a successively encountered rewarding item is retained. Bumble bees learn spatial positions of flower clumps and flight paths between clumps (Heinrich, 1976). *Phyllotreta* flea beetles are known to move more readily between crucifer clumps along the length of cultivated strips than across diverse vegetation separating adjacent strips (Kareiva, 1982). If, through learning, a pest foraged more efficiently by moving up or down the length of a row rather than across rows, any potential beneficial effect of vegetational diversity might be compromised if alternate crop plants or weeds were to be planted between rows rather than both within and between rows.

The above examples would most likely result from sensitization or associative learning. Habituation as a form of learning could also play a role in shaping pest response in monocultures v. polycultures. For example, in cases in which continual exposure to the odor of neighboring nonresource plants masks response of herbivorous pests to host odor (e.g., Thiery and Visser, 1986), part of the cause might lie in habituation to nonresource and resource odor stimuli that are perceived by the insect as being similar. The net effect, however, might parallel that resulting from loss of memory of hosts through encounter with nonhost vegetation. Conversely, frequent contact with alternate crops or nonhost vegetation could result in pest habituation to physical or chemical stimuli associated with such vegetation (Traynier, 1987; but see Blaney and Simmonds, 1985), thereby negating the potential disruptive effect that such contact might have on the memory of host stimuli.

From the foregoing, it appears that were a pest insect to undergo habituation, sensitization, or associative learning in relation to one or more traits affiliated with crop or noncrop vegetation, the outcome in terms of pest injury might be beneficial, adverse, or neutral. The outcome might well depend on the ability of the herbivore in question to learn and remember plant characters and the degree of similarity (as perceived by the insect—Papaj et al., 1988) between stimuli of the principal crop and stimuli of intermixed crop or noncrop plants. Variation in interplay of these factors may in part explain some of the variation in outcome of experiments on pest abundance in monocultures versus diverse agrohabitats.

Many factors having little to do with the nature of pest/plant interactions often have a dominant influence on a grower's decision as to how to structure the composition of an agriculturally cropped field (Prokopy, 1986). Thus, even if pests learned to be more (or less) efficient in exploiting plants in monocultures than in vegetationally diverse cultures, the sorts of modification in composition and spatial arrangements of cropped fields required to take advantage of such knowledge for pest management purposes might not be agriculturally practical.

There are two sorts of cultural practices not directly connected with manipulating vegetational diversity that may be affected by pest learning. One of these involves planting a trap crop (diversionary crop) around the entire margin of a cultivated field to intercept immigrating pest adults before they move into the field (Miller and Cowles, 1990; Hokkanen, 1991). This has been achieved in the case of *Meligethes* beetles, where a border of 3–5 hectares of Chinese cabbage was used to protect interior cauliflower fields of 40–45 hectares from beetle attacks (Hokkanen et al. 1986). Beetles that accumulated in the trap crop were destroyed by pesticide. Under some circumstances, the outcome of such a practice might vary substantially according to the prior experience of the immigrating beetles. If they had prior experience with plants similar to Chinese cabbage, the practice might be even more effective than if the beetles were naïve. If they had prior experience with plants similar to cauliflower, the beetles might bypass the Chinese cabbage and the practice might be less effective than if the beetles were naïve. The second cultural practice involves sudden removal of resources from a locale inhabited by pests. This is a common technique applied in eradication programs against introduced Mediterranean fruit flies, where adults are denied future access to fruit (especially in gardens of homeowners) by stripping all fruit from a host tree (Scribner, 1983). Because medflies exhibit a strong ability to learn and remember for several days characters of host fruit on which they have recently oviposited (Papaj et al., 1987), they may engage in considerable ranging behavior in search of fruit of the type that was removed, bypassing alternate host types. Such movement would be highly counterproductive to efforts to prevent spread of medflies to yet-uninfested areas.

Control Through Host Plant Resistance

Crop plants may possess a broad range of genetically determined characteristics that confer resistance to economic damage by herbivorous insects. Three principal modalities of genetic-based resistance have been identified: antixenosis, antibiosis, and tolerance (Painter, 1951; Kogan, 1986). *Antixenosis* is manifested through plant traits that adversely affect plant finding, examining, and feeding or oviposition behavior of potential pests. Antixenosis may arise from plant physical or chemical properties that provide suboptimal host plant stimuli for a foraging herbivore or repel or deter potential consumers. *Antibiosis* encompasses adverse physiological effects that result from ingestion of plant material by a herbivore. These effects may stem from mild to acute "poisoning" of the herbivore or from some degree of nutritional inadequacy of the plant (Slanksy, 1990). *Tolerance* is the ability of a plant to withstand injury and grow adequately despite supporting a population of herbivorous insects at a density that would cause economic damage to a more susceptible plant. The extent to which each of these forms of plant resistance succeeds in allaying injury from pest herbivores may depend not only on a variety of environmental state factors and on the genetical and physiological state of members of a pest population (Kennedy et al., 1987) but also on the prior experience of the pest with plants, either to the disadvantage or advantage of pest management.

With respect to antixenosis, we believe that under certain crop structure conditions, pests that have utilized and subsequently emigrated from areas planted to susceptible cultivars of a host crop may, at least in part as a consequence of sensitization or associative learning, be unresponsive to neighboring areas of resistant cultivars. We focus, however, on that aspect of antixenosis which involves pest response to plant feeding deterrents (Jermy, 1983). Food that is unpalatable to an insect often, perhaps even in a majority of cases, is not toxic or otherwise harmful (Bernays and Chapman, 1987). In a classic study, Szentesi and Bernays (1984) allowed nymphs of the desert locust, *Schistocerca gregaria*, to feed on sorghum leaves treated with nicotine hydrogen tartrate, a slightly modified plant secondary compound that has no apparent adverse effect on *S. gregaria* (Jermy et al., 1982). Following strong rejection of nicotine-treated leaves on the first day of exposure, the nymphs soon thereafter began to feed on treated leaves at a rate approaching that on untreated leaves. Through elegant experimental design, Szentesi and Bernays (1984) were able to demonstrate that centrally mediated habituation to this feeding deterrent had occurred. Depending on concentration of deterrent substance and environmental context, other herbivorous insects likewise are believed to habituate to feeding deterrents associated with plants (Szentesi and Jermy, 1990).

In cases where (1) antixenosis via presence of a single feeding deterrent compound is the principal or sole form of plant resistance, (2) the deterrent is present in all plant parts fed upon by a pest, and (3) the crop is planted in a monoculture, it is conceivable that pest insects might habituate to the deterrent and thereby overcome plant resistance. To forestall potential negative effects of pest habituation to crop plant deterrents, an approach could be taken analogous to that of forestalling development of genetically-based resistance of pests to plant antixenotic or antibiotic factors (Kennedy et al., 1987; Gould, 1988). This might consist of employing a blend of deterrent compounds, creating heterogeneity of deterrent concentration among various plant tissues through genetic engineering, or creating heterogeneity of plant composition within a cropped area by interplanting ("pyramiding") cultivars possessing different profiles of deterrents.

With respect to antibiosis, there is a growing number of insects, particularly polyphagous species, in which successive contacts with initially acceptable plant food have been found to culminate in sudden or eventual rejection of that food through apparent aversion learning (e.g., Dethier, 1980; Blaney and Simmonds, 1985; Raffa, 1987; Bernays and Lee, 1988). Food aversion learning may involve either sensitization or true associative learning (Szentesi and Jermy, 1990). It is thought to occur as a consequence of post-ingestional metabolic upset (malaise) that arises either through poisoning of the insect by one or more ingested plant secondary compounds or through nutritional inadequacy of the ingested plant material (Bernays and Lee, 1988; Szentesi and Jermy, 1990; Waldbauer and Friedman, 1991). In either case, pests that exhibit food aversion learning may be more easily prevented from causing economic injury when the host crop is planted in monoculture than in a vegetationally diverse agrohabitat that might furnish needed alternative food for the pest to relieve malaise. This could be particularly true in cases where aversion learning leads to total rejection of initially acceptable plant material (Lee and Bernays, 1988; Bernays, this volume).

Plant tolerance of herbivorous pests is apt to be greatest when plants are provided an adequate but not overly abundant supply of nutrients (Mattson and Scriber, 1987). With some insect pests, high levels of plant nutrients such as nitrogen lead to rapid pest population increase and little or no increase in plant growth (Kortisas and Garsed, 1985). To our knowledge, there is to date only one report in which prior experience of an insect with a high level of a particular plant nutrient affects future behavior of the insect in a way interpretable as learning. Minkenberg and Fredrix (1989) found that *Liriomyza* flies that had been exposed previously to tomato plants of high nitrogen content subsequently preferred to feed and oviposit on high-nitrogen plants compared with moderate or low-nitrogen plants. Under horticultural conditions, such phenomena could accentuate population buildup of *Liriomyza* on highly fertilized tomato plants to such a point that genetically-based plant ability to tolerate injury would be compromised.

We conclude from the preceding examples that one or more kinds of learning may affect the response of an insect species to a plant resistance trait and that the consequences of learning may alter the value of that resistance trait to the benefit or detriment of managing a pest that learns. Given present knowledge, however, it would be unwise to modify current practices of employing resistant cultivars in pest-management programs on these bases alone. Genetical- and physiological-state factors may override the importance of the informational state of a foraging pest in shaping response to the agrohabitat. Altering plant resistance traits or protocols for using plants with resistant traits (e.g., monoculture vs. heteroculture) could enhance prospects for managing one type of pest (possibly by taking advantage of learning behavior) but be counterproductive to managing another type of pest (Kogan, 1986). Finally, such alteration could have a positive or negative effect on pest mortality through the action of predators, parasites, or pathogens (Kennedy et al., 1987; Slansky, 1990).

Pesticidal Control

Pesticide applied to crop plants may impinge directly on the exoskeleton of an insect, be absorbed by tarsi or other body parts in contact with treated surfaces, or be ingested during feeding. In addition to toxic effects, several sorts of sublethal effects on insect behavior are known to occur as a result of pesticide absorption or uptake (Pluthero and Singh, 1984; Haynes, 1988). These include sensory perception of the pesticide and behavioral consequences of biochemical or physiological processes generated by pesticide uptake. These effects parallel those associated with natural host plant resistance.

Given that pesticides constitute the dominant pest-management tactic in international agriculture, and given the attention currently directed at developing protocols for slowing the progress of pesticide resistance (Tabashnik, 1989), we are surprised by the lack of attention given to learning as a component of behaviorally based resistance to pesticides. Admittedly, biochemical and physiological mechanisms of resistance are much better understood than behavioral mechanisms (Sparks et al., 1989). Even so, attention to date has been almost exclusively on the role of genetical-state factors that shape behavioral resistance patterns, with little or no regard for insect physiological- and informational-state factors.

Apparent repellency or deterrency of synthetic pesticides such as pyrethroid insecticides to a variety of pest insects (Sparks et al., 1989) could stem not only from genetically-based mechanisms for sensing and avoiding related natural plant toxins (pyrethrums) but also from learning. Tarsal contact of the insect with a pyrethroid-treated surface could be followed by malaise, recovery, and aversion learning. This in fact occurred in a study on German cockroaches by Ebling et al. (1966). Roaches that had been in tarsal contact with organophosphate-treated surfaces learned by associating pesticide with location of the treated surfaces and thus were able to avoid future contact with pesticide. Application of pyrethroid insecticide in orchards is commonplace and is known to stimulate pests such as spider mites to seek out areas of orchards receiving low doses of pyrethroid or none at all (Hall, 1979). Such refuges would be quite suitable for population buildup of mites that would subsequently recolonize the entire orchard as pyrethroid residues decreased. Spider mite outbreaks in orchards following treatment with pyrethroids have been well documented (Penman and Chapman, 1988). It would not be surprising if aversion learning followed by emigration from treated sites were at least in part responsible. Although to our knowledge spider mites have yet to be shown capable of learning, predatory mites are in fact able to learn characters of plants on which spider mite prey have fed (Dicke et al., 1990). Apple maggot flies, *Rhagoletis pomonella*, are capable of learning to refrain from exploiting novel but potentially highly rewarding, resources (Prokopy et al., 1986) and are known to be repelled or deterred by contact with even very low amounts of insecticide (azinphosmethyl) on treated fruit (Reissig et al., 1983). Conceivably, the repellency or deterrency might involve aversion learning (though this has yet to be shown) and could lead to fly emigration from treated trees. Besides repellency or deterrency, ingestion of pesticide, as with ingestion of plant secondary compounds that give rise to malaise followed by recovery, similarly could result in aversion learning (Raffa, 1987).

As we suggested in relation to the possible relevance of learning to insect response to host plant resistance characters, it seems premature to recommend that current patterns of pesticide use be altered to take into account possible effects of insect learning. Nevertheless, we ought to be aware that insect learning could be contributing negatively or positively to the outcome of pesticide treatments.

Behavioral Control

During the process of finding, examining, and ultimately accepting or rejecting a potential resource, a foraging herbivorous insect may be attracted toward, arrested by, repelled from, or deterred by stimuli emanating from resource or nonresource sites (Miller and Strickler, 1984). The stimuli involved could be chemical, physical, or both.

Behavioral methods of managing pest insects involve primarily use of synthetic equivalents or analogues of natural stimuli to (1) attracts pests to or arrest pests at sites where they can be trapped or otherwise eliminated (Prokopy and Owens, 1983; Vité and Baader, 1990; Lanier, 1990), (2) repel or deter pests from finding or using a resource (Bernays, 1983; Prokopy and Owens, 1983; Economopoulos, 1989; Vité and Baader, 1990), or (3) otherwise disrupt one or more stages of the resource-acquisition process, as in the case of pheromonal disruption of mating behavior (Cardé, 1990). Occasionally, behavioral management methods may build upon pest response to stimuli that are unrelated to resource acquisition, such as use of synthetic analogues of alarm pheromone given off by aphids (under attack from predators) to disrupt aphid behavior (Dawson et al., 1988).

Pest learning is perhaps more likely to have a direct impact on the success or failure of a behavioral approach to control than any other control approach. We will focus on two examples in which a phytophagous pest has been demonstrated capable of learning and in which a behavioral method of pest management is appropriate for agricultural conditions. The first concerns use of attractant stimuli to trap the pest. The second concerns use of deterrents to discourage pest buildup.

R. pomonella flies in search of oviposition sites are known to respond positively to volatile compounds of host fruit (Prokopy et al., 1973), particularly to the ester, butyl hexanote (Carle et al., 1987), a constituent of all known attractive host fruit. After moving to host trees in response to host fruit odor (Aluja and Prokopy, 1992), *R. pomonella* flies find individual fruit primarily by vision (Prokopy, 1968; Aluja et al., 1989). Fruit shape, size, and color all serve as important visual cues to a fruit-searching fly (Prokopy, 1968). Recently, host fruit odor and visual stimuli have been employed successfully as a method of controlling *R. pomonella* flies in several commercial orchards (Prokopy et al., 1990b). This approach involved ringing the perimeter of an orchard with sticky-coated red spheres 8 cm in diameter placed 5 m apart. Each sphere was baited with a vial that emitted butyl hexanote. Flies immigrating into an orchard from wild host trees growing in the vicinity (highly probable in eastern North American orchards) are captured by spheres before penetrating into the orchard interior. In most cases, immigration occurs when fruit on wild trees no longer is suitable for egglaying.

R. pomonella flies that have had ovipositional experience with fruit of a particular species or cultivar have greater propensity to oviposit in fruit of similar chemical character and size than fruit of different chemical character and size as a consequence of learning (Papaj and Prokopy, 1986, 1988; Prokopy and Papaj, 1988).

Depending on various environmental- and physiological-state factors, learning of host fruit properties by *R. pomonella* could have a major influence on the efficiency of traps designed to intercept immigrating adults. If, for example, wild host trees from which females emigrated were different from the type planted in a commercial orchard, and if females arriving in the orchard remembered ovipositional experience on wild trees (memory may last at least 3 days—Prokopy et al., unpublished data), then arriving females might reject most or all orchard fruit they visited. This would enhance the likelihood of their visiting a sticky-coated sphere and being captured before ovipositing. If, on the other hand, wild host trees from which females emigrated were the same type as in the commercial orchard, or if immigrating mature females had no prior experience with fruit (possibly owing to absence of fruit that year on wild trees), then arriving females should accept readily orchard fruit that they visited. This would lessen the likelihood of their visiting a sticky-coated sphere before oviposition. Fortunately, circumstances of this sort seem to be uncommon for *R. pomonella* but may be dominant in analogous approaches to control of other insect species.

In a more general vein, as we discussed earlier in relation to effects of learning on sampling populations of insects, it may be important to adjust trap characteristics or trap use patterns according to the specific stimuli of natural resources most recently used by the foraging insect.

S. gregaria locusts are polyphagous insects that assess the quality of potential food plants on the basis of plant chemicals perceived during palpation or biting following alighting (Blaney et al., 1985; Chapman and Bernays, 1989). Feeding deterrents, whether of natural or synthetic origin, offer promise as a behavioral method of insect pest management (Jermy,

1983; 1990; Bernays, 1983). *S. gregaria* are not particularly sensitive to natural concentrations of some secondary plant compounds that are feeding deterrents to other insects, but they are strongly deterred by azadirachtin, an extract of neem plants, when applied to palatable foliage (Bernays, 1983). Of all environmentally safe compounds of natural origin known to deter insect feeding or oviposition behavior, azadirachtin currently offers the greatest potential for widespread effective agriculture use (Olkowski, 1987). Its presence deters feeding by swarms of *S. gregaria* under field conditions (Schmutterer and Ascher, 1987). Just as *S. gregaria* habituated to foliage treated with nicotine hydrogen tartrate (Szentesi and Bernays, 1984), so *S. gregaria* habituates to azadirachtin (Gill, 1972; *vide* Bernays, 1983). Through pest habituation, the efficacy of a compound such as aza-dirachtin applied to crop plants could be significantly compromised. This may be especially likely in polyphagous insects (Jermy, 1990). Ways to circumvent occurrence of habituation could include (1) adjusting the con-centration of deterrent that is applied (Raffa and Frazier, 1988), (2) com-bining application of two or more deterrents (Gould, 1991), (3) combining a deterrent with a low (sublethal) amount of toxicant to neutralize the effect of habituation (Gould, 1991), or, in the case of deterrents that are not absorbed systemically by plants, (4) application of deterrent in het-erogeneous fashion to most but not all of a crop.

Interestingly, there is one behavioral approach to pest management wherein insect habituation may be a distinct advantage to effective control. Bartell (1982) and Cardé (1990) described three ways in which inundating a locale with a natural blend of synthetic sex pheromone might disrupt pest mating behavior: by creating "false" trails of synthetic pheromone, camouflaging trails leading to pheromone-emitting insects, or raising the threshold level of responsiveness of mate-seeking individuals through receptor adaptation or central habituation to pheromone. The degree to which habituation is involved in the mating disruption process remains uncertain, however (Cardé, 1990).

As pointed out by Prokopy (1972), Jermy (1983), Miller and Cowles (1990), Gould (1991), and others, it is doubtful that any one approach to managing pest insects behaviorally through use of natural or synthetic stimuli that attract, arrest, repel, deter, or disrupt individuals will succeed if applied across entire agroecosystems year after year. Through one or another form of learning in concert with selection for resistance and with change in physiological or environmental state, the effectiveness of any one approach will almost certainly diminish over time. Integration of ap-proaches becomes a virtual necessity. Under a framework that includes consideration of pest density, plant acceptability, plant suitability, and time of interaction of pests with plants, Miller and Cowles (1990) have created an elegant integrative stimulo-deterrent diversion concept for stable behavioral-based pest management. The concept involves joint application of an attractant/arrestant with a repellent/deterrent to circumvent potential adverse effects of resource deprivation and perhaps learning. Integration of the tactics of this concept with tactics of habitat management, host plant resistance, and sublethal insecticide use offers a strong potential for mitigating the influence of learning and other state-variable effects leading to an environmentally safe prospect for pest-management programs.

Genetic Control Through Release of Sterile Males

Thus far, the tactics of pest management that we have described have centered largely on management of the state of insect resources or the nonresource environment within or adjacent to a crop. There is one tactic, genetic manipulation of pests, that involves direct management apart from the confines of a cropped area. Genetic manipulation is insect-species-specific and may consist of several approaches, including release of sterile males, release of chromosomally transformed individuals, or release of individuals carrying dominant conditional lethal genetic mutations (Horn, 1988). Only the sterile male approach has received substantial use and will be considered here.

In principle, the sterile-male–release technique is simple. Large numbers of male insects are reared in laboratory cages, sterilized, and released into nature to mate with wild females, which consequently lay only unfertilized eggs, provided the ratio of sterile to fertile males is sufficiently great and provided sterile males are sufficiently competitive with wild males. To date, the greatest success from use of this technique has been in eliminating screwworm flies from the United States and northern Mexico. It has also been used widely with frequent success against the Mediterranean fruit fly, which has been shown capable of learning.

If released sterile medfly males are to be competitive with wild males in acquiring mates, they ought to possess equal ability to locate sites at which mating occurs in nature and to court females at such sites. Wild medfly males form leks on foliage of host plants or nearby nonhost plants, where an initial male releases pheromone and is joined subsequently by other males. It is within leks that the great majority of copulations occurs (Prokopy and Hendrichs, 1979; Hendrichs and Hendrichs, 1990). Whether or not released sterilized medfly males are as able as wild males to find high-quality lekking sites is uncertain. Characteristics of lekking sites appear to include sufficient foliar density to furnish protection against potential predators, presence of nearby fruiting host tree odor, and favorable temperature, illumination, wind, and possibly other microhabitat properties (Hendrichs and Hendrichs, 1990).

As mentioned earlier in regard to methods for estimating absolute population density, insects reared under laboratory conditions and subsequently released into nature might have learned stimuli associated with resources present in laboratory cages but absent from nature, whereas wild individuals might have learned stimuli associated with the natural habitat (Wardle and Borden, 1991). Thus far, learning of habitat stimuli in medflies has been examined only in females. Laboratory-cultured medfly females are known to forage for resources in a manner that is qualitatively similar to but quantitatively different from wild females (Prokopy et al., 1989b). Part of the difference may have a genetic- or physiological-state basis. Part may stem from a difference in learning ability, which is expressed to a lesser degree in laboratory-cultured females than in wild females (Papaj et al., 1987; Prokopy et al., 1990a). Were medfly males to exhibit learning traits that parallel those of medfly females, as is true in *R. pomonella* (Prokopy et al., 1989c), then learning could play a role in the outcome of sterile male releases. For example, if populations of released sterile males had been in laboratory culture for only a few generations and were phenotypically similar to wild males in learning ability, then learning of stimuli or characteristics associated with laboratory conditions could have an adverse effect on the likelihood of released males responding rapidly to natural stimuli associated with lekking sites, provided memory of laboratory stimuli did not fade quickly.

Numerous factors (sociological, political, organizational, as well as biological) determine the success or failure of a program of sterile male releases to suppress or eradicate a population of insects (Scribner, 1983). In some cases, the genetic and physiological state, together with the informational state, of laboratory-cultured released individuals may play a decisive role. There has been much attention to designing protocols for assessing the quality of released sterile individuals (Leppla and Ashley, 1989). To date, the potential influence of learning on individual quality has not been considered. In the next section on augmentative and inundative releases of natural enemies for biological control, we will discuss the importance of exposing individuals before release to stimuli equivalent to those emanating from the forthcoming host habitat and forthcoming hosts so that an appropriate level of prerelease learning might occur. However desirable this might be in some sterile-male–release programs, the sheer quantity of released individuals (often tens of millions per day) undoubtedly renders this approach impractical.

Biological Control Through Beneficial Natural Enemies

Biological control through the action of beneficial natural enemies (or entomophages) is perhaps the most ancient of all forms of pest insect control. In the absence of human intervention for pest-control purposes, entomophages by themselves are often able to maintain populations of potential pests below damaging levels. Indeed, biological control can be considered the foundation upon which all other pest-management tactics ought to be built within an integrative framework. Here, we will focus on entomophagous parasitoids and predators as agents of biological control.

Many factors, including those surrounding learning, have overlapping influences on pest insects and entomophages. Therefore, a solid understanding of entomophage/pest/habitat interactions is important if conflicting effects of IPM tactics on entomophages and pests are to be avoided. Such an understanding is still in its infancy but is developing rapidly.

As with pest insects, only suggestive real agricultural world data are available regarding the relevance of learning by entomophages to pest-management tactics. The principles are similar to those discussed for pest insects, except with a reverse desired effect in the case of entomophages. Substantial laboratory and field-plot data are available regarding the role and importance of learning in the foraging success of entomophages. Many of these data, at least for parasitoids, are discussed in a fundamental way by Turlings and co-authors (this volume). Here we examine learning as a source of variation in the quality and performance of propagated and released entomophages as well as the importance of learning in designing environmental manipulation procedures to enhance entomophage populations.

Two basic features of entomophagous insects are essential to their effectiveness: retention in target areas, and efficient location and attack of hosts or prey. To meet these requirements, we must understand and manage factors that influence entomophage foraging behavior. Learning is a major component but only one of several interwoven sources of behavioral variation in entomophages. Predictably effective performance of entomophages as biological control agents requires matching intrinsic variation in informational and other conditions of the foraging entomophage with conditions of the target environment (Fig. 12.1). Techniques for managing the environmental component of the interaction would be important in all approaches to biological control, be they enhancement of the performance of feral entomophages or laboratory-reared and released populations. On the other hand, management of intrinsic variation in informational conditions of entomophages is more applicable in situations where entomophages are laboratory-reared and released.

Managing the Intrinsic Component of Entomophage Performance

As discussed at the beginning of this chapter, genotypic and phenotypic variables may substantially modify the behavior of a foraging insect. During

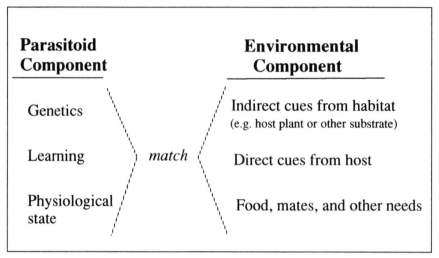

Figure 12.1. The parasitoid and environmental components that must be matched for effective foraging behavior.

laboratory colonization, entomophages are removed from the context of natural selection, thereby allowing changes in desired genotypic and phenotypic characteristics, including changes due to learning (Wardle and Borden, 1986, 1991; Herard et al., 1988; Noldus, 1989). Such changes are of particular concern in the case of inundative releases of entomophages and seasonal inoculative programs where propagation and release are continually artificial (Lewis et al., 1990).

Quality-control procedures are an important part of propagation-and-release programs involving entomophagous insects. Although there has been considerable speculation about desired traits in entomophages (Noldus, 1989), little has been done to develop techniques to measure and monitor these traits throughout propagation and release. Figure 12.2 depicts potential problems that can arise at different points in the propagation-and-release process and the technology needed to address these problems. Our lack of knowledge of specific features essential for entomophage foraging success and of methods for monitoring these features limits our ability to provide a prescription for quality control. However, the conceptual framework for acquiring this knowledge, including ways to monitor and manage changes due to learning, is developing rapidly (Lewis et al., 1990; Lewis and Martin, 1990; Vet et al., 1990).

Recent studies show that learning clearly shapes the foraging behavior of entomophages and is an important consideration for use of entomophages in biological control (Vinson, 1984; Vet et al., 1990; Vet and Groenewold, 1990; Lewis et al., 1990). The response potential of an ento-

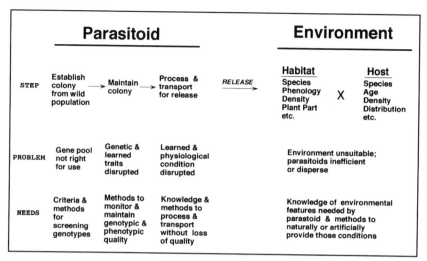

Figure 12.2. Chart of steps, problems, and needed technology for effective propagation and release of parasitoids for pest control. (The parasitoid's genotypic and phenotypic behavioral profile must match the environmental situation.)

mophage can result from experience in preadult as well as adult stages (Vet et al., 1990). Without care, insectary conditions can generate distorted behavioral profiles. For example, Wardle and Borden (1986, 1991) showed that prior experience of wasps (*Exoristis roborator*) with an artificial rearing system (hosts in plastic egg cups) significantly reduced response to hosts in natural situations. By understanding the dynamics of entomophage learning, we may be able to avoid the insect's learning of inappropriate stimuli and provide an appropriate level of experience. Thus, key semiochemicals and other important stimuli could be incorporated artificially into the diet of an entomophage and on the rearing host or prey. This approach may be particularly useful in cases where important learning experiences occur in an immature or early adult stage (Wardle and Borden, 1985; Herard et al., 1988).

Several studies demonstrate improved responsiveness of entomophages to target stimuli merely by providing brief exposure to host-derived products prior to release (Gross et al., 1975; Vinson et al., 1977; Loke and Ashley, 1983; Lewis and Tumlinson, 1988; Ding et al., 1989; Turlings et al., 1989). However, ovipositional experience with a host has been shown to provide even stronger response to target stimuli (Vet and Groenewold, 1990; Papaj and Vet, 1990; Lewis and Martin, 1990), though such prerelease experience may be less feasible. Vet et al. (1990) point out that predictability of behavioral response of parasitoids is increased through learning in two ways: by decreasing variability of response pattern and by increasing mean response level.

Beyond entomophage experience solely with host-derived products, Lewis and Takasu (1990) demonstrated that *Microplitis croceipes* females learn odors associated with food as well as hosts and subsequently make choices between odors linked to the competing needs of feeding and ovipositing based on current state of hunger. This finding highlights the general importance of learning in shaping the foraging behavior of entomophages as well as entomophage physiological state relative to needs other than hosts.

Thus, physiological well-being of entomophages is important to effective foraging behavior. Because shortcomings in neither learning conditions nor physiological state may readily be apparent from general observation, various response evaluation techniques should be employed to monitor the quality of phenotypic traits of a colony.

Managing the Environmental Component of Entomophage Performance

As stated above, effective biological control involves retention of the entomophages in the target area and efficient host search-and-attack behavior. Because profitability of foraging activities of entomophages is determined to a significant degree by learning, there must be sufficient encounters with hosts, food, or other reinforcing stimuli for learning to occur. Otherwise, the entomophages may not remain in the area. For example, the economic damage threshold of a pest may sometimes be at a level below that needed to sustain effective entomophage foraging behavior. Creative use of semiochemicals and other foraging cues could offer a means of providing necessary stimuli and memory reinforcements for manipulating entomophage behavior independent of host density and other variables.

Five sets of indicators can be identified as important in entomophage assessment of the environment and pursuit of hosts or prey: (1) quality, quantity, and mix of cues from a plant or other substrate of the host or prey as indirect indicators of profitability; (2) quality, quantity, and distribution of cues emanating directly from hosts or prey; (3) intraspecific or interspecific marker cues or other indicators of competing entomophages; (4) availability of food, mates, shelter, and other resource needs; and (5) appropriate macro- and microclimatic conditions. Several studies have revealed substantial prospects for application of appropriate indicator cues/conditions to the habitat to enhance entomophage abundance and effectiveness. However, effective use of such indicators for manipulating the behavior of entomophages depends on proper matching of applied cues with the informational state of the entomophage (Lewis et al., 1990). Further, adequate reinforcers such as hosts, prey, or their direct products may be needed to prevent habituation to a static condition and may even strengthen the response (Vinson, 1984; Lewis and Tumlinson, 1988; Turlings et al., 1989; Vet et al., 1990). A substantial understanding of learning mechanisms is important to the success of such manipulation technology.

Studies with *Chrysoperla* predators and *Trichogramma* parasitoids are representative of the limited number of actual field manipulation experiments with entomophages. Hagen et al. (1970) and Hagen and Bishop (1979) used an artificial honeydew to attract adults of the predator *Chrysoperla carnea*. The mixture used provided a kairomone and food supplement, both of which served to increase predator density. Indoleacetaldehyde, a breakdown product of tryptophan present in the yeast hydrolysate of the artificial honeydew, attracted adult lacewings into target fields (van Emden and Hagen, 1976). Other components of the artificial honeydew (sugar, water, whey-yeast hydrolysate) arrested movement of the lacewings and served as a nutritional supplement, thereby promoting oviposition. One week following application of such a mixture to cotton, *Chrysoperla* egg density increased from one to three per plant, and both the density of prey bollworm eggs and the number of damaged bolls declined (Hagen et al., 1970).

Firm evidence of learning in *Chrysoperla* has yet to be shown. Nonetheless, experience by *Chrysoperla* with the artificial honeydew might have shaped subsequent response patterns to honeydew and prey. Studies on possible learning effects would be useful for obtaining consistently effective biocontrol of prey in this as well as other systems artificially manipulating predator behavior.

A predator attacking two or more prey may switch its attack pattern according to abundance and preference for prey species (Murdoch, 1969). Learning is thought to be involved in prey switching behavior and may affect functional responses of predators to prey (Murdoch and Oaten, 1975). Along this line, Houck (1986) demonstrated that switching in preferences of *Stethorus* lady beetle predators between two species of spider mite prey was somewhat modified by the nature of prior experience with these prey.

Use of semiochemicals from both plants and an insect host, *Helicoverpa zea*, has been shown to increase rates of egg parasitism by *Trichogramma* in the field. For example, parasitism of *H. zea* eggs by *Trichogramma* spp. increased from 13% in control plots to 22% in soybeans treated with an extract of scales collected from *H. zea* (Lewis et al., 1975a,b). Similarly, release of a synthetic blend of the sex pheromone of *H. zea* in cotton increased *Trichogramma* parasitism of eggs from 21% in control plots to 36% in treated plots (Lewis et al., 1982). Altieri et al. (1981) demonstrated that spraying various plant extracts onto crops can stimulate increased rates of parasitization. Thus, parasitism of eggs of *H. zea* by *Trichogramma* spp.

was 21% on soybeans treated with an extract of *Amaranthus* compared to 13% on plants sprayed with water. The behaviors which lead to increased parasitism by *Trichogramma* in the presence of either plant extracts or sex pheromones are not yet fully understood, but could well involve learning. As revealed by van Lenteren (1981) and Noldus (1989), *Trichogramma* adults are in fact capable of learning during host-foraging.

From these and other studies, it is apparent that improved knowledge and eventual channeling of learning by entomophages may play a key role in effective use of entomophages as biological control agents. To effectively promote or retain entomophages in desired habitats without interfering with foraging efficiency will require better understanding of the various cues associated with foraging. An improved understanding of learning by entomophages as well as pests within a tritrophic entomophage/pest/habitat interaction framework should enhance future prospects for more effective biocontrol with entomophages.

Conclusion

As others have before us, we have suggested that knowledge of learning, as a component of the informational state of an insect, ought to be integrated with knowledge of physiological and genetical state of the insect if we are to have a robust understanding of the nature of insect behavioral response to environmental stimuli. We have attempted to point out here how insect learning by itself might shape insect responses to environmental stimuli, particularly in ways that could have an impact upon pest-management practices. The practices we have considered include the sampling of pest and beneficial arthropod populations in crops as well as use of several sorts of tactics to prevent arthropod pests from reaching levels causing injury to crops.

Because researchers are only beginning to understand how pest behavior might affect the outcome of current or future pest-management strategies, and because researchers are only beginning to ascertain the degree to which pest and beneficial arthropods are capable of learning environmental stimuli, our review is primarily suggestive. Nonetheless, it is apparent that certain currently used tactics of pest management, particularly behavioral control through use of attractants, arrestants, repellents, or deterrents and biological control through release of beneficial natural enemies have developed to a point that demands immediate fuller understanding of the effects of learning on the success of the tactic. It is also probable that for some cropping situations, learning will eventually be shown conclusively to have a measurable influence on the success of certain pest-management tactics, including cultural control through habitat management, control through host plant resistance, pesticidal control, and genetic control through release of sterile males. We are particularly ignorant of the potential impact of learning by insects on the reliability of current methods of sampling pest and beneficial arthropod abundance in crops.

In the short term, improved knowledge of insect learning ought to bring about substantial positive adjustment in protocols associated with implementation of a given pest-management tactic. In the long term, a stable level of control of all agriculturally important pests in a given crop will require true integration of multiple management tactics.

In conclusion, we should not be dismayed over the gulf that currently exists in the application of knowledge of insect learning to pest management. It is perhaps little greater than that between ecological theory and pest-management practice (Kogan, 1986). Unraveling complex biological phenomena such as the true extent to which learning modifies behavioral patterns requires much time and effort. Turning theory and knowledge into practice is no less demanding. The future challenge to behaviorists conducting research on insect learning and to pest managers will be to determine what learning-based modifications of approaches to pest management ought to be introduced to attain long-term stability of environmentally sound crop-production practices. We are confident this challenge can eventually be met.

Acknowledgments

We are most grateful to Julia Connelly and Amity Lee-Bradley for typing the manuscript.

References

Altieri, M.A. 1987. Agroecology. Westview Press, Boulder, CO.

Altieri, M., Lewis, W.J., Nordlund, D.A., Gueldner, R.C., and Todd, J.W. 1981. Chemical interactions between plants and *Trichogramma* wasps in Georgia soybean fields. Prot. Ecol. **3**:259–263.

Aluja, M., and Prokopy, R.J. 1992. Host search behavior by *Rhagoletis pomonella* flies: Inter-tree movement patterns in response to wind-borne fruit volatiles under field conditions. Physiol. Entomol. **17**:1–8.

Aluja, M., Prokopy, R.J., Elkinton, J.S., and Laurence, F. 1989. Novel approach for tracking and quantifying the movement patterns of insects in three dimensions under seminatural conditions. Environ. Entomol. **18**:1–7.

Andow, D.A. 1991. Vegetational diversity and arthropod population response. Annu. Rev. Entomol. **36**:561–586.

Bartell, R.J. 1982. Mechanisms of communication disruption by pheromone in the control of Lepidoptera: A review. Physiol. Entomol. **7**:353–364.

Bateman, M.A., and Sonleitner, F.J. 1967. The ecology of a natural population of the Queensland fruit fly, *Dacus tryoni*. I. The parameters of the pupal and adult populations during a single season. Aust. J. Zool. **15**:303–335.

Begon, M. 1979. Investigating Animal Abundance: Capture-Recapture for Biologists. University Park Press, Baltimore, 104 pp.

Bell, W.J. 1990. Searching behavior patterns of insects. Annu. Rev. Entomol. **35**:447–467.

Bernays, E.A. 1983. Antifeedants in crop pest management. In D.L. Whitehead and W.S. Bowers (eds.), Natural Products for Innovative Pest Management. Pergamon Press, New York, pp. 259–271.

Bernays, E., and Chapman, R. 1987. The evolution of deterrent responses in plant-feeding insects. In R.F. Chapman, E.A. Bernays, and J.G. Stoffolano (eds.), Perspectives in Chemoreception and Behavior. Springer-Verlag, New York, pp. 159–170.

Bernays, E.A., and Lee, J.C. 1988. Food aversion learning in the polyphagous grasshopper *Schistocerca americana*. Physiol. Entomol. **13**:131–137.

Bernays, E.A., and Wrubel, R.P. 1985. Learning by grasshoppers: Association of colour and light intensity with food. Physiol. Entomol. **10**:359–369.

Blaney, W.M., and Simmonds, M.S.J. 1985. Food selection by locusts: The role of learning in rejection behavior. Entomol. Exp. Appl. **39**:273–278.

Blaney, W.M., Winstanley, C., and Simmonds, M.S.J. 1985. Food selection by locusts: An analysis of rejection behavior. Entomol. Exp. Appl. **38**:35–40.

Blois, C., and Cloarec, A. 1985. Influence of experience on prey selection by *Anax imperator* larvae (Aeschnidae-Odonata). Z. Tierpsychol. **68**:303–312.

Cardé R.T. 1990. Principles of mating disruption. In R.L. Ridgway, R.M. Silverstein, and M.N. Inscoe (eds.), Behavior-Modifying Chemicals for Pest Management. Marcel-Dekkar, New York, pp. 47–71.

Carle, S.A., Averill, A.L., Rule, G.S., Reissig, W.H., and Roelofs, W.L. 1987. Variation in fruit volatiles attractive to apple maggot fly, *Rhagoletis pomonella*. J. Chem. Ecol. **13**:795–805.

Chapman, R.F., and Bernays, E.A. 1989. Insect behavior at the leaf surface and learning as aspects of host plant selection. Experientia **45**:215–222.

Dawson, G.W., Griffiths, D.C., Pickett, J.A., Plumb, R.T., Woodcock, C.M., and Zhong-Ning, Z. 1988. Structure/activity studies on aphid alarm pheromone derivatives and their field use against transmission of barley yellow dwarf virus. Pestic. Sci. **22**:17–30.

Dethier, V.G. 1976. Man's Plague? Insects and Agriculture. Darwin Press, Princeton, NJ.

Dethier, V.G. 1980. Food-aversion learning in two phytophagous caterpillars, *Diacrisia virginica* and *Extigmene congrua*. Physiol. Entomol. **5**:321–325.

Dicke, M., van der Mass, K.J., Takabayashi, J., and Vet, L.E.M. 1990. Learning affects response to volatile allelochemicals by predatory mites. Proc. Exp. Appl. Entomol. **1**:31–36.

Ding, D., Swedenborg, P.D., and Jones, R.L. 1989. Chemical stimuli in host-seeking behavior of *Macrocentrus grandii* (Hymenoptera: Braconidae). Ann. Entomol. Soc. Am. **82**:232–236.

Ebling, W., Wagner, R.E., and Reierson, D.A. 1966. Influence of repellency on the efficacy of blatticides. I. Learned modification of behavior of the German cockroach. J. Econ. Entomol. **59**:1374–1387.

Economopoulos, A.P. 1989. Use of traps based on color and/or shape. In A.S. Robinson and G. Hooper (eds.), Fruit Flies, Their Biology, Natural Enemies and Control. Elsevier, Amsterdam, pp. 315–328.

Emden, H.F. van, and Hagen, K.S. 1976. Olfactory reactions of the green lacewing, *Chrysopa carnea*, to tryptophan and certain breakdown products. Environ. Entomol. **5**:469–473.

Gill, J.S. 1972. Studies on insect feeding deterrents with special reference to the fruit extracts of the neem tree, *Azadirachta indica*. Ph.D. dissertation, University of London.

Gould, F. 1984. Role of behavior in the evolution of insect adaptation to insecticides and resistant host plants. Bull. Entomol. Soc. Am. **30**(4):34–41.

Gould, F. 1988. Evolutionary biology and genetically engineered crops. Bioscience **38**:26–33.

Gould, F. 1991. Arthropod behavior and the efficacy of plant protectants. Annu. Rev. Entomol. **36**:305–330.

Gould, J.L. 1986. The biology of learning. Annu. Rev. Psychol. **37**:163–192.

Gross, H.R., Lewis, W.J., Jones, R.L., and Nordlund, D.A. 1975. Kairomones and their use for management of entomophagous insects: III. Stimulation of *Trichogramma achaeae*, *T. pretiosum* and *Microplitis croceipes* with host-seeking stimuli at time of release to improve their efficiency. J. Chem. Ecol. **1**:431–438.

Hagen, K.S., and Bishop, G.W. 1979. Use of supplemental foods and behavioral chemicals to increase the effectiveness of natural enemies. In D.W. Davis, J.A. McMurtry, and S.C. Hoyt (eds.), Biological Control and Insect Management. California Agricultural Experimental Station Publication 4096, Berkeley, pp. 49–60.

Hagen, K.S., Sawall, E.F., Jr., and Tassan, R.L. 1970. The use of food sprays to increase effectiveness of entomophagous insects. Proc. Tall Timbers Conf. Ecol. Anim. Control Habitat Manage. **2**:59–81.

Hall, F.R. 1979. Effects of synthetic pyrethroids on major insect and mite pests of apple. J. Econ. Entomol. **72**:441–446.

Harrison, J.M., and Breed, M.D. 1987. Temporal learning in the giant tropical ant, *Paraponera clavata*. Physiol. Entomol. **12**:317–320.

Haynes, K.F. 1988. Sublethal effects of neurotoxic insecticides on insect behavior. Annu. Rev. Entomol. **33**:149–168.

Heinrich, B. 1976. The foraging specializations of individual bumblebees. Ecol. Monogr. **42**:105–128.

Hendrichs, J., and Hendrichs, M.A. 1990. Mediterranean fruit fly in nature: Location and diel pattern of feeding and other activities on fruiting and nonfruiting hosts and nonhosts. Ann. Entomol. Soc. Am. **83**:632–641.

Herard, F.H., Keller, M.A., Lewis, W.J., and Tumlinson, J.H. 1988. Beneficial arthropod behavior mediated by airborne semiochemicals. IV. Influence of host diet on host-oriented flight chamber responses of *Microplitis demolitor*. J. Chem. Ecol. **14**:1597–1606.

Herzog, D.C., and Funderburk, J.E. 1986. Ecological basis for habitat management and pest cultural control. In M. Kogan (ed.), Ecological Theory and Integrated Pest Management Practice. John Wiley, New York, pp. 217–250.

Hokkanen, H.M.T. 1991. Trap cropping in pest management. Annu. Rev. Entomol. **36**:119–138.

Hokkanen, H., Granlund, H., Husberg, G.B., and Markkula, M. 1986. Trap crops used successfully to control *Meligethes aeneus*, the rape blossom beetle. Ann. Entomol. Fennici **52**:115–120.

Horn, D.J. 1988. Ecological Approach to Pest Management. Guilford Press, New York, 285 pp.

Houck, M.A. 1986. Prey preference in *Stethorus punctum* (Coleoptera: Coccinellidae). Environ. Entomol. **15**:967–970.

Jermy, T. 1983. Multiplicity of insect antifeedants in plants. In D.L. Whitehead and W.S. Bowers (eds.), Natural Products for Innovative Pest Management. Pergamon Press, New York, pp. 223–236.

Jermy, T. 1987. The role of experience in the host selection of phytophagous insects. In R.F. Chapman, E.A. Bernays, and J.G. Stoffolano (eds.), Perspectives in Chemoreception. Springer-Verlag, New York, pp. 143–157.

Jermy, T. 1990. Prospects of antifeedant approach to pest control—A critical review. J. Chem. Ecol. **16**:3151–3166.

Jermy, T., Bernays, E.A., and Szentesi, A. 1982. The effect of repeated exposure to feeding deterrents on their acceptibility to phytophagous insects. In J.H. Visser and A.K. Minks (eds.), Proceedings of the 5th International Symposium on Insect–Plant Relationships. Pudoc, Wageningen, pp. 25–32.

Jones, R.E. 1987. Behavioral evolution in the cabbage butterfly (*Pieris rapae*). Oecologia **72**:69–76.

Jones, R.E., and Ives, P.M. 1979. The adaptiveness of searching and host selection behavior in *Pieris rapae*. Aust. J. Ecol. **4**:75–86.

Kareiva, P. 1982. Experimental and mathematical analysis of herbivore movement: quantifying the influence of plant spacing and quality on foraging discrimination. Ecol. Monogr. **52**:261–282.

Kareiva, P. 1983. Influence of vegetation texture on herbivore populations: resource concentrations and herbivore movement. In R.F. Denno and M.S. McClure (eds.), Variable Plants and Herbivores in Natural and Managed Systems. Academic Press, New York, pp. 259–289.

Katsoyannos, B.I. 1987. Effect of color properties of spheres on their attractiveness for *Ceratitis capitata* flies in the field. J. Appl. Entomol. **104**:79–85.

Kennedy, G.C., Gould, F., Deponti, O.M.B., and Stinner, R.E. 1987. Ecological, agricultural, genetic and commercial considerations in the deployment of insect-resistant germplasm. Environ. Entomol. **16**:327–338.

Kogan, M. 1986. Plant defense strategies and host-plant resistance. In M. Kogan (ed.), Ecological Theory and Integrated Pest Management Practice. John Wiley, New York, pp. 83–134.

Kortisas, V.M., and Garsed, S.G. 1985. The effects of nitrogen and sulfur nutrition on the response of brussel sprout plants to infestation by the aphid *Brevicoryne brassicae*. Ann. Appl. Biol. **106**:1–15.

Lanier, G.N. 1990. Principles of attraction-annihilation: Mass trapping and other means. In R.L. Ridgway, R.M. Silverstein, and M.N. Inscoe (eds.), Behavior-Modifying Chemicals for Insect Management. Marcel-Dekker, New York, pp. 25–46.

Lee, J.C., and Bernays, E.A. 1988. Declining acceptability of a food plant for the polyphagous grasshopper, *Schistocerca americana*: The role of food aversion learning. Physiol. Entomol. **13**:291–301.

Lenteren, J.C. van 1981. Host discrimination by parasitoids. In D.A. Nordlund, R.L. Jones, and W.J. Lewis (eds.), Semiochemicals: Their Role in Pest Control. John Wiley, New York, pp. 153–179.

Leppla, N.C., and Ashley, T.R. 1989. Quality control in insect mass production: A review and model. Bull. Entomol. Soc. Am. **35**:33–44.

Lewis, A.C. 1986. Memory constraints and flower choice in *Pieris rapae*. Science, **232**:863–865.

Lewis, A.C., and Lipani, G.A. 1990. Learning and flower use in butterflies: Hypotheses from honey bees. In E.A. Bernays (ed.), Insect–Plant Interactions. CRC Press, Boca Raton, FL, pp. 95–110.

Lewis, W.J., Jones, R.L., Nordlund, D.A., and Sparks, A.N. 1975a. Kairomones and their use for management of entomophagous insects: I. Evaluation for increasing rates of parasitization by *Trichogramma* spp. in the field. J. Chem. Ecol. **1**:343–347.

Lewis, W.J., Jones, R.L., Nordlund, D.A., and Gross, H.R. 1975b. Kairomones and their use for management of entomophagous insects: II. Mechanism causing increase in rate of parasitization by *Trichogramma* spp. J. Chem. Ecol. **1**:349–360.

Lewis, W.J., Nordlund, D.A., Gueldner, R.C., Teal, P.E.A., and Tumlinson, J.H. 1982. Kairomones and their use for management of entomophagous insects: XIII. Kairomonal activity for *Trichogramma* spp. of abdominal tips, excretions and a synthetic sex pheromone blend of *Heliothis zea* (Boddie) moths. J. Chem. Ecol. **8**:1323–1331.

Lewis, W.J., and Tumlinson, J.H. 1988. Host detection by chemically mediated associative learning in a parasitic wasp. Nature **331**:257–259.

Lewis, W.J., and Martin, W.R., Jr. 1990. Semiochemicals for use with parasitoids: Status and future. J. Chem. Ecol. **16**:3067–3089.

Lewis, W.J., and Takasu, K. 1990. Use of learned odors by a parasitic wasp in accordance with host and food needs. Nature **348**:635–636.

Lewis, W.J., Vet, L.E.M., Tumlinson, J.H., van Lenteren, J.C., and Papaj, D.R. 1990. Variations in parasitoid foraging behavior: Essential element of a sound biological control theory. Environ. Entomol. **19**:1183–1193.

Loke, W.H., Ashley, T.R., and Sailer, R.I. 1983. Influence of fall armyworm, *Spodoptera frugiperda* (Lepidoptera: Noctuidae) larvae, and corn plant damage on host finding in *Apanteles marginiventris*. Environ. Entomol. **12**:911–915.

Maldonado, H., Jaffe, K., and Balderrama, N. 1979. The dynamics of learning in the praying mantis (*Stagmatoptera biocellata*). J. Insect Physiol. **25**:525–533.

Mallet, J., Longino, J.T., Murawski, D., Murawski, A., and Gamoa, A.S. 1987. Handling effects in *Heliconius*: Where do all the butterflies go? J. Anim. Ecol. **56**:377–386.

Marler, P., and Terrace, H.S. 1984. The Biology of Learning. Dahlem Konf. Life Sci. Res. Rept. 29, Springer-Verlag, Berlin.

Mattson, W.J., and Scriber, J.M. 1987. Feeding ecology of insect folivores of woody plants: water, nitrogen, fiber, and mineral considerations. In F. Slanksy and J.G. Rodriguez (eds.), The Nutritional Ecology of Insects, Mites and Spiders. John Wiley, New York, pp. 105–146.

McInnis, D.O. 1989. Artificial oviposition sphere for Mediterranean fruit flies in field cages. J. Econ. Entomol. **82**:1382–1385.

Miller, J.R., and Strickler, K.S. 1984. Finding and accepting host plants. In W.J. Bell and R.T. Cardé (eds.), Chemical Ecology of Insects. Chapman and Hall, London, pp. 127–157.

Miller, J.R., and Cowles, R.S. 1990. Stimulo-deterrent diversion: A concept and its possible application to onion maggot control. J. Chem. Ecol. **16**:3197–3212.

Minkenberg, O.P.J.M., and Fredrix, M.J.J. 1989. Preference and performance of an herbivorous fly, *Liriomyza trifolii* (Diptera: Agromyzidae), on tomato plants differing in leaf nitrogen. Ann. Entomol. Soc. Am. **82**:350–354.

Murdoch, W.W. 1969. Switching in general predators: Experiments on predator specificity and stability of prey populations. Ecol. Monogr. **39**:335–354.

Murdoch, W.W., and Oaten, A. 1975. Predation and population stability. Adv. Ecol. Res. **9**:2–131.

Nakagawa, S., Prokopy, R.J., Wong, T.T.Y., Ziegler, J.R., Mitchell, S.M., Urago, T., and Harris, E.J. 1978. Visual orientation of *Ceratitis capitata* flies to fruit models. Entomol. Exp. Appl. **24**:193–198.

Noldus, L.P.J.J. 1989. Semiochemicals and quality of beneficial arthropods: general considerations with special reference to *Trichogramma* spp. J. Appl. Entomol. **108**:425–451.

Nottingham, S.F. 1988. Host plant finding for oviposition by adult cabbage root fly, *Delia radicum*. J. Insect Physiol. **34**:227–234.

Olkowski, W. 1987. Update: neem—a new era in pest control products? IPM Pract. **9**(10):1–8.

Painter, R.H. 1951. Insect Resistance in Crop Plants. Macmillan, New York, 520 pp.

Papaj, D.R. 1984. Causes of variation of host discrimination behavior in the butterfly, *Battus philenor*. Ph.D. dissertation, Duke University, Durham, NC.

Papaj, D.R. 1986. Shifts in foraging behavior by a *Battus philenor* population: Field evidence for switching by individual butterflies. Behav. Ecol. Sociobiol. **19**:31–39.

Papaj, D.R., and Prokopy, R.J. 1986. Phytochemical basis of learning in *Rhagoletis pomonella* and other herbivorous insects. J. Chem. Ecol. **12**:1125–1143.

Papaj, D.R., and Prokopy, R.J. 1988. The effect of prior adult experience on components of habitat preference in the apple maggot fly (*Rhagoletis pomonella*). Oecologia **76**:538–543.

Papaj, D.R., and Prokopy, R.J. 1989. Ecological and evolutionary aspects of learning in phytophagous insects. Annu. Rev. Entomol. **34**:315–350.

Papaj, D.R., Prokopy, R.J., McDonald, P.T., and Wong, T.T.Y. 1987. Differences in learning between wild and laboratory *Ceratitis capitata* flies. Entomol. Exp. Appl. **45**:65–72.

Papaj, D.R., Opp, S.B., Prokopy, R.J., and Wong, T.T.Y. 1988. Cross-induction of fruit acceptance by the medfly *Ceratitis capitata*: The role of fruit size and chemistry. J. Insect Behav. **2**:241–254.

Papaj, D.R., and Rausher, M. 1983. Individual variation in host location by phytophagous insects. In S. Ahmad (ed.), Herbivorous Insects: Host-Seeking Behavior and Mechanisms. Academic Press, New York, pp. 77–124.

Papaj, D.R., and Vet, L.E.M. 1990. Odor learning and foraging success in the parasitoid, *Leptopilina heterotoma*. J. Chem. Ecol. **16**:3137–3150.

Pasteels, J.M., and Gregoire, J.C. 1984. Selective predation on chemically defended chrysomelid larvae: a conditioning process. J. Chem. Ecol. **10**:1693–1700.

Pearson, D.L. 1985. The function of multiple anti-predator mechanisms in adult tiger beetles. Ecol. Entomol. **10**:65–72.

Penman, D.R., and Chapman, R.B. 1988. Pesticide-induced mite outbreaks: pyrethroids and spider mites. Exp. Appl. Acarol. **4**:265–276.

Plotkin, H.C. 1979. Learning in a carabid beetle (*Pterostichus melanerius*). Anim. Behav. **27**:567–575.

Pluthero, F.G., and Singh, R.S. 1984. Insect behavioral responses to toxins: practical and evolutionary considerations. Can. Entomol. **116**:57–68.

Prokopy, R.J. 1968. Visual responses of apple maggot flies, *Rhagoletis pomonella*: Orchard studies. Entomol. Exp. Appl. **11**:403–422.

Prokopy, R.J. 1972. Evidence for a marking pheromone deterring repeated oviposition in apple maggot flies. Environ. Entomol. **1**:326–332.

Prokopy, R.J. 1986. Toward a world of less pesticide. Mass. Agr. Exp. Sta. Res. Bull. **710**:1–22.

Prokopy, R.J., and Hendrichs, J. 1979. Mating behavior of *Ceratitis capitata* on a field-caged host tree. Ann. Entomol. Soc. Am. **72:**642–648.

Prokopy, R.J., and Owens, E.D. 1983. Visual detection of plants by herbivorous insects. Annu. Rev. Entomol. **28:**337–364.

Prokopy, R.J., and Papaj, D.R. 1988. Learning of apple fruit biotypes by apple maggot flies. J. Insect Behav. **1:**67–74.

Prokopy, R.J., Moericke, V., and Bush, G.L. 1973. Attraction of apple maggot flies to odor of apples. Environ. Entomol. **2:**743–749.

Prokopy, R.J., Papaj, D.R., Cooley, S.S., and Kallet, C. 1986. On the nature of learning in oviposition site acceptance by apple maggot flies. Anim. Behav. **34:**98–107.

Prokopy, R.J., Green, T.A., and Wong, T.T.Y. 1989a. Learning to find fruit in *Ceratitis capitata* flies. Entomol. Exp. Appl. **53:**65–72.

Prokopy, R.J., Aluja, M., and Wong, T.T.Y. 1989b. Foraging behavior of laboratory cultured Mediterranean fruit flies on field-caged host trees. Proc. Hawaiian Entomol. Soc. **29:**103–109.

Prokopy, R.J., Cooley, S.S., and Opp, S.B. 1989c. Prior experience influences the fruit residence of male apple maggot flies, *Rhagoletis pomonella*. J. Insect Behav. **2:**39–48.

Prokopy, R.J., Green, T., Wong, T.T.Y., and McInnis, D.O. 1990a. Influence of experience on acceptance of artificial oviposition substrates in *Ceratitis capitata*. Proc. Hawaiian Entomol. Soc. **30:**91–95.

Prokopy, R.J., Johnson, S.A., and O'Brien, M.T. 1990b. Second-stage integrated management of apple arthropod pests. Entomol. Exp. Appl. **54:**9–19.

Raffa, K.F. 1987. Maintenance of innate feeding preferences by a polyphagous insect despite ingestion of applied deleterious chemicals. Entomol. Exp. Appl. **44:**221–227.

Raffa, K.F., and Frazier, J.L. 1988. A generalized model for quantifying behavioral de-sensitization to antifeedants. Entomol. Exp. Appl. **46:**93–100.

Reissig, W.H., Stanely, B.H., Valla, M.E., Seem, R.C., and Bourke, J.B. 1983. Effects of surface residues of azinphosmethyl on apple maggot behavior, oviposition and mortality. Environ. Entomol. **12:**815–822.

Risch, S.J., Andow, D.A., and Altieri, M.A. 1983. Agroecosystem diversity and pest control: Data, tentative conclusions, and new research directions. Environ. Entomol. **12:**625–629.

Roitberg, B.D., and Prokopy, R.J. 1981. Experience required for pheromone recognition by the apple maggot fly. Nature **292:**540–541.

Root, R.B. 1973. Organization of plant-arthropod association in simple and diverse habitats: the fauna of collards (*Brassica oleracea*). Ecol. Monogr. **43:**95–124.

Root, R.B., and Kareiva, P.M. 1984. The search for resources by cabbage butterflies (*Pieris rapae*): Ecological consequences and adaptive significance of Markovian movements in a patchy environment. Ecology **65:**147–165.

Schmutterer, H., and Ascher, K.R.S. (eds.). 1987. Natural pesticides from the neem tree (*Azadirachta indica*) and other tropical plants. Proceedings of the 3rd International Neem Conference, Nairobi (1986). GTZ Grub H., Eschborn, 703 pp.

Scribner, J. 1983. The medfly in California: Organization of the eradication program and public policy. Hortscience **18**:47–52.

Sheehan, W., and Shelton, A.M. 1989. The role of experience in plant foraging by the aphid parasitoid *Diaretiella rapae*. J. Insect Behav. **2**:743–759.

Slansky, F. 1990. Insect nutritional ecology as a basis for studying host plant resistance. Florida Entomol. **73**:359–378.

Southwood, T.R.E. 1978. Ecological Methods, 2nd ed. Chapman and Hall, London, 524 pp.

Sparks, T.C., Lockwood, J.A., Byford, R.L., Graves, J.B., and Leonard, B.R. 1989. The role of behavior in insecticide resistance. Pestic. Sci. **26**:383–399.

Srygley, R.B., and Chai, P. 1990. Flight morphology of neotropical butterflies: Palatability and distribution of mass to the thorax and abdomen. Oecologia **84**:491–499.

Stanton, M.L. 1983. Spatial patterns in the plant community and their effects upon insect search. In S. Ahmad (ed.), Herbivorous Insects: Host-Seeking Behavior and Mechanisms. Academic Press, New York, pp. 125–157.

Stanton, M.L. 1984. Short-term learning and the searching accuracy of egglaying butterflies. Anim. Behav. **32**:33–40.

Steiner, A.L. 1981. Anti-predator strategies. II. Grasshoppers attacked by *Prionyx parkeri* and some *Tachysphex* wasps. Psyche **88**:1–24.

Szentesi, A., and Bernays, E.A. 1984. A study of behavioral habituation to a feeding deterrent in nymphs of *Schistocerca gregaria*. Physiol. Entomol. **9**:329–349.

Szentesi, A., and Jermy, T. 1990. The role of experience in host plant choice by phytophagus insects. In E.A. Bernays (ed.), Insect/Plant Interactions, Vol. II. CRC Press, Boca Raton, FL, pp. 39–74.

Tabashnik, B. 1989. Managing resistance with multiple pesticide tactics: theory, evidence and recommendations. J. Econ. Entomol. **82**:1263–1269.

Thorpe, W.H., and Jones, F.G.W. 1937. Olfactory conditioning and its relation to the problem of host selection. Proc. R. Soc. Lond. B **134**:56–81.

Traynier, R.M.M. 1987. Learning without neurosis in host finding and oviposition by the cabbage butterfly, *Pieris rapae*. In V. Labeyrie, G. Fabres, and D. Lachaise (eds.), Insects-Plants. Junk Dordrecht, Netherlands, pp. 243–247.

Trimble, R.M., Pree, D.J., and Vickers, D.M. 1990. Survey for insecticide resistance in some Ontario populations of the apple leafminer parasite *Pholetesor ornigis*. Can. Entomol. **122**:969–973.

Turlings, T.C.J., Tumlinson, J.H., Lewis, W.J., and Vet, L.E.M. 1989. Beneficial arthropod behavior mediated by airborne semiochemicals. VII. Learning of host-related odors induced by a brief contact experience with host by-products in

Cotesia marginiventris, (Cresson), a generalist larval parasitoid. J. Insect Behav. **2**:217–225.

Turlings, T.C., Tumlinson, J.H., and Lewis, W.J. 1990. Exploitation of herbivore-induced plant odors by host-seeking parasitoid wasps. Science **250**:1251–1253.

Vet, L.E.M., and Groenewold, A.W. 1990. Semiochemicals and learning in parasitoids. J. Chem. Ecol. **16**:3119–3135.

Vet, L., Lewis, W.J., Papaj, D.R., and van Lenteren, J.C. 1990. A variable-response model for parasitoid foraging behavior. J. Insect Behav. **3**:471–490.

Vinson, S.B. 1984. The behavior of parasitoids. In G.A. Kerkut and L.I. Gilbert (eds.), Comprehensive Insect Physiology, Biochemistry and Pharmacology. Pergamon, New York, pp. 417–469.

Vinson, S.B., Barfield, C.S., and Henson, R.D. 1977. Ovioposition behavior of *Bracon melitor*, a parasitoid of the boll weevil (*Anthonomus grandis*). II. Associative learning. Physiol. Entomol. **2**:157–164.

Visser, J.H., and Thiery, D. 1986. Effects of feeding experience on the odour-conditioned anemotaxes of Colorado potato beetles. Entomol. Exp. Appl. **42**:198–200.

Vité, J.P., and Baader, E. 1990. Present and future use of semiochemicals in pest management of bark beetles. J. Chem. Ecol. **16**:3031–3041.

Waldbauer, G.P., and Friedman, S. 1991. Self-selection of optimal diets by insects. Annu. Rev. Entomol. **36**:43–63.

Wardle, A.R. 1990. Learning of host microhabitat colour by *Exeristes roborator* (Hymenoptera: Ichneumonidae). Anim. Behav. **39**:914–923.

Wardle, A.R., and Borden, J.H. 1985. Age-dependent associative learning by *Exeristes roborator* (F.) (Hymenoptera: Ichneumonidae). Can. Entomol. **117**:605–616.

Wardle, A.R., and Borden, J.H. 1986. Detrimental effect of prior conditioning on host habitat location by *Exeristes roborator*. Naturwissenschaften **73**:509–560.

Wardle, A.R., and Borden, J.H. 1991. Effect of prior experience on the response of *Exeristes roborator* to a natural host and microhabitat in a seminatural environment. Environ. Entomol. **20**:889–898.

13

Cognition in Bees: From Stimulus Reception to Behavioral Change

Reuven Dukas and Leslie A. Real

Introduction

Recent foraging models split the problems animals face into two stages: (1) the updating of information about alternative food sources and (2) the determination of behavior subject to this information (reviewed by Kacelnik and Krebs, 1985; Kacelnik et al., 1987; Stephens and Krebs, 1986). The cognitive abilities of different foragers are crucial at both stages. Species with limited cognitive abilities may rely heavily on relatively restrictive and inflexible decision rules. On the other hand, species with superior abilities to acquire information about the distribution, availability, and relative profitability of different food types may show superior and more flexible performance in foraging tasks (Pyke et al., 1977; Orians, 1981; Kamil and Mauldin, 1987; Schoener, 1987). In spite of the crucial importance of cognition for foraging performance, foraging studies, with few exceptions, rarely include information about the actual cognitive abilities of the animals under investigation. Nevertheless, all studies of foraging behavior make explicit or implicit assumptions about what animals perceive and know.

Definitions of cognition often vary from one authority to the next. Throughout our discussion we adopt what might as easily be called the "computational-representational" approach to animal learning and behavior (Gallistel, 1990). This approach suggests that behavior is influenced by three underlying processes. The first process encodes and translates information from the environment into a form that can be acted upon through mental operations. The second process involves the application of computational algorithms to a given set of encoded information. These mental operations that are specified by computational rules give rise to representations of the environment that comprise the third cognitive process. The emphasis on internal mental states that arise from computational operations is the characteristic stamp of the cognitive approach.

Cognition is often divided into a number of interdependent stages: reception, attention, representation, memory, problem solving, and in some cases, communication and language (Eysenck and Keane, 1990; Simon and Kaplan, 1989). Many cognitive studies using wild animals focus only on aspects of learning and memory. Learning is commonly defined as the modification of behavior through practice, training, or experience; the modification of behavior is almost always the only criterion used to test experimentally for learning (Papaj and Prokopy, 1989). Here we argue that discussing learning and memory alone is not sufficient and that detailed studies of *all* cognitive stages are necessary in order to understand and predict how animals modify their behavior in response to perceived stimuli and experience.

For two reasons, we concentrate on bees in our discussion of the different cognitive stages. First, extensive studies on bees, especially the social honey bees and bumble bees, have revealed details of their cognitive abilities (von Frisch, 1967; Menzel et al., 1974; Heinrich, 1979; Pyke, 1984; Seely, 1985; Gould and Gould, 1988; Bitterman, 1988; Real, 1991). Second, we are most familiar with bees from our own research. We do not attempt to summarize all that is known about cognition in bees. Rather, we try to identify relevant topics that are of importance for research on cognition and behavior. We begin with a critical review of cognition in the Apoidea and then discuss several ecological and evolutionary factors that may have significantly affected the structure of cognitive systems.

Reception

Reception is the acquisition of information about the environment through the sense organs. In the discussion of reception, we follow most cognitive scientists and artificial-intelligence researchers and focus on vision. This is because more is known about the visual system and its interrelatedness to other cognitive processes than about any other sensory system (Fischler and Firschein, 1987; Bruce and Green, 1990).

Visual reception, or seeing, is the physical recording of the pattern of light energy received from the environment through the eye. The interpretation of this physical recording is done in the brain and may involve the representation of information in the memory system, and the matching of that information with existing representations (Bruce and Green, 1990; Eysenck and Keane, 1990); these topics will be discussed below. The physical process of seeing is much better understood than the cognitive process of interpretation. All eyes must meet two requirements: first, they must divide the environment into an appropriate number of separately resolved sectors; this is termed the resolving power of the eye. Second, the eye must ensure that enough light is received from each sector for differences between sectors to be detectable; this is the sensitivity of the eye (Land, 1981).

Almost all the information needed in order to know both the resolving power and the absolute sensitivity of eyes is available from simple anatomical measurements of the eye and its receptors. Given knowledge of the optical principle on which the eye is based, the eyes of different species can then be compared, using as criteria their resolving power and sensitivity. A comparison between the compound eye of a worker honey bee and the simple human eye reveals a large difference in resolving power, 1.9° for the bee, compared with 0.014° for man. The resolution of the human eye is thus more than two orders of magnitude greater than that of the bee eye. In contrast to resolution, the absolute sensitivity of the bee's eye is $0.3 \ \mu m^2$, which falls within the range of sensitivity of the human eye—$0.02 \ \mu m^2$ in bright light and $37 \ \mu m^2$ in dark (Land, 1981).

The very low resolution of the compound eye imposes severe limits on the ability of insects to identify and discriminate between objects visually. Proximity can reduce this problem (Land, 1981; Gould and Gould, 1988) but is energetically expensive. A possible solution lies in the use of other senses. While the sense of smell is highly developed in insects, it is ineffective at long-distance discrimination, because odors from different sources get mixed together over a short distance. The low resolution of the compound eye may therefore make detection of a certain flower type much harder for bees than it would be for humans. The degree of difficulty in a detection task may affect the amount of attention devoted to this task. In addition, low visual resolution may significantly reduce the ability of individual bees to acquire information about the availability of alternative food sources. Therefore, visual resolution may be a key factor in determining floral-choice behavior in bees.

Given the relatively low resolving power of the insect eye, floral species may have evolved systems to increase their detectability in the natural environment. For example, increasing contrast between flower and background, while not affecting the resolving power of the eye, will increase detectability. In some cases adaptation in plants for increasing contrast can be quite dramatic. Most studies indicate that flowers reflect ultraviolet light and contrast sharply with foliar backgrounds that are UV-absorbing. However, many dune plants have reversed patterns in which flowers are UV-absorbing. The background in this case, sand, is UV-reflecting. In some species of *Heliotropium* found in Mexico, the flowers are unusual in being UV-absorbing against a foliar background. The leaves of these species are, however, glaucous, which increases their UV reflectance (Frolich, 1976).

The representation of floral types influenced by degrees of contrast is implicitly computational since the visual image is determined in part by the magnitudes of differences in excitation between adjacent areas of the visual field. Differential patterns of excitation and inhibition associated with contrast patterns have been used to explain some aspects of human visual perception, especially optical illusions, and may prove important in explaining aspects of bee vision and floral choice.

Attention

In order to process information effectively, animals must attend at any one time to only a limited amount of internal or external information. The study of attention may be defined as the analysis of the effectiveness of information processing when there are multiple simultaneous stimuli (Kahneman,1973; Shiffrin and Schenider, 1977; Navon and Gopher, 1979; Roitblat, 1987; Allport, 1989; Posner and Peterson, 1990; Eysenck and Keane, 1990). For example, the ability of human subjects to correctly detect camouflaged target types is inversely related to the number of target types: the more different types for which the subject is looking, the worse his or her efficiency at detecting any given type (e.g., Lindsey, 1970; Schneider and Shiffrin, 1977). Similarly, the reader of this chapter may find it impossible to listen simultaneously to, for example, the news on the radio. In this case, the attention demanded for reading prevents the reader from processing additional information effectively.

Attention may have its principal effects either through limiting what enters short-term memory or through affecting the capacity to retrieve what is already in long-term memory. At times it may be difficult to determine on phenomenological grounds which of these two processes— storage or retrieval—accounts for any observed pattern of behavior. Mechanistic neurobiological studies may help resolve some of these difficulties. For example, some recent studies provide a neurological basis for trade-offs in attention. These studies suggest that increasing the amount of attention directed toward a stimulus enhances the responsiveness and selectivity of neurons that process the stimulus in the brain. Consequently, the processing of information associated with this stimulus is enhanced, enabling an improved performance on the associated task (Spitzer et al., 1988; Corbetta et al., 1990; Poster and Peterson, 1990). For example, Corbetta et al. (1990) measured changes in regional cerebral blood flow of human subjects that were attending to either one or three attributes of the same set of visual stimuli. They found that selective attention to a single attribute was associated with a higher neural enhancement in the attended region of the visual cortex. Correspondingly, subjects showed a higher sensitivity for discriminating subtle stimulus changes when they were focusing attention on one attribute than when dividing attention among these attributes.

Studies on attention have focused mostly on humans. However, attentional limitations may have significant effects on the behavior of other animals (Milinski, 1989; Papaj, 1990; Dukas and Real, in review, a). Dukas and Ellner (in press) examined how attentional limitations may affect the optimal diet of foragers. Their model is based on Gendron and Staddon's (1983) extension of the basic optimal diet model (Stephens and Krebs, 1986). The capture rate of a given prey type is a function of prey density, the search rate of the forager, and the probability of detecting this prey type. The probability of detecting a prey type is a decreasing function of the forager's search rate and an increasing function of the amount of attention it devotes to this prey type.

Dukas and Ellner's (in press) model predicts that when prey are hard to detect, foragers should search for a single prey type in order to maximize their net rate of energy intake (Fig. 13.1, example a). In this case, division of attention among several prey types results in lower probabilities of detecting each prey type and smaller overall prey capture. On the other hand, if prey are easy to detect, foragers should divide their attention among several prey types. In this case, the optimal allocation of attention among prey types depends on their relative conspicuousness. When prey types differ in their conspicuousness, but all are relatively inconspicuous, foragers should devote more attention to the most conspicuous prey type (Fig. 13.1, examples b, c). Here a marginal increase in attention to the more conspicuous prey type results in a higher increase in the probability that this prey will be detected than in the less conspicuous one. However, if all prey types are relatively more conspicuous, foragers should devote more attention to the least conspicuous prey type (Fig. 13.1, example e). In this case, a marginal increase in attention to the less conspicuous prey type results in a higher increase in the probability of detecting this prey than in the more conspicuous one.

Attentional constraint may explain why bees tend to restrict their visits to the flowers of a single species of plant while bypassing other equally rewarding plant species, a behavior usually termed "flower constancy" (see Waser, 1986). Dukas and Real (in review, a) tested whether bumble bees reduce their ability to discriminate rewarding flowers from nonrewarding ones when foraging on an increasing number of rewarding floral types. They allowed bumble bees (*Bombus impatiens*) to forage in an enclosure containing equal proportions of randomly distributed rewarding and nonrewarding flowers of *Abelia floribunda*. Each of the rewarding flowers always contained 1 µl of 30% sugar water, and each of the nonrewarding flowers contained 5 µl water. To create the four floral types used in the experiments, they painted two of the five petals of each flower with either yellow,

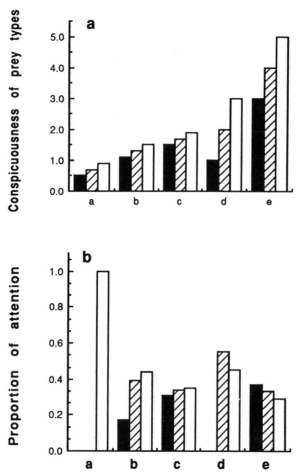

Figure 13.1. Five examples of the conspicuousness of prey types 1 (black bar), 2 (hatched bar), and 3 (white bar). The optimal proportion of attention given to each of the three prey types for each of the five examples. In all five examples, all three prey types have the same caloric content of 30 calories/prey, the same density of 4 prey/m^2, and handling time of 0.01 minutes/prey. The values of the constants for energy expenditure are $f = 20$ and $b = 1.6$. From Dukas and Ellner (in press).

white, purple, or blue paint. In one set of experiments the enclosure always contained 50 nonrewarding flowers of a single type and 50 rewarding flowers. In sessions of experiment 1, all 50 rewarding flowers were of a single type. In sessions of experiments 2 and 3, there were equal proportions of two and three floral types, respectively.

The bees' learning rates were higher in experiments with fewer numbers or rewarding floral types (Fig. 13.2). In addition, the asymptotic performance of bees, representing their maximum capacity for discrimination between rewarding and nonrewarding flowers at the end of each experimental session, was higher with fewer numbers of rewarding floral types (Fig. 13.2). These results suggest that when bees divide attention among an increasing number of rewarding floral types, their ability to detect these types among nonrewarding flowers is reduced considerably.

Representation

The information received by the sensory organs must be transformed into a code the brain uses for processing, storing and interpretation (Holland et al., 1986; Roitblat, 1987; Boden, 1988; Gallistel, 1989, 1990; Pylyshyn, 1989; Bruce and Green, 1990; Eysenck and Keane, 1990). While the concept of representation is central to cognitive science, it is defined and used in different ways. We follow here the definition of Roitblat (1987) and Eysenck and Keane (1990): a list of codes corresponding to certain features of an object. The manner in which information is represented is crucial to its interpretation. For example, many of us find it much easier to travel with a map, i.e., an external pictorial representation of an area, than with a long list of directions (Fischler and Firschein, 1987).

Earlier studies suggested that honey bees represent visual information as a list of relevant parameter values (e.g., spatial frequency and the proportion of different colors) (Anderson, 1972; Gould, 1984). Gould (1986a) tested honey bees for pictorial representation of visual stimuli. He allowed bees to choose between two alternative artificial flowers that looked different but had the same spatial frequency and color area (e.g., one flower was a mirror image of the other). Bees easily distinguished between the alternative flowers in a way that suggested that they form pictorial representations.

Gould (1986a) also calculated the resolution of the visual representation in honey bees by measuring the minimum apparent size of each element in flowers which bees recognized as different. He suggested that the resolution of the visual representation in honey bees is only about 8°, compared with the 1–2° resolution of honey bee vision. Note that the lower resolution of the visual representation reduces even more the ability of

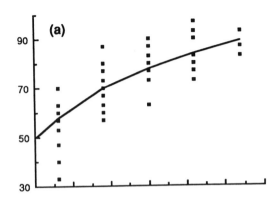

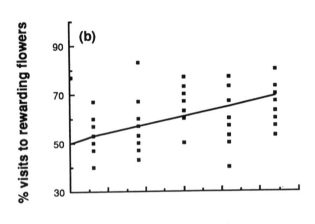

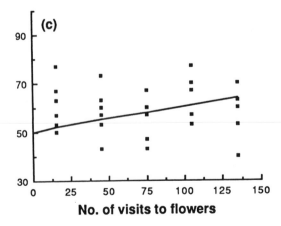

honey bees to detect certain floral types. This is because object recognition involves matching the information extracted from the visual stimulus with the represented information stored within the memory system (Hildreth and Ullman, 1989; Eysenck and Keane, 1990; Bruce and Green, 1990).

Honey bees apparently maintain pictorial representation of landmarks as well as object representations. They encode the size, shape, and color of landmarks near food sources and around their hive and use these representations to locate their destination (von Frisch, 1967; Gould, 1987a). Interestingly, the resolution of these representations (ca. 3.5°) is higher than the resolution of flower representations (ca. 8°) (Gould and Gould, 1988). Possibly, honey bees must have better resolution for landmark representations in order to correctly identify landmarks from a longer flight distance. The resolution of pictorial representations is probably an increasing function of the amount of neural circuitry needed to store each picture. A limited memory capacity may therefore be a crucial factor determining the degree of this resolution. The ability to form pictorial representations probably enhances honey bees' ability to correctly identify flowers and landmarks. However, this ability may be constrained by the low visual resolution and even lower resolution of representations in bees.

Spatial representations are often attributed to the formation of "cognitive maps." By definition, a cognitive map is an internal representation of the geometric relations among noticeable points in the animal's environment (Tolman, 1948; Wehner and Menzel, 1990; Gallistel, 1989). While cognitive maps were originally believed to be a property of vertebrates only, Gould (1986b) suggested that honey bees have mental maps as well. Gould (1986b)

Figure 13.2. Changes in the proportions of visits to rewarding flowers with experience. Panels a, b, and c present the results for experiments 1, 2, and 3, involving one, two, and three rewarding floral types and 11, 11, and eight bees, respectively. Each data point represents the proportion of visits to rewarding flowers by a bee in a trial of 30 visits. Data points for each experiments are for all sessions of this experiment. (Some overlapping data points are hidden.) The regression lines are calculated from the model $n = 30 - 15e^{-\delta v}$, where n is the number of visits to rewarding flowers per trial of 30 visits, 15 is the expected number of visits to nonrewarding flowers per trial prior to learning, δ is the learning rate, and v is the cumulative total number of visits. Learning rate in experiment 1 was significantly higher than learning rate in experiment 2 ($F = 69$, $p < 0.0001$, df = 1,101). Learning rate in experiment 3 was not significantly different than zero ($r^2 = 0.01$, $n = 40$, $p > 0.5$). The number of visits to rewarding flowers in the last learning trial was significantly greater in experiment 1 than in experiment 2 and greater in experiment 2 than in experiment 3 (Fig. 13.4a, $F = 29.3$, $p < 0.0001$, df = 2,72, Ryan-Einot-Gabriel-Welch Multiple Range Test $p < 0.05$).

displaced bees departing the hive to an unfamiliar site and observed them flying along a novel route to the unseen feeding station.

Dyer (1991) posed two alternative explanations for Gould's results. First, the route may not have been novel. In order to demonstrate the existence of a cognitive map in a certain species, one must show unequivocally that this species computes a short route between two points without having traveled along this route before (Wehner and Menzel, 1990). Second, the route may have been novel but bees may have used something other than a cognitive map (e.g., the panorama of major landmarks—pastures, forest, hills—surrounding the unseen feeding station) to return to the original feeding station.

Dyer (1991) attempted to distinguish among these alternative explanations by displacing bees trained to a feeding station A to a feeding station B from which bees could not use the panorama of major landmarks to return to the unseen feeding station A (the landmarks being obscured by forest surrounding station B). The cognitive map hypothesis predicts that bees can compute their way from feeding station B back to feeding station A, even if they have not previously traveled along this route and even if no panorama of major landmarks surrounding station A is available. Bees did not fly back to feeding station A; instead, most flew to the hive along the route with which they were familiar from previous training. This behavior does not support the cognitive map hypothesis (Dyer, 1991). Other displacement experiments with honey bees have led to a similar conclusion—namely, that honey bees probably do not have large-scale cognitive maps (Wehner and Menzel, 1990). While evidence for large-scale cognitive maps remains controversial, there is substantial evidence that honey bees maintain at least local metric spatial representations. Cartwright and Collett (1983) trained individual worker honey bees to fly from their colony through a gap in a curtain into a featureless matte white room. The room contained a small feeder and one or more upright black cylinders that acted as landmarks. The position of the upright black landmarks and feeder within the room varied across experiments. However, the positions of the landmarks and feeder with respect to each other were held constant across experiments. During test trials the feeders were removed and the searching behavior of workers was monitored. In almost every experiment, searching behavior was concentrated in the area where the feeder should have been if bees were using relative position with respect to the landmarks as a cue. The concentration of searching effort to the predicted area was observed even when there was only a single landmark. If bees were assessing only the distance from the landmark as their representation of the location of the feeder, then foraging should have been distributed in a ring around the landmark. That the bees search only along a single compass bearing suggests that use of the landmark depends on a geocentric map with metric information. The compass bearing of a landmark from a given point is a geocentric angular distance relation. Consequently, honey bees must represent angular and distance relationships at least over relatively small scales.

The mechanisms that allow bees to locate their positions while moving across landscapes may involve a variety of navigational techniques. Wehner and Menzel (1990) suggest that honey bees use two basic navigational systems. First, they may use path integration. This means that during flight, honey bees continuously monitor their speed and the angles turned during movement and integrate these data so as to track current position relative to their home. Path integration has been suggested for directional movement in other social hymenoptera, especially ants (Wehner and Srinivasan, 1981). Second, bees may use the pictorial representation of landmarks surrounding their goal in order to pinpoint its exact location (Wehner and Menzel, 1990). Honey bees may also follow familiar landmarks on their way between the hive and food sources and use this pictorial representation of landmarks as a backup system on overcast days when they cannot see the sun (Dyer and Gould, 1981). In short, honey bees have an effective navigational system that enables them to find the shortest route between food sources and their hive under a variety of environmental conditions.

Memory

While the structure of representations is significant for performance, the retention of the represented information is no less important. Experiments on honey bees and bumble bees show that they can remember information about landmarks and flowers for a considerable period of time. For example, honey bees showed no reduction in the correct choice of rewarding flowers after 2 weeks of separation from the test floral types (Menzel et al., 1974). Bumble bees also showed good memory retention and no reduction in the correct choice of rewarding flowers after a 2-day separation (Dukas, personal observation). This length of deprivation constitutes the longest time period tested for bumble bees.

Prior studies on memory retention in bees generally kept individual subjects in isolation between the training and test trials. However, studies on memory clearly show that retention is not a passive process. Rather, forgetting is a result of the performance of other activities that interfere with the retention of previously learned information (Alloway, 1972). Therefore, isolating bees probably facilitates memory retention compared to natural conditions. More realistic studies of memory capacity in bees must explore the consequences of intervening activity on the retention of previously learned tasks (Menzel et al., this volume).

Limited memory capacity in bees has been suggested as a reason for the bees' tendency to restrict foraging to single floral species while bypassing equally rewarding species, i.e., the tendency to show "flower constancy." If learning to recognize or manipulate one floral type interferes with the bee's memory about a previously visited floral type, then being constant to one floral type may increase the efficiency of resource use (Darwin, 1876; Waser, 1983, 1986; Lewis, 1986; Lewis, this volume). Waser (1986) found that bees foraging on arrays of randomly distributed flowers of two dissimilar species showed a higher degree of flower constancy than bees foraging on two similar species. This result is in agreement with the memory constraint hypothesis. Experiments on butterflies also suggest that performance of one activity interferes with memory about other activities (Lewis, 1986; Stanton, 1983). However, the capacity of butterfly memory relative to that of social bees remains unclear.

Two studies suggest that honey bees can remember simultaneously to recognize and manipulate at least two floral types. Koltermann (1974) trained honey bees to visit an artificial flower scented with geraniol in the morning and a flower scented with thyme oil 6 hours later in the afternoon. The two scented flowers were presented to the bees simultaneously on the following day. Bees mostly visited the geraniol-scented flower in the morning and the thyme-oil-scented flower in the afternoon.

Gould (1987b) trained honey bees to land on the lower right petal of a blue artificial flower from 0930 to 1100 and on the lower left petal of a yellow flower from 1100 to 1230. When tested on a later day, bees significantly selected the correct petal of the correct flower more than 80% of the time (compared with the 9% for new recruits). The above two studies, as well as others, suggest that honey bees represent information about flowers on a time linked basis (von Frisch, 1967; Koltermann, 1974; Gould, 1987b). However, these studies do not seem to agree with the memory constraint hypothesis.

Problem Solving

Information represented in the bee brain is used to solve problems or learn, i.e., to change its behavior in response to past experience. This represented information already reflects the bee's receptional, attentional, and memory limitations. Therefore, one must also study the cognitive stages discussed above in order to understand the structure of decision rules bees use to solve problems.

The most common problems faced by worker social bees involve foraging tasks. Bees must constantly decide what plant species to visit, how many flowers to visit on an inflorescence or in a flower patch, and how far to fly between flowers and patches. Worker social bees are among the most convenient organisms for testing foraging theories, since they engage almost exclusively in foraging activity and are not "distracted" by other activities such as mating and nest construction (Pyke, 1984; Dukas and Real, 1991; Real, 1991). Studies on bumble bee foraging have focused primarily on problems of patch exploitation and floral choice. Tests of the optimal residence time of bees on inflorescenses and patches suggest that bees maximize the expected long-term net rate of energy intake from flowers (Pyke, 1980, 1982, 1984; Stephens and Krebs, 1986; Schoener, 1987). Pleasants (1989) proposed a rule based on the marginal-value theorem (Charnov, 1976) for foragers making decisions about when to leave an inflorescence. The rule states that an optimal forager should leave an inflorescence if the amount of nectar it has just received indicates that its expected rate of energy intake for staying will be less than its expected rate of energy intake for leaving. The predictions of this rule were tested with data from studies of bumble bees foraging on larkspur (*Delphinium nelsonii*). The behavior of simulated bees using this rule was very similar to actual bumble bee behavior.

Experiments on floral choice by bees have primarily focused on how the distribution of nectar rewards within different floral types affects visitation rate to that type. In one of the first studies of this type, Waddington and Holden (1979) allowed honey bees to forage for nectar among two types of randomly distributed artificial flowers that differed in shape and color. They varied the average nectar volume per flower of each type by varying the proportions of flowers containing 2 μl nectar. In the first three experiments, equal proportions of each of the two floral types contained nectar. In the other four experiments, a higher proportion of flowers of type 1 contained nectar. Waddington and Holden (1979) predicted that bees would visit equal proportions of the floral types in the first three experiments. For the other four experiments, they predicted that bees would visit higher proportions of flowers of type 1, in relation to the expected higher caloric intake from flowers of this type. The observed proportion of bee visits were consistent with the predictions of Waddington and Holden's (1979) optimal foraging model, based on maximizing the net rate of energy gain.

Real and his colleagues (Real, 1981, 1991; Real et al., 1982, 1990; Harder and Real, 1987) have performed a series of experiments to determine if bumble bees exhibit choices that are consistent with expected utility theory—the dominant model for human decision-making under uncertainty (French, 1986). Individual bumble bees were allowed to forage for nectar among two types of randomly distributed artificial flowers that differed in color. The densities of each flower type were equal, and the value of each type was varied by changing the probability distribution of the nectar volume per flower. Bumble bees showed flower preferences that increased with increases in the average nectar volume per flower but decreased with increases in the nectar variance. In other words, bumble bees showed a trade-off between the arithmetic mean reward size and the variance in reward size in the formation of floral preferences.

This pattern of behavior is consistent with the expected utility model for choice. The expected utility model is governed by two factors: how nectar rewards are translated into some measure of utility to the bee and the probabilities of different reward outcomes upon visiting individual flowers. Harder and Real (1987) have suggested that the appropriate utility function for bees is a positive, diminishing function that translates nectar volumes into rates of net energetic gain (Fig. 13.3). This nonlinearity is sufficient to account for the bumble bees' response to variance in nectar reward, and simulations of bumble bee foraging based on maximizing the expected value of this utility model have proven quite robust in predicting forager preferences.

The biomechanical utility model suggested by Harder and Real (1987) assumes that bumble bees maximize the expected short-term net rate of energy gain, $E(R/T)$, where R is the reward and T is the time required to

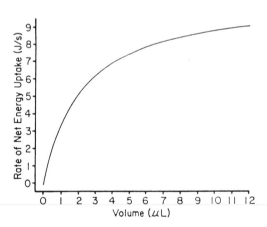

Figure 13.3. A plot of the model "utility" function based on empirically established biomechanics of nectar extraction. The model relates rate of net energetic gain in individual bumble bees to nectar volume in visited flowers. Under conditions in which bees maximize the expected rate of net energy gain, the nonlinear relation accounts for the bees' sensitivity to variance in nectar rewards per flower.

get this reward from the flower. Short-term rate of gain is not the usual currency employed in foraging models (Stephens and Krebs, 1986), and there is some question of what constitutes the appropriate time scale for evaluating choices. Possingham et al. (1990), for example, have argued that the long-term rate of energy gain is the appropriate currency. However, the foraging dynamics of simulated bees using short-term energy maximization is in better agreement with the observed pattern of bumble bee behavior (Real et al., 1990) compared with the pattern predicted from long-term rate maximization. The discrepancy could reflect a constraint. All of the potential decision rules used by bees, short or long term, are subject to cognitive constraints (Cheverton et al., 1985; Stephens and Krebs, 1986; Schoener, 1987; Real, 1991). For example, Real et al. (1990) suggested that if bees have a limited memory capacity, calculations based on short-term rate maximization may be more accurate and more efficient than calculations based on long-term rate maximization.

Alternatively, short-term rate maximization could have functional value. In some cases, the use of short-term or long-term characterizations of the expected rate of energetic gain may depend on ecological context. For example, the high degree of spatial autocorrelation that exists in natural populations of flowers have been used to explain the short-term averaging scheme apparently employed by bumble bees (Real, 1991). Cuthill et al. (1990) similarly have shown that starlings will switch between using short-term and long-term calculations of benefit depending on the correlation in reward delay. When there is a high degree of temporal autocorrelation then starlings only exhibit short-term memory. Uncorrelated reward schedules lead to more long-term averaging behavior. Short-term rate maximization may also prove advantageous under certain forms of correlation between temporal delays in access to reward and reward size (Caraco et al., 1991).

Context dependence in adaptive computational rules is suggested by a number of other foraging studies. For example, Carter and Dill (1990) showed that individual worker bumble bees switched their risk preferences as the energetic state of the colony was changed. When Carter and Dill experimentally depleted the colony's energetic reserves (by draining honey pots), workers switched from being risk-averse to being risk-seeking in their floral choices. The exact mechanisms that trigger a switch in the computational rules remain unclear.

It is still unknown how many past visits bees consider in order to make foraging decisions. Cresswell (1990) suggested that bumble bees (*B. bimaculatus*) foraging on wild bergamot (*Monarda fistulosa*) considered only the reward in the last flower visited in order to decide whether to leave an inflorescence. On the other hand, Dukas and Real (in review, b) found that bumble bees (*B. bimaculatus*) considered the nectar volumes in at least three flowers visited in determining subsequent flight distance. Dukas and Real (1991c) also found that visiting at least three empty flowers of a rewarding floral type did not alter bumble bees' subsequent proportions of visits to this floral type. This suggests that foraging decisions in bumble bees are based on cumulative experience in at least the last three floral visits.

The limited memory and truncated sampling that seem to characterize bumble bee foraging can have a profound influence on the representation of the parameters in the environment that govern choice and decisions. For example, small sequential samples from a highly skewed nectar distribution of rewards can lead to misrepresentation of the probability of rewards associated with different floral types. Limited sampling may account for the underestimation of rare events by individual bees and the underexploitation of relatively rare floral types in functional response experiments (Real, 1990, 1991).

In conclusion, it seems fair to assume that bees have a limited cognitive and computational capacity and that this limitation may lead to biases in behavior of great ecological importance (Real, 1991). However, more studies are needed in order to quantify the exact cognitive limitations of bees and how these limitations influence the structure of the decision rules bees use in solving foraging problems. Clearly, we need to know much more about how bees perceive and process information about the reward characteristics of flowers in order to produce a synthetic and predictive theory of choice.

Communication and Language

Cognitive science has developed, for the most part, from two often-independent histories (Simon and Kaplan, 1989). One branch has been primarily concerned with how organisms, especially humans, acquire skills, learn tasks, and solve problems. This branch has been of major interest to psychologists, decision theorists, and adherents of artificial intelligence. The second branch of cognitive science has focused primarily on language and its unique place in human cognition. The degree to which language is a unique feature of human communication systems and the degree to which all other cognitive functions are tied to language is one of the most important current issues in cognition, drawing on the disciplines of philosophy, neuroscience, linguistics, and psycholinguistics. Having explored some of the cognitive aspects of problem solving in bees, we now turn to issues of communication and language.

Communication can be defined as an action by one individual (the sender) that alters the behavior of another individual (the receiver) (Wilson, 1971).

In most solitary bees, communication is restricted to species-specific messages indicating the sex and reproductive readiness of the sender (Eickwort and Ginsberg, 1980; Hefetz, 1983; Gould and Gould, 1982; Gerling et al., 1989).ˑCarpenter bees, *Xylocopa virginica*, in addition, deposit a pheromone secreted by the Dufour's gland on flowers they visit. Flowers "marked" with this pheromone are avoided by conspecifics for about 10 minutes (Frankie and Vinson, 1977). A similar "marking" of flowers seems to exist in other species of carpenter bees (Gerling et al., 1989) and possibly bumble bees (Cameron, 1981; Schmitt and Bertch, 1990). The ultimate function of the scent a bee leaves on a flower is still unclear. Leaving scent may be akin to "leaving footprints" rather than active marking. Additional studies are needed in order to establish the commonness of scent "marking" and its functional significance among foraging bees under natural conditions.

Communication has several important functions in social species, where individuals may communicate their status, alert other members of the colony for predators, and recruit for food or other resources (Wilson, 1971; Gould and Gould, 1982). The ability to communicate can significantly facilitate social coordination and control. All social bees have some form of communication between the queen and workers. In honey bees, decanoic acid released by the queen has at least three effects on the colony. First, it represses the ovaries of the workers, preventing the development of laying workers. Second, it attracts nurse bees to the queen and elicits their care. Third, it inhibits the production of new queens (Wilson, 1971; Gould and Gould, 1988). Through the release of pheromones the queen controls the reproductive state of the colony. However, the queen is not responsible for either the allocation of workers among different tasks or the recruitment of foragers to different food sources. The control of such colony-level activities is a result of the integration of information across individual workers (Wilson, 1962, 1971; Oster and Wilson 1978; Seely, 1985; Seely et al., 1991; Wilson and Hölldobler, 1988; Gordon, 1989; Franks, 1989).

It should be noted that communication about food sources is unknown in primitively social bees and bumble bees (Wilson, 1971; Heinrich, 1979; Waddington, 1981). Such communication is probably restricted to three other genera of the Apoidea, *Melipona*, *Trigona*, and *Apis*. Species of *Melipona* only alert nestmates to the presence of a rich food source. Several *Trigona* species use a pheromone secreted from the mandibular glands to mark a scent trail between the food source and the nest. The recruited workers then follow this trail on their way to the food source. This method of communication about food is similar to that common among ant species (Lindauer and Kerr, 1960; von Frisch, 1967; Wilson, 1971).

Karl von Frisch (reviewed in von Frisch, 1967) discovered the waggle dance of bees and proposed that scouts use this dance to communicate about food sources. Wenner (1967) and Johnson (1967) challenged the idea of dance language in honey bees (see also Wenner and Wells, 1990). However, experiments by Gould (1975) and preliminary experiments by Michelsen et al. (1989) with a mechanical model of a dancing bee further suggest that honey bees use the dance language to communicate the direction and distance to a food source.

The dance language of honey bees is probably the only form of communication in invertebrates that has been regarded as a primitive "language." However, a closer examination reveals that honey bee communication falls far short of at least one definition of true language. Anderson (1985) defined language according to four criteria: arbitrariness of symbolic units, displacement, semanticity, and productivity. Human language satisfies all of these requirements. Honey bee communication fails to meet at least one and more likely two of these criteria.

The arbitrariness-of-units criterion establishes that the particular signals used to communicate are not inherent in the goal of the communication but are arbitrarily assigned their meanings. For example, baring teeth or raising fists are not arbitrary signals for aggression. However, saying, "I'm going to get you" employs an arbitrary set of signals. Honey bee communication does employ an arbitrary set of symbols by which one bee informs others about the direction, distance, and relative quality of distant objects. This first criterion seems to be satisfied by the bees' dance.

The second criterion for language, displacement, requires that signals maintain their semantic referents when displayed away from the immediate vicinity of the referent. Since the dance of incoming foragers is performed at some distance away from the spatial location of the area represented in the dance, this criterion seems to be satisfied by the bees.

Semanticity, the third criterion, requires that a signal activate a representation of the event to which it relates (Pearce, 1987). For example, there is clear evidence that chimpanzees learn the semantic reference for at least some of the words they learn. One chimp, "Sarah," was trained to know the lexigraphic set of signals for "brown color of chocolate" in the absence of any chocolate. When presented with four discs of different colors she responded correctly to the question, "What color brown?" The lexigraphic symbol for chocolate must have been able to elicit a representation of chocolate in order for Sarah to know what color the lexigram "brown" referred to (Premack, 1976).

Gould (1984) trained honey bees to respond to artificial feeders at different locations around a lake. When the feeder was on the edge of the lake workers would respond and recruit to the resource. However, when the feeder was located in the center of the lake, bees could not be recruited to this station. Obviously, bees are responding to more than the fixed signal, but must be interpreting signals as well. Failure to recruit indicates that the sender's message was essentially "meaningless." Such results may indicate that the bee dance has a semantic as well as a purely procedural component.

The fourth criterion, corresponding to productivity, is the property of language which appears to be the most unique to humans. Productivity in language depends on the existence of syntactic rules and relations that promote the generation of new semantically significant signals through the combination of other semantically significant signals. The honey bee's dance is not easy to interpret with respect to this criterion. To some degree the dance shows an infinite capacity for production since there is an infinite number of combinations of angular and distance signals. However, information about angle and distance may not be separable within the bee's dance and therefore location may not be the product of combining semantically significant signals through any kind of syntactic relation. Instead, location may be a single signal with an infinite degree of modulation in expression. There is no substantial evidence that any animal besides humans can combine signals into new meanings.

Consequently, bee language seems to meet the first two criteria and possibly the third, but fails to meet the last criterion requiring syntax. While bees come up short in this regard, their capacity for communicating complex information in perhaps semantically significant fashion is impressive. This complex capacity may have evolved to meet the demands of collective decision-making prerequisite to the emergence of social organization in colonies of bees.

One of the better-studied examples of the role of communication in collective decision-making is the way foragers in honey bee colonies choose among nectar sources over space and time (Visscher and Seely, 1982; Seely, 1985; Seely et al., 1991). The basis for this colony-level decision-making is the modulation of the rates of bees' recruitment and abandonment from different nectar sources in relation to their profitability (Seely et al., 1991). Each forager seems to estimate the absolute profitability of the nectar source based mostly on the abundance and quality of nectar and the distance from the hive. Foragers that encounter a highly profitable nectar source continue to visit this source, work quickly, and actively recruit additional bees through dancing. On the other hand, foragers that encounter low-quality nectar sources tend to abandon those sources, or if they continue to visit them, they are less excited and do not perform recruitment dances (Seely et al., 1991). Honey bees seem to make reasonable colony-level foraging decisions based on the integrated behavior of individuals.

Population Genetic and Comparative Aspects of Cognition

Cognition and Evolution

There is at present little direct evidence that cognition has undergone adaptive evolutionary change. Still, it is worth asking whether specific cognitive functions can evolve in an adaptive manner. In order to show that natural selection can alter cognitive functions, we must demonstrate (1) that genetically based variation in a given cognitive function exists and (2) that differences in the cognitive function are associated with differences in fitness (Papaj and Prokopy, 1989).

Genetic studies on several cognitive abilities in humans and other mammals report relatively high heritabilities (Plomin, 1990; Plomin et al., 1990; Wahlsten, 1972). For example, in more than 30 twin studies involving more than 10,000 pairs of twins, identical and fraternal twin correlations for general cognitive ability (IQ) averaged 0.85 and 0.60, respectively. These results are consistent with heritability values of about 50% (Plomin, 1990). Brandes (1988) examined the heritability of learning in honey bees using conditioning of the proboscis extension reflex (Bitterman et al., 1983). He found heritability values (h^2) of 0.4–0.5, which are consistent with heritability values for learning ability in other species (Plomin et al., 1990).

While genetically based variation in learning clearly exists, almost nothing is known about the association between learning and fitness (Orians, 1981; Johnston, 1982; Papaj and Prokopy, 1989). It is relatively easy to measure the effect of learning on the efficiency of flower handling by bees (Strickler, 1978; Laverty, 1980; Laverty and Plowright, 1988) and the consequent effect on bee fitness or at least a component of fitness such as energy intake (Strickler, 1978). Therefore, bees may be a good group for conducting a critical test of the association between learning and fitness.

Species Differences in Cognitive Abilities

Several studies suggest adaptive evolutionary change in cognitive abilities based on comparisons across species. For example, food-storing birds show longer-lasting spatial memory compared to closely related nonstoring species (Sherry, 1988; Balda and Kamil, 1989; Krebs et al., 1990). At least one study suggests similar differences in cognitive abilities in bees. Geographic races of honey bee differ considerably in colony size and foraging range; these differences are associated with a difference in the indication of distance in the honey bee dance (Gould, 1982). The racial differences in foraging range may also be associated with differences in cognitive capacities. This is because the longer-distance foragers probably encounter a wider range of sensory stimuli. A correlation between foraging range and relative brain size is known in three orders of mammals: bats, rodents, and primates (Eisenberg and Wilson, 1978; Harvey and Krebs, 1990).

If the range of sensory stimuli in an animal's environment relates to selection for increased cognitive abilities, one might expect generalist species to show higher cognitive abilities than specialist species (Papaj and Prokopy, 1989; Dukas and Real, 1991). There is still no clear evidence for this notion in either phytophagous insects (Papaj and Prokopy, 1989) or bees.

The Costs and Benefits of Cognitive Functions

Higher levels of cognitive ability are uncommon among animals and are mostly restricted to several groups of mammals and birds (Mayr, 1974; Johnston, 1982). Among insects, the social hymenoptera seem to be exceptional in their relatively higher cognitive abilities (Menzel et al., 1974; Heinrich, 1984; Real, 1991, 1992; Dukas and Real, 1991). The relatively high learning and memory capabilities of social hymenoptera call into question two common generalizations about cognition. The first is that cognitive abilities should be limited in short-lived species, because these species have only a limited time to exploit the knowledge they acquire. The second is that small body size limits neural complexity and consequent cognitive ability (Mayr, 1974; Johnston, 1982; Staddon, 1983).

Life may be perceived as a trade-off between spending time and energy learning new things, and exploiting things already known (Staddon, 1983). However, from this it does not follow that short-lived species should not spend time learning. Assuming no constraints on cognitive abilities, the *relative* time spent on learning should reflect the potential benefits and costs of learning. As already stated, we simply do not know the evolutionary benefits of learning in any well-studied group of organisms.

Spending time and energy on learning may sometimes prove a good investment regardless of life expectancy, especially if learning can significantly improve individual performance. For example, when inexperienced worker bumble bees (*B. vagans*) begin foraging on jewelweed (*Impatiens biflora*), about 50% land on and probe in the wrong place for nectar. Overall, inexperienced bees visit one to three flowers per minute. However, experienced bees visit 11 flowers per minute. Thus, the cost of learning to handle the flowers is about 110 calories/minutes, which is the reward available from eight flowers (Heinrich, 1984). However, it takes bumble bees only about 100 visits to become proficient at handling the flowers that offer much higher energy intake compared to other easy-to-handle flowers (Heinrich, 1984; see also Laverty, 1980; Laverty and Plowright, 1988).

In addition to time and energy, another cost of learning is the vulnerability of young inexperienced individuals. During the period when experience is gained, young animals may be an easy target for predators, or may simply starve to death (Morse, 1980; Johnston, 1982). Evidence for the severe cost of youth and inexperience exists for several bird species (Lack, 1966; Sullivan, 1988). For example, the major mortality factor for recently fledged juvenile juncos (*Junco phaenotus*) is starvation. Juveniles fail to forage effectively, resulting in loss of body weight which is highly correlated with risk of mortality (Sullivan, 1988; Weathers and Sullivan, 1989).

Parental Care and Sociality

Parental care can significantly reduce the vulnerability of inexperienced young. Indeed, it seems that higher cognitive abilities in birds and mammals are always associated with an extensive period of parental care (Mayr, 1974; Morse, 1980; Johnston, 1982).

The care of young in the social hymenoptera may be one of the main reasons for the higher cognitive abilities observed in this group. Emerging solitary bees must have a fixed genetically based representation of flowers, because they have to locate flowers independently. On the other hand, in honey bees, the scouts provide the inexperienced foragers with information about the direction, and distance to, and odor of certain flowers. Therefore, honey bees need not have a fixed genetically based representation of flower appearance, and they may be more open to learning about food sources. It is interesting to note that honey bees are probably the only temperate bee species that visit food sources other than flowers. Honey bees may forage on the juice of fruits, and even on leftovers of man-made soft-drinks in garbage cans, items that clearly do not resemble flowers.

Another aspect of sociality could be associated with increased cognitive abilities. Social behavior requires recognition of and communication among colony members. This type of interaction may select for a general increase in learning and memory capabilities in highly social species (Johnston, 1982; Essock-Vitals and Seyfarth, 1986; Dukas and Real, 1991). Dukas and Real (1991) tested this prediction by comparing the performance of a social species of bumble bee (*B. bimaculatus*) and a solitary species of carpenter bee *Xylocopa virginica* in three experiments in which bees had to learn to discriminate between rewarding and nonrewarding flowers differing only in color. Discriminating rewarding from nonrewarding flowers is a simple learning task that generalist bees, such as carpenter bees and bumble bees, always face while foraging under natural conditions. In all three experiments, bumble bees showed higher learning rates compared to carpenter bees (Fig. 13.4). These results support our hypothesis that social bees show higher learning capabilities than solitary bees.

Conclusions

Using the cognitive approach for studying animal behavior in general and learning in particular seems very useful. Looking only at changes in behavior due to past experience is clearly not sufficient for understanding and explaining complex behavioral patterns of animals. First, the nature

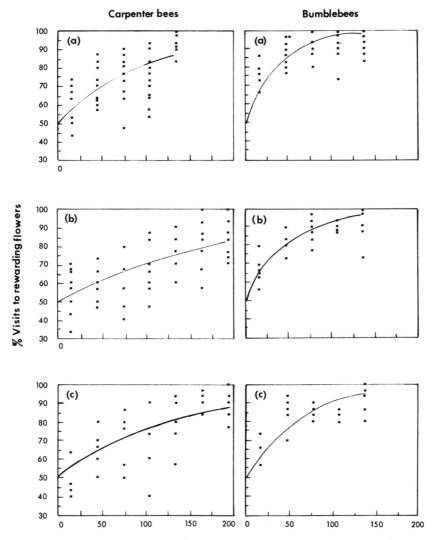

Figure 13.4. Learning rates, or changes in proportions of visits to rewarding flowers by carpenter bees (left) and bumble bees (right). a, Rewarding flowers are painted white and nonrewarding flowers are painted blue. The nonlinear regression equation is given by $p = 1 - 0.5 * e^{-0.01 * V}$ ($R^2 = 0.46$, $p < 0.001$, df $= 76$) for carpenter bees and $p = 1 - 0.5 * e^{-0.028 * V}$ ($R^2 = 0.64$, $p < 0.001$, df $= 58$) for bumble bees. b, Rewarding flowers are painted white and nonrewarding flowers are painted yellow. The nonlinear regression equation is given by $p = 1 - 0.5 * e^{-0.005 * V}$ ($R^2 = 0.48$, $p < 0.001$, df $= 62$) for carpenter bees and $p = 1 - 0.5 * e^{-0.02 * V}$ ($R^2 = 0.76$, $p < 0.001$, df $= 38$) for bumble bees c, Rewarding flowers are painted yellow and nonrewarding flowers are white. The nonlinear regression equation is given by $p = 1 - 0.5 * e^{-0.007 * V}$ ($R^2 = 0.6$, $p < 0.001$, df $= 38$) for carpenter bees and $p = 1 - 0.5 * e^{-0.0168 * V}$ ($R^2 = 0.8$, $p < 0.001$, df $= 28$) for bumble bees. From Dukas and Real (1991).

of an animal's sensory system strongly affects the way it perceives its environment. Second, the animal's attentional capacity determines the amount of sensory information that can be processed simultaneously and effectively. Third, usually only a portion of the information perceived is encoded and represented in the brain. The amount of the information represented and the structure of the representation determines the effectiveness of using this information in the future. Fourth, memory, or the retention of representations over time, determines how long an animal can use its past experience to solve certain problems. Fifth, the structure of an animal's decision rules shapes the way the animal's behavior changes in response to stimuli perceived from the environment. Such decision rules are subject to computational limitations that differ among species. Finally, populations do not consist of independently acting individuals. Interactions among individuals, most notably communication, may directly affect the individual's perception and cognitive function.

While some of the above cognitive stages have been extensively studied in bees and other animals, others, such as attention and decision rules, require further exploration. We must also construct critical tests of the association between different cognitive functions and the reproductive success of individuals. Such studies will ultimately lead to a better understanding of the evolutionary origins of animal cognition and intelligence.

Acknowledgments

We thank J. Walters, S. Ellner, N. Waser, and B. Zielinski for comments on the manuscript. This research was supported through NSF grants BNS 8719292 and BNS 8900292 to L.A.R.

References

Alloway, T.M. 1972. Learning in insects except Apoidea. In W.C. Corning, J.A. Dyal, A.O.D. Willows (eds.), Invertebrate Learning. Plenum, New York, pp. 131–171.

Allport, A. 1989. Visual attention. In M.I. Posner (ed.), Foundations of Cognitive Science, MIT Press, Cambridge, MA, pp. 631–682.

Anderson, A.M. 1972. The ability of honey bees to generalize visual stimuli. In R. Wehner (ed.), Information Processing in the Visual System of Arthropods, Springer-Verlag, Berlin, pp. 207–212.

Anderson, JR 1985. Cognitive Psychology and Its Implications. Freeman, San Fransisco.

Balda, R.P., and Kamil, A.C. 1989. A comparative study of cache recovery by three corvid species. Anim. Behav. **38**:486–495.

Bitterman, M.E., Menzel, R., Fietz, A., and Schafer, S. 1983. Classical conditioning of proboscis extension in honey bees (*Apis mellifera*). J. Comp. Psychol. **97**:107–119.

Bitterman, M.E. 1988. Vertebrate-invertebrate comparisons. In H.J. Jerison and I. Jerison (eds.), Intelligence and Evolutionary Biology. Springer-Verlag, Berlin, pp. 251–275.

Boden, M.A. 1988. Artificial intelligence and biological intelligence. In H.J. Jerison and I. Jerison (eds.), Intelligence and Evolutionary Biology. Springer-Verlag, Berlin, pp. 45–71.

Brandes, C. 1988. Estimation of heritability of learning behavior in honey bees (*Apis mellifera capensis*). Behav. Genet. **18**:119–132.

Bruce, V., and Green, P.R. 1990. Visual Perception: Physiology, Psychology and Ecology. Lawrence Erlbaum Associates, Hillside, NJ.

Cameron, S.A. 1981. Chemical signals in bumble bee foraging. Behav. Ecol. Sociobiol. **9**:257–260.

Caraco, T., Kacelnik, A., Mesnick, N., and Smulewitz, M. 1992. Short-term rate maximization when rewards and delays covary. Anim. Behav., in press.

Carter, R., and Dill, L. 1990. Why are bumble bees risk-sensitive foragers? Behav. Ecol. Sociobiol. **26**:121–127.

Cartwright, B.A., and Collett, T.S. 1983. Landmark learning in bees: Experiments and models. J. Comp. Physiol. **151**:521–543.

Charnov, E.L. 1976. Optimal foraging: The marginal value theorem. Theor. Popul. Biol. **9**:129–136.

Cheverton, J., Kacelnik, A. and Krebs, J.R. 1985. Optimal foraging: Constraints and currencies. In Holldobler and Lindauer (eds.), Experimental Behavioral Ecology. G. Fischer Verlag, Stuttgart, pp. 110–126.

Corbetta, M., Miezin, F.M., Dobmeyer, S., Shulman, G.L. and Petersen, S.E. 1990. Attentional modulation of neural processing of shape, colour, and velocity in humans. Science **248**:1556–1559.

Cresswell, J.E. 1990. How and why do nectar-foraging bumble bees initiate movements between inflorescences of wild bergamot *Monarda fistulosa* (Lamiaceae). Oecologia, **82**:450–460.

Cuthill, I., Kacelnik, A., Krebs, J.R., Haccou, P., and Iwasa, Y. 1990. Starlings exploiting patches: The effect of recent experience on foraging decisions. Anim. Behav. **40**:625–640.

Darwin, C. 1876. On the Effects of Cross and Self Fertilization in the Vegetable Kingdom. John Murray, London.

Dukas, R., and Ellner, S. Information processing and prey detection. Ecology, in press.

Dukas, R., and Real, L.A. 1991. Learning foraging tasks by bees: A comparison between solitary and social species. Anim. Behav. **42**:269–276.

Dukas, R., and Real, L.A.. Learning constraints and floral choice behaviour in bumble bees. Anim. Behav., in review, a.

Dukas, R., and Real, L.A. Effects of recent experience on foraging decisions in bumble bees. Oecologia, in review, b.

Dyer, F.C. 1991. Bees acquire route-based memories but not cognitive maps in a familiar landscape. Anim. Behav. **41**:239–246.

Dyer, F.C., and Gould, J.L. 1981. Honey bee orientation: A backup system for cloudy days. Science **214**:1041–1042.

Eickwort, G.C., and Ginsberg, H.S. 1980. Foraging and mating behaviour in Apoidea. Annu. Rev. Entomol. **25**:421–446.

Eisenberg, J.F., and Wilson, D.E., 1978. Relative brain size and feeding strategies in the Chiroptera. Evolution **32**:740–751.

Essock–Vitale, S., and Seyfarth, R.M. 1986. Intelligence and social cognition. In B.B. Smuts, R.W. Wrangham, D.L., Cheney, T.T. Struhsaper, and R.M. Seyfarth (eds.), Primate Societies. University of Chicago Press, Chicago, pp. 452–461.

Eysenck, M.W., and Keane, M.T. 1990. Cognitive Psychology. Lawrence Erlbaum Associates, London.

Fischler, M.A., and Firschein, O. 1987. Intelligence: the Eye, the Brain, and the Computer. Addison–Wesley, Reading, PA.

Frankie, G.W., and Vinson, S.B. 1977. Scent marking of passion flowers in Texas by females of *Xylocopa virginica texana* (Hymenoptera: Anthophoridae). J. Kansas Entomol. Soc. **50**:613–625.

Franks, N.R. 1989. Army ants: a collective intelligence. Am. Sci. **77**:139–145.

French, S. 1986. Decision Theory. John Wiley, New York.

Frisch, K. von. 1967. The Dance Language and Orientation of Bees. Harvard University Press, Cambridge, MA.

Frohlich, M.W. 1976. Appearance of vegetation in ultraviolet light: Absorbing flowers, reflecting backgrounds. Science **194**:839–846.

Gallistel, C.R. 1989. Animal cognition: The representation of space, time and number. Annu. Rev. Psychol. **40**:155–189.

Gallistel, C.R. 1990. The Organization of Learning. Cambridge, MA, MIT Press.

Gendron, R.P., and Staddon, J.E.R. 1983. Searching for cryptic prey: The effects of search rate. Am. Nat. **121**:172–186.

Gerling, D., Velthuis, H.H.W., and Hefetz, A. 1989. Bionomics of the large carpenter bees of the genus *Xylocopa*. Annu. Rev. Entomol. **34**:163–190.

Gordon, D. 1989. Caste and change in social insects. Oxford Surveys Evol. Biol. **6**:55–72.

Gould, J.L. 1975. Honey bee recruitment: The dance language controversy. Science **189**:685–692.

Gould, J.L. 1982. Why do honey bees have dialects? Behav. Ecol. Sociobiol. **10**:53–56.

Gould, J.L. 1984. The natural history of honey bee learning. In P. Marler and H. Terrace (eds.), The Biology of Learning. Springer–Verlag, Berlin, pp. 149–180.

Gould, J.L. 1986a. Pattern learning by honey bees. Anim. Behav. **34**:990–997.

Gould, J.L. 1986b. The local map of honey bees: Do insects have cognitive maps? Science **232**:861–863.

Gould, J.L. 1987a. Landmark learning by honey bees. Anim. Behav. **35**:26–34.

Gould, J.L. 1987b. Honey bees store learned flower-landing behavior according to time of day. Anim. Behav. 35:1579–1581.

Gould, J.L., and Gould, C.G. 1982. The insect mind: Physics or metaphysics? In D.R. Griffin (ed.), Animal Mind – Human Mind Springer-Verlag, Berlin, pp. 269–298.

Gould, J.L, and Gould, C.G. 1988. The honey bee. Scientific American, New York.

Harder, L.D., and Real, L.A. 1987. Why are bumblebees risk aversive? Ecology **68**:1104–1108.

Harvey, P.H., and Krebs, J.R. 1990. Comparing brains. Science **249**:140–146.

Hefetz, A. 1983. Function of secretion of mandibular gland of male in territorial behavior of *Xylocopa sulcatipes* (Hymenoptera: Anthophoridae). J. Chem. Ecol. **9**:923–931.

Heinrich, B. 1979. Bumblebee Economics. Harvard University Press, Cambridge, MA.

Heinrich, B. 1984. Learning in invertebrates. In P. Marler and H. Terrace (eds.), The Biology of Learning, Springer-Verlag, Berlin, pp. 135–147.

Hildreth, H.C., and Ullman, S. 1989. The computational study of vision. In M.I. Posner (ed.), Foundations of Cognitive Science, MIT Press, Cambridge, MA, pp. 581–630.

Holland, J.H., Holyoak, K.J., Nisbett, R.E., and Thagard, P.R. 1986. Induction: Processes of Inference, Learning, and Discovery. MIT Press, Cambridge, MA.

Johnson, D.L. 1967. Honey bees: Do they use the direction information contained in their dance maneuver? Science **155**:847–849.

Johnston, T.D. 1982. Selective costs and benefits in the evolution of learning. Adv. Study Behav. **12**:65–106.

Kacelnik, A., and Krebs, J.R. 1985. Learning to exploit patchily distributed food. In R.M. Sibly and R.H. Smith (eds.), Behavioural Ecology. Blackwell's Scientific Publications, Oxford, England, pp. 189–205.

Kacelnik, A., Krebs, J.R., and Ens, B. 1987. Foraging in a changing environment: An experiment with starlings (*Sturnus vulgaris*). In M.L. Commons, A. Kacelnik, and S.J. Shettleworth (eds.), Quantitative Analyses of Behavior. Lawrence Erbaum Associates, Hillsdale, NJ, pp. 63–87.

Kahneman, D. 1973. Attention and effort. Prentice Hall, Englewood Cliffs, NJ.

Kamil, A.C., and Mauldin, J.E. 1987. A comparative-ecological approach to the study of learning. In R.C. Bolles and M.D. Beecher (eds.), Evolution and Learning. Lawrence Erlbaum Associates, Hillsdale, NJ, pp. 117–133.

Koltermann, R. 1974. Periodicity in the activity and learning performance of the honeybee. In L.B. Browne (ed.), Experimental Analysis of Insect Behavior. Berlin: Springer-Verlag, Berlin, pp. 218–227.

Krebs, J.R., Healy, S.D., and Shettleworth, S.J. 1990. Spatial memory of Paridae: Comparison of a storing and non-storing species, the coal tit, *Parus ater*, and the great tit, *P. major*. Anim. Behav. **36**:733–740.

Lack, D. 1966. Population Studies of Birds. Clarendon Press, Oxford, England.

Land, M.F., 1981. Optics and vision in invertebrates. In H. Autrum (ed.), Comparative Physiology and Evolution of Vision in Invertebrates, Vol. B. Springer-Verlag, Berlin, pp. 471–592.

Laverty, T.M. 1980. Bumble bee foraging: Floral complexity and learning. *Can. J. Zool.* **58**:1324–1335.

Laverty, T.M., and Plowright, C. 1988. Flower handling by bumblebees: A comparison of specialists and generalists. Anim. Behav., **36**:733–740.

Lewis, A.C. 1986. Memory constraints and flower choice in *Pieris rapae*. Science **232**:863–865.

Lindauer, M., and Kerr, W.E. 1960. Communication between the workers of stingless bees. Bee World **41**:65–71.

Lindsey, P.H. 1970. Multiple processing in perception. In D.I. Mostofsky (ed.), Attention: Contemporary Theory and Analysis. Meredith Corp., New York, pp. 149–171.

Mayr, E. 1974. Behavior programs and evolutionary strategies. Am. Sci. **62**:650–659.

Menzel, R., Erber J., and Masuhr, T. 1974. Learning and memory in the honeybee. In L. Barton-Browne (ed.), Experimental Analysis of Insect Behavior. Berlin: Springer-Verlag, Berlin, pp. 195–217.

Michelsen, A., Anderson, B.B., Kirchner, W.H., and Lindauer, M. 1989. Honey bees can be recruited by a mechanical model of a dancing bee. Naturwissenschaften **76**:277–280.

Milinski, M. 1989. Information overload and food selection. In R.H. Hughes (ed.), Behavioural Mechanisms of Food Selection. NATO ASI Series. Springer-Verlag, Berlin, pp. 721–737.

Morse, D.H. 1980. Behavioural Mechanisms in Ecology. Harvard University Press, Cambridge MA.

Navon, D., and Gopher, D., 1979. On the economy of the human processing system. Psychological Rev. **86**:214–255.

Orians, G.H. 1981. Foraging behavior and the evolution of discrimination abilities. In A.C. Kamil and T.D. Sargent (eds.), Foraging Behavior, Garland, New York, pp. 389–405.

Oster, G., and Wilson, E.O. 1978. Caste and Ecology in the Social Insects. Princeton, NJ, Princeton University Press.

Papaj, D.R., 1990. Interference with learning in pipevine swallowtail butterflies: Behavioral constraint or possible adaptation? Symp. Biol. Hung. **39**:89–101.

Papaj, D.R. and Prokopy, R.J. 1989. Ecological and evolutionary aspects of learning in phytophagous insects. Annu. Rev. Entomol., **34**:315–350.

Pearce, J.M. 1987. An Introduction to Animal Cognition. Lawrence Erlbaum Associates, Hillsdale, NJ.

Pleasants, J.M. 1989. Optimal foraging by nectarivores: A test of the marginal value theorem. Am. Nat. **134**:51–71.

Plomin, R. 1990. The role of inheritance in behavior. Science **248**:183–188.

Plomin, R., DeFries, J.C., and McClearn, G.E. 1990. Behavioral Genetics. W. H. Freeman, New York.

Posner, M.I., and Peterson, S.E. 1990. The attention system of the human brain. Annu. Rev. Neurosci. **13**:25–42.

Possingham, H.P., Houston, A.I., and McNamara, J.M. 1990. Risk-averse foraging in bees: Comments on the model of Harder and Real. Ecology **71**:1622–1624.

Premack, D. 1976. Intelligence in Ape and Man. Lawrence Erlbaum Associates, Hillsdale, NJ.

Pyke, G.H. 1980. Optimal foraging in bumblebees: Calculation of net rate of energy intake and optimal patch choice. Theor. Popul. biol. **17**:232–246.

Pyke, G.H. 1982. Foraging in bumblebees: Rule of departure from an inflorescence. Can. J. Zool. **60**:417–428.

Pyke, G.H. 1984. Optimal foraging theory: A critical review. Annu. Rev. Ecol. Syst. **15**:523–575.

Pyke, G.H., H.R. Pulliam, and Charnov, E.L. 1977. Optimal foraging: A selective review of theory and tests. Q. Rev. Biol. **52**:137–154.

Pylyshyn, Z.W. 1989. Computing in cognitive science. In M.I. Posner (ed.), Foundations of Cognitive Science. MIT Press, Cambridge, MA, pp. 51–91.

Real, L. 1981. Uncertainty and pollinator-plant interactions: The foraging behavior of bees and wasps on artificial flowers. Ecology, **62**:20–26.

Real, L. 1990. Predator switching and the interpretation of animal choice behavior: the case for constrained optimization. In R.N. Hughes (ed.), Behavioural Mechanisms of Food Selection. Springer-Verlag, Heidelberg, pp. 1–21.

Real, L.A. 1991. Animal choice behaviour and the evolution of cognitive architecture. Science **253**:980–986.

Real, L.A. 1992. Biological information processing and the evolutionary ecology of cognition. Am. Nat., in press.

Real, L.A., Ott, J., and Silverfine, E. 1982. On the tradeoff between the mean and the variance in foraging: Effects of spatial distribution and color preference. Ecology **63**:1617–1623.

Real, L., Ellner, S., and Harder, L.D. 1990. Short-term energy maximization and risk-aversion in bumble bees: A reply to Possingham et al. Ecology **71**:1625–1628.

Roitblat, H.L. 1987. Introduction to Comparative Cognition. W.H. Freeman, New York.

Schmitt, U. and Bertsch, A. 1990. Do foraging bumblebees scent-mark food sources and does it matter? Oecologia **82**:137–144.

Schneider, W., and Shiffrin, R.M. 1977. Controlled and automatic human information processing: I. Detection, search, and attention. Psychol. Rev. 84:1–66.

Schoener, T.W. 1987. A brief history of optimal foraging ecology. In A.C. Kamil, J.R. Krebs, and H.R. Pulliam (eds.), Foraging Behavior. Plenum Press, New York, pp. 5–67.

Seely, T.D. 1985. Honeybee Ecology. Princeton University Press, Princeton, NJ.

Seely, T.D., Camazine, S., and Sneyd, J. 1991. Collective decision-making in honey bees: How colonies choose among nectar sources. Behav. Ecol. Sociobiol. **28**:277–290.

Sherry, D.F. 1988. Learning and adaptation in food-storing birds. In R.C. Bolles and M.D. Beecher (eds.), Evolution and Learning. Lawrence Erlbaum Associates, Hillsdale, NJ.

Shiffrin, R.M., and Schneider, W. 1977. Controlled and automatic human information processing: II. Perceptual learning, automatic attending, and a general theory. Psychol. Rev. **84**:127–190.

Simon, H.A., and Kaplan, C.R. 1989. Foundations of cognitive science. In M.I. Posner (ed.), Foundations of Cognitive Science. MIT Press, Cambridge, MA, pp. 1–47.

Spitzer, H., Desimone, R., and Moran, J. 1988. Increased attention enhances both behavioral and neuronal performance. Science **240**:338–340.

Staddon, J.E.R. 1983. Adaptive Behaviour and Learning. Cambridge University Press, Cambridge, England.

Stanton, L.S. 1983. Short-term learning and the searching accuracy of egg-laying butterflies. Anim. Behav., **31**:33–40.

Stephens, D.W., and Krebs, J.R. 1986. Foraging Theory. Princeton University Press, Princeton, NJ.

Strickler, K. 1978. Specialization and foraging efficiency of solitary bees. Ecology **60**:998–1009.

Sullivan, K.A. 1988. Ontogeny of time budgets in yellow-eyed Juncos: adaptation to ecological constraints. Ecology **69**:118–124.

Tolman, E.C. 1948. Cognitive maps in rats and men. Psychol. Rev. **55**:189–208.

Visscher, P.K., and Seeley, T.D. 1982. Foraging strategy of honey bee colonies in a temperate deciduous forest. Ecology **63**:1790–1801.

Waddington, K.D. 1981. Patterns of size variation in bees and evolution of communication systems. Evolution **35**:813–814.

Waddington, K.D., and Holden, L.R. 1979. Optimal foraging: On flower selection by bees. Am. Nat. **114**:179–196.

Wahlsten, D. 1972. Genetic experiments with animal learning. Behav. Biol. **7**:143–182.

Waser, N.M. 1983. The adaptive nature of flower traits: Ideas and evidence. In L. Real (ed.), Pollination Biology. Academic Press, New York, pp. 241–185.

Waser, N.M. 1986. Flower constancy: Definition, cause, and measurement. Am. Nat. **127:**593–603.

Weathers, W.W., and Sullivan, K.A. 1989. Juvenile foraging proficiency, parental effort, and avian reproductive success. Ecol. Monog. **59:**223–246.

Wehner, R., and Menzel, R. 1990. Do insects have cognitive maps? Annu. Rev. Neurosci. **13:**403–414.

Wehner, R., and Strinivasan, M.V. 1981. Searching behavior of desert ants, genus *Cataglyphis* (Formicidae, Hymenopters). J. Comp. Physiol. **142:**315–338.

Wenner, A.M. 1967. Honey bees: Do they use the distance information contained in their dance maneuver? Science **155:**847–849.

Wenner, A.M., and Wells, P.H. 1990. Anatomy of a Controversy. Columbia University Press, New York.

Wilson, E.O. 1962. Chemical communication among workers of the fire ant *Solenopsis saevissima* (Fr. Smith). 1. The organization of mass-foraging. Anim. Behav. **10:**134–147.

Wilson, E.O. 1971. The Insect Societies. Harvard University Press, Cambridge, MA.

Wilson, E.O., and Hölldobler, B. 1988. Dense heterarchies and mass communication as the basis of organization in ant colonies. Trends Ecol. Evol. **3:**65–68.

14

Afterword: Learning, Adaptation, and the Lessons of O

Daniel R. Papaj

Learning Defined as Adaptive

In 1956, the insect ethologist W.H. Thorpe in a book entitled *Learning and Instinct in Animals* defined learning as "that process which manifests itself by adaptive change in behavior as a result of experience." The definition was both widely accepted and widely criticized. On one hand, it was embraced by ethologists as prominent as Konrad Lorenz (cf. Lorenz, 1965) and, until the most recent edition, was the definition of choice in Alcock's deservedly best-selling textbook on animal behavior. On the other hand, the definition was criticized by anthropologists and behavioral biologists alike on the grounds that defining learning as adaptive behavioral change undermined efforts to evaluate the evolution of learning (Stenhouse, 1973; Hailman, 1985; Goodall, 1986; Papaj and Prokopy, 1989). While it is possible that learned behavior is the result of adaptive evolutionary change, these critics argued, Thorpe's definition amounted to presuming the hypothesis of learning as adaptation to be true before it was validated.

The objection is a powerful one, so long as the term "adaptive" is taken to refer to a trait "that serves a definable function and has evolved under the action of natural selection" (Alcock, 1989). This is the sense usually intended by contemporary biologists. However, Lorenz and possibly Thorpe used the term "adaptive" in another sense. Lorenz meant only that individual behavior after learning had higher survival value in the current environment than did behavior before learning (cf. Lorenz, 1965). In other words, animals generally learned to "do the right thing."

The chapters in this book provide ample evidence of behavioral changes which are adaptive in this sense. Honey bees learn to orient *toward* (not away from) odor or visual stimuli associated with a sugar reward and *away from* (not toward) stimuli associated with an electric shock (Gould, Chapter 2, Menzel et al., Chapter 4; Smith, Chapter 5). Butterflies learn to handle *more effectively* (not less effectively) a flower associated with successful nectar extraction (Lewis, Chapter 9). Grasshoppers learn to *avoid* (not prefer) a food item associated with a postingestive malaise (Bernays, Chapter 1). Parasitoids learn to orient *toward* (not away from) the odor or color of a host microhabitat associated with oviposition or contact with a host chemical (Turlings et al., Chapter 3).

The adapting nature of learning is not always so obvious. My own chapter (Chapter 10) illustrates a subtle and little-studied form of learning, learning to be consistent, that was key to the Lamarckian perspective on the evolution of instinct. Like other forms of learning, this one can be shown to be adaptive in the sense used by Lorenz. As a female parasitic wasp learns, she becomes more consistent in her movements up an odor plume. While it might not be immediately obvious how consistency would benefit the foraging parasitoid, a simple simulation model (Chapter 10) suggests that consistent movement increases the female wasp's effectiveness at finding the odor source and presumably her effectiveness at finding host larvae in which to lay eggs.

At first glance, some behavioral changes seem clearly *not* to be adaptive in the Lorenzian sense. Smith (Chapter 5) describes a learning phenomenon, stimulus generalization, which appears at first to be maladaptive. Honey bees which learn the odor of a particular compound (e.g., geraniol) in association with a sucrose reward also show heightened responses to compounds similar in structure (e.g., nerol), even though these compounds have never been associated with a reward. Smith proposes that stimulus generalization may be a way in which animals cope with stochastic variability in the stimuli emitted by their biotic resources. Where a forager is likely not to encounter exactly the complex of stimuli associated with a reward, it might pay to respond to stimuli which resemble to some degree the stimuli rewarded. Even learning phenomena that look for all the world like design flaws may permit an animal to adapt its behavior to its current environment.

Learning and Evolutionary Change

Saying that learning generally increases the survival value of current and future behavior is not the same as saying that learning evolved and is maintained through natural selection. The conditions under which learning will be favored by selection is the focus of chapters by Mangel (Chapter 6), Roitberg et al. (Chapter 7), and Stephens (Chapter 8). The difference between "adaptive" as used by Lorenz and "adaptive" as used by these authors can be summarized as follows: Lorenz would have us compare the current fitness of an animal before and after experience. Mangel, Roitberg, and Stephens, by contrast, would have us compare the overall fitness of an animal which learns with that of an animal which does not. Learning can be both adapt*ing* and, like all behavior, adapt*ive*.

The alternative notions of behavior as adaptive and learning as adapting are not independent of one another. As I argue in Chapter 10, adaptive evolutionary change in behavior may be influenced greatly by the degree to which learning is adapting. Adapting or optimizing learning, by mimicking effects of selection, might sometimes retard adaptive evolutionary change in behavior (Chapter 10; Boyd and Richerson, 1985). This capacity of learning to put adaptive evolutionary change in check prompts an important question: when we observe behavior in diverse species to be adaptive or even optimal, is such behavior a consequence of the incessant and recent action of natural selection or is it the result of preexisting adapting, or optimizing, mechanisms such as learning?

Learning and the Evolution of Optimal Behavior

This question is a meaningful one because, despite plausible reasons for thinking that natural selection will not produce optimal behavior (Gould and Lewontin, 1979; Pierce and Ollason, 1987), animals—including insects—seem to exhibit optimal or nearly optimal behavior much of the time. Certainly many people believe that they do (Boyd and Richerson, 1985; Krebs and Davies, 1987). Moreover, widespread patterns of adaptive or optimal behavior have commonly been taken as an indication of the ongoing and prevalent action of selection on behavior (cf. Alcock, 1989). In what appears to be the prevailing view (widely implied if not widely expressed), optimal behavior is presumed to reflect adaptation and, given that it is expressed by a diversity of animal species engaged in a variety of activities, relatively recent and rampant adaptation at that.

There exists an alternative explanation for widespread patterns of optimal behavior. It is conceivable that extant species possess adapting or even optimizing behavioral mechanisms that arose early on in the history of a particular group. It is possible that these mechanisms have been commandeered in more or less their original form to solve various problems as environments have changed and taxa have diversified. If so, the expression of optimal or nearly optimal behavior in new species or by old species in new environments may have required little or even no genetic change.

These alternative perspectives on the evolution of optimal behavior are diagrammed in a pair of imaginary phylogenies shown in Figure 14.1. Each phylogeny shows the evolution of an optimizing mechanism called *O*. *O*

A.

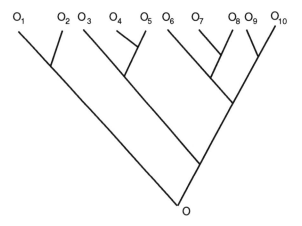

B.

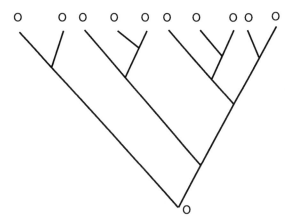

Figure 14.1. A. Hypothetical phylogeny onto which a character denoted as O, an optimizing mechanism, has been mapped. In this scheme, different O's ($O_1 \ldots O_{10}$) have evolved in recent taxa. Here selection for optimal behavior is ongoing. B. An alternative scheme using the same hypothetical phylogeny as in A. Here one O has evolved early on in the group and is retained in extant species. Despite the conservatism of O to evolutionary change, all recent species exhibit optimal or nearly optimal behavior.

permits an animal to adopt optimal or nearly optimal behavior in the environment in which it currently finds itself. O is, in short, a problem-solving mechanism (see Dukas and Real, Chapter 13, for a beautiful summary on aspects of cognition related to problem solving).[1]

Figure 14.1A portrays a phylogeny onto which the character O has been mapped. Here distinct optimizing mechanisms $O_1 \ldots O_{10}$ have arisen independently in recently evolved species. As taxa diversified and entered new ecological niches, optimizing mechanisms arose or, if already present, were refined greatly by selection. In this case, a close fit between the "structure" of behavior and its function (i.e., optimal or nearly optimal behavior) reflects the recent action of natural selection on behavior in the species under study.

Figure 14.1B portrays an alternative diagram of evolutionary change in O. Here O arose early on in the group. As taxa diversified and species entered new ecological niches, O was implemented with essentially no genetic change to solve new problems successfully, even optimally. Under this scenario, optimal behavior differs in form among species simply because the problems which species are asked to solve differ themselves in form. The optimizing mechanism O is the same in all species. We might ask if, in this instance, optimal behavior in modern species reflects the recent action of optimizing selection. Even though all ten species exhibit optimal behavior of somewhat differing form, the answer is no. The ability to express optimal behavior evolved under selection early on and was simply retained by descendant species. Were we to infer that natural selection acted consistently across species to optimize behavior we would, as Rosenheim points out in Chapter 11, be guilty of pseudoreplication. As in Rosenheim's example of learning and patterns of diet specialization, the behavior of the species shown in Figure 14.1B constitutes in essence just a single data point (or actually the historical residue of a data point that no longer exists).

If the latter scenario is the more correct one, it would challenge the perspective that broad patterns of optimal behavior observed in extant species are evidence of the prevalent, enduring action of optimizing selection. In the extreme, optimal behavior in current circumstances would instead be nothing more than the manifestation of an optimizing mechanism which evolved under selection long ago. While perhaps distressing to a few behavioral ecologists, this perspective might actually comfort field ecologists and ecological geneticists who have been hard-pressed to find com-

---

[1]O represents the system of neurological and hormonal processes that permits an animal to express the behavior appropriate to a particular environment. O includes mechanisms related to learning and also to motivation (cf. Mangel, Chapter 6). However, O is more than this. As Lorenz (1981) noted in his review of Baerend's famous work on digger wasps, a great deal of behavioral plasticity can be achieved without learning.

pelling evidence for optimizing selection in nature (Endler, 1986; Travis, 1989).

Note that nowhere in these arguments am I appealing to examples of nonadaptive behavior or invoking reasons to think that behavior will not be perfectly or nearly perfectly adapted (although such arguments are persuasive ones; cf. Gould and Lewontin, 1979; Pierce and Ollason, 1987). I am actually adopting an adaptationist perspective in postulating the evolution of an optimizing mechanism, O, designed to maximize individual fitness. However, I am also speculating that O arose relatively long ago and that modern species employ essentially the same O in ways that always seem to maximize fitness. If true, it would mean that we behavioral ecologists are frequently fleshing out the adapting properties of O without saying a great deal about how behavior has changed under selection. We behavioral ecologists may be engaged more in study of proximate causes of a special mechanism of behavioral flexibility (i.e., the workings of O) than of ultimate evolutionary causes of behavior. This last point is an ironic one in that many, if not most, behavioral ecologists typically aim to do just the opposite (i.e., study ultimate causes and not proximate ones) (cf. Krebs and Davies, 1987; Alcock, 1989).

Implications and Evidence

The two scenarios for the evolution of optimal behavior outlined above are not mutually exclusive and the truth could fall somewhere between them. Evidence as to which scenario is more correct is lacking. However, the second scenario carries two implications worthy of discussion. These are addressed briefly below.

1. *O or elements of O are relatively ancient.* The perspective that optimal behavior in extant species has been achieved with little or no genetic change implies that behavioral plasticity of an adapting form predates the taxon or taxa whose optimal behavior is under scrutiny. If we were, for example, to suggest that optimal behavior across all animal phyla has been achieved with little or no genetic change, then O must be very ancient.

Despite a long history of interest in the subject (Morgan, 1986; Bitterman, 1965, 1988), data bearing on phylogenetic aspects of learning, one component of O, are sparse. Certainly such information is lacking in this volume. This deficiency is nowhere more evident than in Rosenheim's excellent contribution (Chapter 11), which necessarily takes a methodological, and not an empirical, tack toward the phylogenetic analysis of learning within the insects.

Accepting that data are limited, there is nevertheless some indication that two forms of learning, habituation and associative learning, are relatively ancient. Figure 14.2 shows a tentative (very tentative!) diagram of the occurrence of habituation and associative learning in major animal phyla. Habituation appears almost as soon as things acquire nerve nets (i.e., in the Cnidaria). The pattern for associative learning is more open to debate. It may or may not occur in the Cnidaria; it was, however, almost certainly present in ancestors of the acoelomates. As Smith (Chapter 5) notes, we must be wary of inferences about homologies in learning among taxa until such time as we can guarantee that the mechanism underlying a given form of learning in, say, a mollusk is the same as that in, say, an

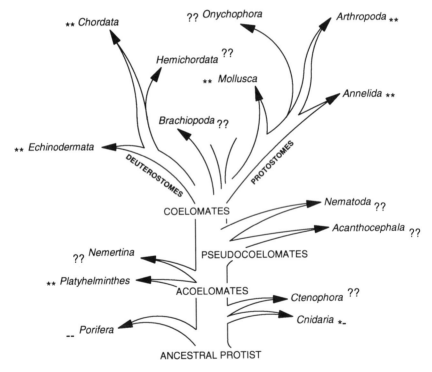

Figure 14.2. Phylogeny of the major animal phyla onto which the occurrence of two forms of learning, habituation and associative learning, has been mapped. The occurrence of habituation is indicated by an "*" in the first member of a pair of symbols; the occurrence of associative learning is indicated by a "*" in the second member of the pair. A "-" indicates that members of a group have been surveyed for a particular form of learning and found to lack that form; a "?" indicates a failure to find information on a particular form of learning in a particular group. Phylogeny from Wessells and Hopson (1988).

insect or a chordate. Still, evidence to date suggests that learning arose almost as soon as did nervous tissue (perhaps even earlier, depending on one's definition of learning) and that even associative learning was present in relatively primitive groups. Tierney (1986) has gone so far as to speculate that plasticity is an inherent property of nervous tissue and, hence, an inherent property of behavior. In short, at least one element of *O* appears to have been in place for a very long time.

2. *O can be used to solve novel problems.* Also implied in the idea that optimal behavior has been achieved in recent groups with little or no genetic change is an ability for existing *O*'s to be applied to the solution of novel problems. As taxa diversify or as old species are thrust into new environments, new problems are encountered. These new problems are presumably solved by extant optimizing mechanisms, else selection would shape these mechanisms as presumed in the conventional scenario. Is there evidence that animals can apply existing learning abilities to solve novel problems?

I believe that the answer to this question is yes, though data are scant once more. The classic example of an animal learning to solve a novel problem is that of blue tits in Europe which learned to remove the caps from milk bottles and extract the cream within (Fisher and Hinde, 1948). Since such birds are adapted to pry open bark from trees and extract insects beneath, the problem represented by caps on milk bottles is admittedly not as novel as was first thought. There are nevertheless numerous other examples among the vertebrates. Anyone who has been to the circus or attended a marine animal show has observed animals solving all sorts of problems which are, to varying degrees, novel ones. Any one of a number of peculiar pet tricks are evidence of animals solving novel problems through learning. (Insofar as a biscuit or some other food reward represents a suitable resource, such solutions are generally adapting in form!) Similarly, the capacity for parrots and other animals (Pepperberg, 1987) to learn human sounds and syntax may reflect an ability to solve novel problems related to communication.

Are there corresponding examples among the insects? Perhaps, although data for insects are even more meager than that for vertebrates. Honey bees thrust by man into alien habitats appear to have solved, probably through learning and perhaps without genetic change, problems associated with finding and extracting nectar from novel flower species. In some cases (e.g., alfalfa), these flowers appear to be different in key respects from ones to which honey bees have been exposed over evolutionary time.

Similarly, herbivorous insects have solved problems associated with exploitation of plants in alien habitats to which insects have been introduced or of plants (usually crops or ornamentals) thrust into the insects' native habitat. Occasionally these plants are in systematic terms very different from native hosts and might thus constitute a novel problem for the herbivore. It is conceivable that these problems have sometimes been solved in part by virtue of behavioral flexibility and with perhaps less genetic change in behavioral traits than might be anticipated (Courtney et al., 1989; Jaenike and Papaj, 1992). Much work needs to be done before we can conclude that this is so. However, it is worth noting that practitioners of integrated pest management are increasingly aware that behavioral flexibility may permit pest insects to circumvent man's best efforts to reduce their fitness and control their numbers (Roitberg and Angerilli, 1986; Prokopy and Lewis, Chapter 12).

In a sense, man poses for insects a novel problem and literally dares them to come up with a solution. Frequently, of course, they do so. I predict that we will find increasingly that they do so with the help of preexisting mechanisms of behavioral flexibility, including learning, and with relatively little genetic change in behavior. The ability of insects to solve novel problems is not necessarily a wholly gloomy prospect for humankind. Practitioners of biological control of pest insects sometimes pit natural enemies against novel host insects in non-native habitats (cf. Prokopy and Lewis, Chapter 12). These efforts may succeed in part because behavioral plasticity, mediated partly through learning, permits natural enemies to adapt to novel hosts in novel habitats. Biological control may be an instance where novel problem solving by insects has been put in the service of humankind.

A final example from the insects (and the only one that does not involve the response of insects to some action by humans) relates to kin recognition in the social hymenoptera. The ability to recognize kin is thought to have facilitated the evolution of altruistic behavior. In the social hymenoptera, such recognition is mediated frequently by discrimination among odors and such discrimination is usually learned at least in part. Hölldobler and Michener (1980) proposed that the ability of advanced eusocial species to learn to discriminate kin from nonkin was a consequence of a preexisting ability in solitary ancestors to learn odors associated with recognition of an individual's nest-site entrance. Novel problem solving may thus have played a key role in the evolution of altruistic behavior in eusocial insects.

Our own species provides the best examples of how existing problem-solving mechanisms have been applied by animals to solve novel problems. In humans, cultural change has greatly outstripped biological evolution. Nobody would argue that human behavior has evolved under natural selection in lockstep with cultural change and yet human behavior remains principally adaptive in form (Boyd and Richerson, 1985). Apparently, learning (insight learning and social learning in particular) helps human beings today solve all sorts of culture-based problems that never confronted us over the course of our biological evolution. Associative learning even helps us to solve effectively problems of a kind that natural selection could not otherwise solve, i.e., problems that arise over the course of a single generation and never appear again. Does it require so much of a leap of faith to suppose that associative learning has been similarly commandeered (though perhaps in less conspicuous ways) in animals other than human beings? Behavioral ecologists place great value in general solutions to specific optimization problems that can be applied to new problems as they occur to us. Would we really be surprised if Nature endowed animals early on in their phylogenetic history with general problem-solving mechanisms that were assigned to novel problems materializing later in their history? If so, it may well be that optimal behavior in extant taxa is in large part a consequence of optimizing mechanisms that evolved long ago.

Summary and Future Directions

In closing, I emphasize that I am not arguing that optimal behavior has arisen entirely without the action of natural selection. I am only proposing that, given adaptive evolution in ancestral taxa of a mechanism by which behavior is adjusted and fitness is maximized or nearly maximized in the current environment, optimal behavior in extant taxa may have been achieved with a minimum of genetic change and in the absence of strong optimizing selection.

What information is needed to evaluate these ideas? First and foremost, we need to document better the actual significance of learning in nature. Is it not cause for wonder that a national laboratory initiated a multimillion dollar effort to decipher the physiological mechanism of learning in *Drosophila*, an animal for which the significance of learning in nature has not been established unambiguously (Jaenike 1982, 1983, 1985; Hoffman, 1985; Hoffman and Turelli, 1985)? Is it not remarkable that, for one intensively studied form of learning in herbivorous insects, i.e., larval induction of food preference in the Lepidoptera [see Papaj and Prokopy (1986) and Jermy (1987) for reviews], there is only a single, rudimentary field study (Dethier, 1988)? Is it not curious that, despite the recent flurry of papers in major journals on the subject of learning in entomophagous parasitoids, there are only two studies bearing on its possible (emphasis on the word "possible") significance in the field (Lewis and Martin, 1990; Papaj and Vet, 1990)? Neither study was designed in such a way as to demonstrate conclusively that experience affects host use by parasitoids in nature.

We also need information on the mechanism and function of learning, as well as its evolutionary history, With respect to function, we need continued effort by theoreticians to define exactly how learning (and behavior in general) ought to be suited to the diversity of niches filled by animals. How well and in what ways does learning adapt the animal to its current environment? At the same time, we need to specify better the physiological mechanisms underlying learning. Menzel et al. (Chapter 4) argue forcefully that the rigor of behavioral ecologists' conclusions about the value of learning depends partly on the extent to which their models are realistic with respect to mechanism. More than that, studies of mechanism (especially when conducted from an ecological perspective) ultimately provide evolutionary biologists with the characters in the "evolutionary play." This is no less true of learning than of any other trait.

Finally, we need more data on learning in more species. Were I to attempt the exercise portrayed in Figure 14.2 at the level of orders within the insects, the ?s would perhaps outnumber the *s. It is imperative that we adopt comparative and especially phylogenetic approaches to learning and behavior. If the ideas outlined above have been understated in the literature, it is principally because phylogenetic analyses of behavior (including learning) have lagged far behind functional ones. The current proliferation of molecular and theoretical techniques for phylogeny reconstruction (Maddison, 1990; Brooks and McLennan, 1991; Rosenheim, Chapter 11) should facilitate greatly the validation or refutation of the opinions expressed here. I can hardly wait.

Acknowledgments

I thank Ann Hedrick, Alcinda Lewis, Bill Mitchell, Peter Smallwood, and Bill Wcislo for comments on an earlier draft.

References

Alcock, J. 1989. Animal Behavior: An Evolutionary Approach, 4th ed. Sinauer, Sunderland, MA, 596 pp.

Bitterman, M.E. 1965. Phyletic differences in learning. Am. Psychol. **20**:396–410.

Bitterman, M.E. 1988. Vertebrate-invertebrate comparisons. In H.J. Jerison and I. Jerison (eds.), Intelligence and Evolutionary Biology. Springer-Verlag, New York, pp. 251–275.

Boyd, R., and Richerson, P.J. 1985. *Culture and the Evolutionary Process*. University of Chicago Press, Chicago.

Brooks, D.R., and McLennan, D.A. 1991. Phylogeny, Ecology and Behavior: A Research Programme in Comparative Biology. University of Chicago Press, Chicago, 434 pp.

Courtney, S.P., Chen, G.K., and Gardner, A. 1989. A general model for individual host selection. Oikos **55**:55–56.

Dethier, V.G. 1988. Induction and aversion-learning in polyphagous arctiid larvae (Lepidoptera) in an ecological setting. Can. Entomol. **120**:125–131.

Endler, J.A. 1986. Natural Selection in the Wild. Princeton University Press, Princeton, NJ.

Fisher, R.A., and Hinde, R.A. 1948. Br. Birds **42**:347–357.

Goodall, J. 1986. The Chimpanzees of Gombe: Patterns of Behavior. Belknap Press of Harvard University Press, Cambridge, MA, 673 pp.

Gould, S.J., and Lewontin, R.C. 1979. The spandrels of San Marcos and the Panglossian paradigm: A critique of the adaptionist programme. Proc. R. Soc. Lond., B **205**:581–598.

Hailman, J.P. 1985. Historical notes on the biology of learning. In Issues in the Ecological Study of Learning. Lawrence Erlbaum Associates, Hillsdale, NJ, pp. 27–57.

Hoffmann, A.A. 1985. Effects of experience on oviposition and attraction in *Drosophila*: Comparing apples and oranges. Am. Nat. **126**:41–51.

Hoffmann, A.A., and Turelli, M. 1985. Distribution of *Drosophila melanogaster* on alternative resources: Effects of experience and starvation. Am. Nat. **126**:662–679.

Hölldobler, B., and Michener, C.D. 1980. Mechanisms of identification and discrimination in social Hymenoptera. In H. Markl (ed.), Evolution of Social Behavior: Hypotheses and Empirical Tests. Dahlem Conferences 1980, Verlag Chemie, Weinheim, Germany, pp. 35–58.

Jaenike, J. 1982. Environmental modification of oviposition behavior in *Drosophila*. Am. Nat. **119**:784–802.

Jaenike, J. 1983. Induction of host preference in *Drosophila melanogaster*. Oecologia **58**:320–325.

Jaenike, J. 1985. Genetic and environmental determinants of food preference in *Drosophila tripunctata*. Evolution **39**:362–369.

Jaenike, J., and Papaj, D.R. 1992. Learning and patterns of host use by insects. In M. Isman and B.D. Roitberg (eds.), Chemical Ecology: An Evolutionary Perspective. Routledge, Chapman and Hall, New York, in press.

Jermy, T. 1987. The role of experience in the host selection of phytophagous insects. In Proceedings of the 5th International Symposium on Plant–Insect Relationships, Pudoc, Wageningen, pp. 25–32.

Krebs, J.R., and Davies, N.B. 1987. Introduction to Behavioural Ecology. Blackwell Scientific Publications, Oxford, England.

Lewis, W.J., and Martin, W.R., Jr. 1990. Semiochemicals for use with parasitoids: Status and future. J. Chem. Ecol., **16**:3067–3090.

Lorenz, K. 1965. Evolution and Modification of Behavior. University of Chicago Press, Chicago.

Lorenz, K. 1981. The Foundations of Ethology. Springer-Verlag, New York.

Maddison, W.P. 1990. A method for testing the correlated evolution of two binary characters: Are gains or losses concentrated on certain branches of a phylogenetic tree? Evolution **44**:539–557.

Morgan, C.L. 1986. Habit and Instinct. Arnold, London.

Papaj, D.R., and Prokopy, R.J. 1989. Ecological and evolutionary aspects of learning in phytophagous insects. Annu. Rev. Entomol. **34**:315–350.

Papaj, D.R., and Vet, L.E.M. 1990. Odor learning and foraging success in the parasitoid, *Leptopilina heterotoma*. J. Chem. Ecol., **16**:3137–3150.

Papaj, D.R., and Prokopy, R.J. 1986. Phytochemical basis of learning in *Rhagoletis pomonella* and other herbivorous insects. J. Chem. Ecol. **12**:1125–1143.

Pepperberg, I.M. 1987. Interspecies communication: A tool for assessing conceptual abilities in the African Grey parrot (*Psittacus erithacus*). In G. Greenberg and E. Tobach (eds.) Language, Cognition, and Consciousness: Integrative Levels. Lawrence Erlbaum Associates, Hillsdale, NJ, pp.31–56.

Pierce, G.J. and Ollason, J.G. 1987. Eight reasons why optimal foraging theory is a complete waste of time. Oikos **49**:111–118.

Roitberg, B.D., and Angerilli, N.P.D. 1986. Management of temperate-zone deciduous fruit pests: Applied behavioural ecology. Agric. Zool. Rev. **1**:137–165.

Stenhouse, D. 1973. The Evolution of Intelligence: A General Theory and Some of Its Implications. George Allen and Unwin, London.

Thorpe, W.H. 1956. Learning and Instinct in Animals. Methuen, London.

Tierney, A.J. 1986. The evolution of learned and innate behavior: Contributions from genetics and neurobiology in a theory of behavioral evolution. Anim. Learn. Behav. **14**:339–348.

Travis, J. 1989. The role of optimizing selection in natural populations. Annu. Rev. Ecol. Syst. **220**:279–296.

Wessels, N.K., and Hopson, J.L. 1988. Biology. Random House, New York.

Index